AutoLISP® to Visual LISP™:
Design Solutions
for AutoCAD®

AutoLISP® to Visual LISP™: Design Solutions for AutoCAD®

Kevin Standiford

THOMSON

LEARNING™

autodesk
press

Australia • Canada • Mexico • Singapore • Spain • United Kingdom • United States

THOMSON

™

LEARNING

AutoLISP® to Visual LISP™:
Design Solutions for AutoCAD®
by Kevin Standiford

Autodesk Press Staff

Executive Director:
Alar Elken

Executive Editor:
Sandy Clark

Acquisitions Editor:
Michael Kopf

Developmental Editor:
John Fisher

Executive Marketing Manager:
Maura Theriault

Executive Production Manager:
Mary Ellen Black

Production Coordinators:
Stacy Masucci and Larry Main

Art and Design Coordinator:
Mary Beth Vought

Marketing Coordinator:
Paula Collins

Technology Project Manager:
Tom Smith

Cover Illustration:
Tom White

COPYRIGHT © 2001 Thomson Learning™.

Printed in Canada
 2 3 4 5 6 7 8 9 10 XXX 05 04 03 02 01

For more information, contact Autodesk Press, 3 Columbia Circle, Box 15-015, Albany, New York, 12212-15015 USA.

Or find us on the World Wide Web at http://www.autodeskpress.com

Library of Congress Cataloging-in-Publication Data

Standiford, Kevin.
  AutoLISP to Visual LISP : solutions for AutoCAD / Kevin Standiford.
    p. cm.
  ISBN 0-7668-1517-X
  1. LISP (Computer program language) 2. AutoCAD (Computer file) I. Title.

  QA76.73.L23 S715 2000
  005.13'3—dc21
                              00-031312

Notice To The Reader

Publisher does not warrant or guarantee any of the products described herein or perform any independent analysis in connection with any of the product information contained herein. Publisher does not assume, and expressly disclaims, any obligation to obtain and include information other than that provided to it by the manufacturer.

The reader is expressly warned to consider and adopt all safety precautions that might be indicated by the activities herein and to avoid all potential hazards. By following the instructions contained herein, the reader willingly assumes all risks in connection with such instructions.

The publisher makes no representation or warranties of any kind, including but not limited to, the warranties of fitness for particular purpose or merchantability, nor are any such representations implied with respect to the material set forth herein, and the publisher takes no responsibility with respect to such material. The publisher shall not be liable for any special, consequential, or exemplary damages resulting, in whole or part, from the readers' use of, or reliance upon, this material. Autodesk does not guarantee the performance of the software and Autodesk assumes no responsibility or liability for the performance of the software or for errors in this manual.

CONTENTS

Quick Contents

FOREWARD XV

PREFACE XIX

CHAPTER 1   INTRODUCTION TO PROGRAMMING 1

CHAPTER 2   INTERFACING WITH THE USER, SAVING
OUTPUT, MATH FUNCTIONS 37

CHAPTER 3   LIST PROCESSING, MANIPULATING AND
CONVERTING STRINGS, MAKING DECISIONS 75

CHAPTER 4   ADVANCED LIST PROCESSING
AND LOOPS 115

CHAPTER 5   EXTENDED ENTITY DATA, XRECORDS,
SYMBOL TABLES, DICTIONARIES, AND
SELECTION SETS 147

CHAPTER 6   DIESEL AND DIALOG BOXES 177

CHAPTER 7   INTERFACING WITH AND MANAGING
DIALOG BOXES 221

CHAPTER 8   INTRODUCTION TO VISUAL LISP 263

CHAPTER 9   OBJECT ORIENTED PROGRAMMING
AND ACTIVEX 355

CHAPTER 10   REACTORS 399

CHAPTER 11   MULTIPLE DOCUMENT INTERFACE 441

CHAPTER 12   WINDOWS 95, 98, AND NT REGISTRY, LOADING
AND EXECUTING VBA APPLICATIONS 467

INDEX 497

CONTENTS

FOREWARD ....**XV**

A WORD FROM THOMSON LEARNING™ XV

A WORD FROM AUTODESK, INC. XV

THE INITIAL IDEA XVI

TOPICS FOR THE PROGRAMMER SERIES XVII

WHO READS PROGRAMMING BOOKS? XVII

THIRST, THEME, AND VARIATION XVIII

PREFACE ....**XIX**

FEATURES OF THIS EDITION XXI

HOW TO USE THIS BOOK XXI

TYPOGRAPHIC CONVENTIONS XXII

WE WANT TO HEAR FROM YOU XXII

ABOUT THE AUTHOR XXIII

ACKNOWLEDGMENTS XXIII

DEDICATION XXIV

CHAPTER 1   INTRODUCTION TO PROGRAMMING......................... 1

INTRODUCTION TO PROGRAMMING 2

WHAT IS AUTOLISP? 6

PROGRAMMING TECHNIQUES 6

    Phase One – Eight Steps to Solving a Design Problem 6

    Phase Two – Planning the Source Code 14

WRITING AN AUTOLISP EXPRESSION 18

    Spacing and Alignment Issues 21

SAVING AN AUTOLISP EXPRESSION 21
    Windows-Based Word Processing Programs 22
DEFINING A FUNCTION 26
    Adding Comments to a Function 29
LOADING AN AUTOLISP PROGRAM 29
    AutoLISP LOAD Function 29
    AutoCAD APPLOAD Command 32
SUMMARY 34
REVIEW QUESTIONS 35

CHAPTER 2   INTERFACING WITH THE USER, SAVING OUTPUT, MATH FUNCTIONS 37

AUTOLISP DATA TYPES 38
    Subrs and External Subrs 38
    Integers 39
    Real or Real Numbers 39
    Strings 40
    Lists 41
    Selection Sets 41
    Entity Names 41
    File Descriptors 42
    VLA 43
    Symbols and Variables 43
LOCAL AND GLOBAL VARIABLES 44
    Declaring Local Variables 44
    Declaring a Variable in AutoLISP 45
    Recalling the Value of a Global Variable at the AutoCAD Command Prompt 46
PROMPTING THE USER FOR INPUT 47
    Requesting String Information 47
    Requesting Numeric Information 48
    Requesting Angular Information 50
    Requesting Coordinates and Other Miscellaneous Information 51
UTILIZING THE RESULTS 55
    Writing the Results to the Screen 55
    Writing the Results to a File 57
    Reading Information from a File 58

PERFORMING MATHEMATICAL OPERATIONS 59
   Basic Mathematical Operations 59
   Equalities and Inequalities 62
PERFORMING COMPLEX MATHEMATICAL OPERATIONS 65
   Exponential and Logarithm Functions 65
   Trigonometry Functions 65
APPLICATION – CALCULATING BEND ALLOWANCES 67
SUMMARY 72
REVIEW QUESTIONS 73

CHAPTER 3   LIST PROCESSING, MANIPULATING AND CONVERTING STRINGS, MAKING DECISIONS 75

INTRODUCTION TO LISTS AND LIST PROCESSING 76
   Lists in AutoCAD 77
   Creating a List 77
   Combining two or More Lists into a Single List 78
   Determining If an Item Is a List 79
   Retrieving Information from a List 80
   Determining the Length of a List 82
PROCESSING THE ELEMENTS OF A LIST 83
   Manipulating and Converting Strings 83
   Converting Numeric Data to String Data 84
   Converting String Data to Numeric Data 85
   Merging Text Strings 85
   Truncating a Text String 87
   Determining the Length of a String 88
   Miscellaneous String Manipulations 90
MAKING DECISIONS 91
   Using the AutoLISP IF Function 93
   Evaluating Multiple Test Expressions Using the COND Function 100
APPLICATION – CONSTRUCTING A SIMPLE HARMONIC MOTION CAM
DISPLACEMENT DIAGRAM 101
SUMMARY 110
REVIEW QUESTIONS AND EXERCISES 111

**CHAPTER 4   ADVANCED LIST PROCESSING
AND LOOPS** 115

INTRODUCTION TO ADVANCED LIST PROCESSING 116
　　Association List 116
　　AutoCAD and Association Lists 124
LOOPS 137
　　Creating a Loop Using the WHILE Function 138
　　Creating a Loop Using the REPEAT Function 140
SUMMARY 142
REVIEW QUESTIONS AND EXERCISES 143

**CHAPTER 5   EXTENDED ENTITY DATA, XRECORDS,
SYMBOL TABLES, DICTIONARIES, AND
SELECTION SETS** 147

INTRODUCTION TO EXTENDED ENTITY DATA 148
　　Using Extended Entity Data 149
　　Xrecords 156
SYMBOL TABLES AND DICTIONARIES 157
　　Symbol Tables 157
　　Dictionary Objects 160
SELECTION SETS 161
　　Building a Selection Set in AutoLISP 161
　　Working with Selection Sets 166
APPLICATION – CONSTRUCTING AND ATTACHING EXTENDED ENTITY
DATA TO A SIMPLE GEAR TRAIN 169
SUMMARY 172
REVIEW QUESTIONS AND EXERCISES 174

CHAPTER 6   DIESEL AND DIALOG BOXES 177

INTRODUCTION TO DIESEL 178
　　Using the MODEMACRO System Variable to Configure the AutoCAD
　　Status Line 178
　　Making the Status Bar Reflect AutoCAD Internal State Using DIESEL 181
　　Using DIESEL Expressions in AutoCAD Menu Files 188
　　DIESEL Functions 190

INTRODUCTION TO DIALOG BOXES 192

Dialog Box File Format 193

Components of a Dialog Box 193

Dialog Box Style and Syntax 195

Child /Parent Relationship 202

Defining a Dialog Box 203

Dialog Box Design Considerations 216

SUMMARY 217

REVIEW QUESTIONS AND EXERCISES 218

CHAPTER 7 INTERFACING WITH AND MANAGING DIALOG BOXES 221

INTRODUCTION TO DIALOG BOX MANAGEMENT 222

Loading a Dialog Box into Memory 222

Initializing the Dialog Box 223

Displaying the Actual Dialog Box 225

Terminating a Dialog Box 225

Terminating All Current Dialog Boxes 226

Unloading a Dialog Box 227

Setting the Value of a Tile 227

Retrieving the Value of a Dialog Box's Tile 228

Action Expressions and Callback Functions 228

Setting the Mode of a Dialog Box 232

List Boxes and Popup Lists 233

Images 235

APPLICATION - DETERMINING SPOT ELEVATIONS 238

SUMMARY 258

REVIEW QUESTIONS 260

CHAPTER 8 INTRODUCTION TO VISUAL LISP 263

INTRODUCTION TO VISUAL LISP 264

Enhancements Made to AutoLISP 265

WORKING WITH VISUAL LISP 265

Launching the Visual LISP IDE 265

Navigating the Visual LISP IDE Menus 266

Navigating the Visual LISP IDE Toolbars 272

The Visual LISP Console Window 275

The Visual LISP Trace Window 285

The Visual LISP Text Editor 286

Opening, Closing, Creating, and Saving Files 298

DEVELOPING APPLICATIONS WITH VISUAL LISP 303

Marking Sections of the Source Code with Bookmarks 303

Advancing the Cursor without Using Bookmarks 305

Using Visual LISP to Complete a Word by Matching 306

The Apropos Feature and Working with Symbols 306

Results Obtained from Apropos 309

Using Apropos to Complete a Word 314

Watch Window and Inspection Window 315

The Visual LISP Trace Stack 319

Symbol Service 325

Browsing the AutoCAD Database 326

Previewing a Portion of a Dialog Box and the Entire Dialog Box 334

Using the Animate Debugging Function 336

VISUAL LISP PROJECTS 337

Creating a Project 338

Opening an Existing Project 349

Creating a Project FAS File 349

Creating a Standalone Application 349

SUMMARY 351

REVIEW QUESTIONS 352

CHAPTER 9   OBJECT ORIENTED PROGRAMMING
              AND ACTIVEX 355

INTRODUCTION TO OBJECT ORIENTED PROGRAMMING 357

Classes, Objects, Methods, and Properties 358

Encapsulation, Polymorphism, and Inheritance 359

INTRODUCTION TO ACTIVEX 359

Object Oriented Programming and AutoCAD 360

ActiveX and AutoCAD 360

Understanding How ActiveX Works with AutoCAD and Visual LISP 360

Using ActiveX in Conjunction with AutoLISP 367

Viewing the Properties of an Object with the Inspection Tool 369
Working With Visual Basic Data Types 372
Using ActiveX to Access an Object's Properties 378
Working with Collections 380
Freeing Memory and Releasing Objects 385
Accessing Applications other than AutoCAD with ActiveX 385

APPLICATION - USING ACTIVEX TO WRITE THE RESULTS FROM AN

AUTOLISP APPLICATION TO WORD 387

SUMMARY 394

REVIEW QUESTIONS AND EXERCISES 396

CHAPTER 10   REACTORS **399**

INTRODUCTION TO REACTORS 400
Creating a Reactor 400
Generating a List of Reactor Types 403
Callback Events 404
Callback Functions 405
Creating an AutoCAD Reactor 408
Creating an Object Reactor 409
Using Reactors with Multiple Drawings 418
Determining a Reactor's Notification Setting 420
Obtaining and Modifying Information about Reactors 421
Disabling and Enabling Reactors 429
Making a Reactor Transient or Persistent 430

APPLICATION – INCORPORATING A REACTOR 433

SUMMARY 437

REVIEW QUESTIONS AND EXERCISES 438

CHAPTER 11   MULTIPLE DOCUMENT INTERFACE **441**

INTRODUCTION 442

MULTIPLE DOCUMENT INTERFACE 442
NameSpace 443
Blackboards 460
Error Messages 462

Limitations of Working within Multiple Document Interface Using
AutoLISP 464

SUMMARY 465

REVIEW QUESTIONS 466

**CHAPTER 12   WINDOWS 95, 98, AND NT REGISTRY, LOADING
AND EXECUTING VBA APPLICATIONS 467**

INTRODUCTION TO WINDOWS REGISTRY 468

The Windows Registry 468

AutoLISP and the Windows Registry 470

APPLICATION - SPOT ELEVATION 477

VISUAL BASIC FOR APPLICATIONS AND VISUAL LISP 491

SUMMARY 495

REVIEW QUESTIONS 496

INDEX **497**

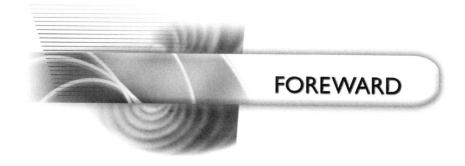

FOREWARD

A WORD FROM THOMSON LEARNING™

Autodesk Press was formed in 1995 as a global strategic alliance between Thomson Learning and Autodesk, Inc. We are pleased to bring you the premier publishing list of Autodesk student software and learning and training materials to support the Autodesk family of products. AutoCAD® is such a powerful product that everyone who uses it would benefit from a mentor to help them unlock its full potential. This is the premise upon which the Programmer Series was conceived. The titles in this series cover the most advanced topics that will help you maximize AutoCAD. Our Programmer Series titles also bring you the best and the brightest authors in the AutoCAD community. Maybe you've read their columns in a CAD journal, maybe you've heard them speak at an Autodesk event, or maybe you're new to these authors—whatever the case may be, we know you'll enjoy and apply what you'll learn from them. We thank you for selecting this title and wish you well on your programming journey.

Sandy Clark
Executive Editor
Autodesk Press

A WORD FROM AUTODESK, INC.

From the birth of AutoCAD onward, there has been a large library of source material on how to use the software; however, relatively little material on customization of AutoCAD has been made available. There have been a few texts written about AutoLISP®, and many general AutoCAD books include a chapter or two on customization. On the whole, however, the number of texts on programming AutoCAD has been inadequate given the amount of open technology, the number of application programming interfaces (APIs), and the sheer volume of opportunity to program the AutoCAD design system.

Four years ago, when I started working with developers as the product manager for AutoCAD APIs, I immediately found an enormous demand for supporting texts about ObjectARX™, AutoLISP, Visual Basic for Applications®, and AutoCAD® OEM. The demand for a "programming series" of books stems from AutoCAD's history of being the most programmable, customizable, and extensible design system on the market. By one measure, over 70% of the AutoCAD customer base "programs" AutoCAD using Visual LISP, VBA, or menu customization, all in an effort to increase productivity. The demand for increasing the designer's productivity extends deeply, creating a demand for new source texts to increase productivity of the programming and customization process itself.

The standard Autodesk documentation for AutoCAD provides the original source of technical material on the AutoCAD API technology. However, it tends toward the clinical, which is a natural result of describing the software while it is being created in the software development lab. Documentation in fully applied depth and breadth is only completed through the collective experiences of hundreds of thousands of developers, customers, and users as they interpret and apply the system in ways specific to their needs.

As a result, demand is high for other interpretations of how to use AutoCAD APIs. This kind of instruction develops the technique required to innovate. It develops programmer instinct by instructing when to use one interface over another and provides direction for interpretive nuances that can only be developed through experience. AutoCAD customers and developers look for shortcuts to learning and for alternative reference material. Customers and developers alike want to accelerate their programming learning experience, thereby shortening the time needed to become expert and enabling them to focus sooner and better on their own specific customization or development projects. Completing a customization or development project sooner, faster, and better means greater productivity during the development project and more rapid deployment of the result.

For the CAD manager, increasing productivity through accelerated learning means increasing his or her CAD department's productivity. For the professional developer, this means bringing applications to market faster and remaining competitive.

THE INITIAL IDEA

An incident that occurred at Autodesk University in Los Angeles in the fall of 1997 illustrated dramatically a dynamic demand for AutoCAD API technology information. Bill Kramer was presenting an overview session on AutoCAD's ObjectARX. The Autodesk University officials had planned on having 20 to 30 registrants for this session and had assigned an appropriately sized room to the session. About three weeks before it took place, I received a telephone call indicating that the registrations for the session had reached the capacity of the room, and we would be moving the session. We moved the session two more times due to increasing registration from customers, CAD Managers, designers, corporate design managers, and even developers attending the Autodesk University customer event. What they all had in common was an interest in seeing how ObjectARX, an AutoCAD API, was going to increase their own or their department's design productivity. When the presentation started, I walked into the back of the room to see what appeared to me to be over 250 people in a room which was now, standing room only. That may have been the moment when I decided to act, or it may have been just when the actual intensity of this demand became apparent to me; I'm not sure which.

The audience was eager to hear what Bill Kramer was going to say about the power of ObjectARX. Nearly an hour of unexpected follow-up questions and answers followed. Bill had successfully evangelized a technical subject to an audience spanning non-technical to technical individuals. This was a revelation for me, and the beginning of my interest in developing new ways to communicate to more people the technical aspects, power, and benefits of AutoCAD technology and APIs.

As a result, I asked Bill if he would write a book on the very topic he just presented. I'm happy to say that Bill's book has become one of the first books written in the new AutoCAD Programmer Series with Autodesk Press.

TOPICS FOR THE PROGRAMMER SERIES

The extent of AutoCAD's open programmable design system made the need for a series of books apparent early on. My team, under the leadership of Cynde Hargrave, Senior AutoCAD Marketing Manager, began working with Autodesk Press to develop this series.

It was an exciting project, with no shortage of interested authors covering a range of topics from AutoCAD's open kernel in ObjectARX to the Windows standard for application programming in VBA. The result is a complete library of references in Autodesk's Programmer Series covering ObjectARX, Visual LISP, AutoLISP, customizing AutoCAD through ActiveX Automationð and Microsoft Visual Basic for Applications, AutoCAD database connectivity, and general customization of AutoCAD.

WHO READS PROGRAMMING BOOKS?

Every AutoCAD user will find books in this series to fit their AutoCAD customization or development interests. The collective goal we had with our team and Autodesk Press for developing this series, identifying titles, and matching them with authors was to provide a broad spectrum of coverage across a wide variety of customization content and a wide range of reader interest and experience.

Collaborating as a team, Autodesk Press and Autodesk developed a programmer series covering all the important APIs and customization topics. In addition, the series provides information that spans use and experience levels from the novice just starting to customize AutoCAD to the professional programmer or developer looking for another interpretive reference to increase his or her experience in developing powerful applications for AutoCAD.

THIRST, THEME, AND VARIATION

I compare this thirst for knowledge with the interest musicians have in listening to music performed in different ways. For me, it is to hear Vivaldi's *The Four Seasons* time and time again. Musicians play from the same notes written on the page, with the identical crescendos and decrescendos and other instructions describing the "technical" aspects of the music.

All of the information to play the piece is there. However, the true creative design and beauty only manifests through the collection of individual musicians, each applying a unique experience and interpretation based on all that he or she has learned before from other mentors in addition to his or her own practice in playing the written notes.

By learning from other interpretations of technically identical music, musicians benefit the most from a new, and unique, interpretation and individual perception. This makes it possible for musicians to amplify their own experience with the technical content in the music. The result is another unique understanding and personalized interpretation of the music.

Similar to musical interpretation is the learning, mentoring, creative processes, and resources required in developing great software, programming AutoCAD applications, and customizing the AutoCAD design system. This process results in books such as Autodesk's Programmer Series, written by industry and AutoCAD experts who truly love working with AutoCAD and personalizing their work through development and customization experience. These authors, through this programmer series, evangelize others, enabling them to gain from their own experiences. For us, the readers, we gain the benefit from their interpretation, and obtain the value through different presentation of the technical information, by this wide spectrum of authors.

Andrew Stein
Senior Manager
Autodesk Business Research, Analysis and Planning

PREFACE

The computer age has touched every aspect of our modern world. Computers control our energy, transportation, communications, manufacturing, agriculture, as well as our personal lives. They are used in every discipline of engineering from aerospace to civil and every phase from conception to final product output (drawings, CNC code, word processing, etc.). Therefore it is critical that the engineering professional become well versed in its control, as well as the usage of this modern tool, which includes the operation as well as the preparation of software. Although, software is becoming more versatile and readily available that will automate the many tedious and even complex task that are performed daily in today's engineering offices, there is still occasions in which specialized functions are still performed using just the basic system. Therefore, any person or can write specialized programs can save a company many valuable hours of production time. This book is designed to give the student the tools necessary to accomplish that task, by introducing the basics of customizing the AutoCAD workstation using AutoLISP, Diesel and Visual LISP for AutoCAD 2000 programming languages.

AutoLISP is a variation of an older programming language (LISP) that has been used for many years in the research and development of artificial intelligence. LISP has three important features that makes it a desirable programming language to base Autodesk's AutoLISP programming on. First, because it is a simple language to learn, the programmer productivity curb is increased. This allows the programmer to spend more time developing useful programs and less time learning all the in and out of the language. Second, the program is very powerful. This is an important aspect of any programming language. Why spend the time learning a language if it is limited in the quality of the programs that can be created with it. Third, it is an excellent language for graphic interface. This feature might be one of the most important of all, sense AutoCAD is a graphics based program. All these features have been carried over into Autodesk's AutoLISP making it an extremely power but yet simple language to master.

Diesel (Direct Interpretively Evaluation String Expression Language) is a recent addition to the Autodesk family of programming languages. It offers the programmer the capability of alter the AutoCAD status line.

Visual LISP for AutoCAD 2000 is an extension of the AutoLISP programming language, in which the capabilities have been extended so that object interface can now be accomplished using the Microsoft ActiveX Automation interface. The package also includes a set of programming tools that now include an integrated development environment, a compiler and debugger. Even though this is a new product developed by Autodesk, it is however an extension of the AutoLISP programming language. Therefore any student wishing to develop skills using Visual LISP as a programming tool must also become well versed in the use of AutoLISP commands.

This textbook is designed to take a little different path in teaching the student the principles and techniques of AutoLISP programming. It's main objective is to teach programming from a design standpoint of view. In other words the student will be learning the theory of programming while applying the concepts learned to developing AutoCAD into a powerful design application, instead of applying them to improve AutoCAD drafting ability. The textbook starts with a basic introduction to problem solving techniques, along with an introduction to programming, in which the student combine these skills when faced with the extensive coverage of the AutoLISP programming language. Once the student has somewhat mastered AutoLISP programming, then the focus is shifted to enhancing their programming abilities by introducing Diesel and DCL in which they learn to incorporate graphics into their programs, making them more user friendly. Finally, the student is exposed to Visual LISP. It is imperative that the student has a firm grasp and understanding of AutoLISP principles and techniques before starting this section. This is because the basic functions and theory that applies to AutoLISP programming are carried over into Visual LISP. Therefore, even though Visual LISP is new, it is really an extension of the AutoLISP programming language, giving the programmer editing, and debugging features along with other tools that can be used to create AutoLISP programs.

FEATURES OF THIS EDITION

- Extensive coverage of the concepts and theories of computer programming using AutoLISP

- Introduction to the Windows 95/98 and NT registry

- Introduction to event driven programming using DCL

- Integrating Visual Basic GUI into an AutoLISP program

- Introduction to Microsoft's ActiveX

- Introduction to the basic concepts and theories of Object Oriented Programming.

- Coverage of the concepts and theories regarding the use of Reactors

- Chapter review questions and exercises designed for all levels

HOW TO USE THIS BOOK

This book is designed to be used in a course ranging in length from eleven weeks to sixteen weeks. It is not intended to teach students the basics of AutoCAD or even design theory, but instead it is designed to aid the student in the development of an understanding of the theory and principles of AutoLISP programming. Upon completion of this text the student will be able to combine the principles of AutoLISP programming and their level of knowledge in design theory to transform AutoCAD into the ultimate design package, tailored to a specific industry. Therefore a basic understanding of design and AutoCAD is necessary before this text can be effectively used. Additional information on the use of AutoCAD can be found at Delmar's web site http://www.autodesk.com. The success of this book is largely dependent upon the amount of time the reader is willing to spend behind the computer working out the example programs and exercises located in the book. The material covered in this textbook, exercises, review questions, and practice problems are designed to work with AutoCAD 2000.

A CD-ROM has been included to help the reader set up the example programs located in the text. On this disk are example drawings, AutoLISP source code files, DCL source code files, Complied AutoLISP applications, Visual Basic forms files, Visual Basic project files and Visual Basic Project Workspace files. All drawings contained on the student CD-ROM are saved in AutoCAD 2000 format.

TYPOGRAPHIC CONVENTIONS

In order to make this text easier to use, we have adopted specific typographic conventions which help to distinquish various elements found throughout the book. The following table lists the each text element and gives an example of how each element would appear in the text.

Text Element	Example
Program code examples are shown in monospaced type.	`(SETQ point (LIST 5 4 3))` `(princ "\nEnter Resistance to start program ")`
Command prompts are indented and shown in a non-serif typeface.	Command: **LINE**
Input to be entered is shown in boldface.	Specify first point: **!point**
Instructions after prompt sequences are enclosed in parentheses.	Select objects: *(Use an object selection method.)*
Keys you press on the keyboard are shown in small caps.	CTRL, F12, ESC, ENTER
The plus mark denotes keys you press simultaneously on the keyboard.	CTRL+C, SHIFT+F1
File and folder names are shown are enclosed in quotes when referred to in a sentence.	Double-click the file name "spot_elevation.lsp". ...the directory is "C:\ProgramFiles\ACAD2000"
AutoLISP function names are shown in small caps.	...argument for the FILL_TILE function.
AutoCAD commands and system variables are shown in small caps.	...using the AutoCAD LINE command. DIMBLK, DWGNAME, LTSCALE

WE WANT TO HEAR FROM YOU

If you have any questions or comments about this text, please contact:

The CADD-Drafting Team
c/o Delmar, Thomson Learning
3 Columbia Circle
P.O. Box 15015
Albany, NY 12212-5015
www.autodeskpress.com
www.cadd-drafting.com

ABOUT THE AUTHOR

Kevin Standiford received his Bachelor of Science in Mechanical Engineering Technology from the University of Arkansas at Little Rock. He has over seven years combined experience consulting in the fields of mechanical, electrical, and civil engineering. He has developed several short courses, seminars, and college classes, including Introduction to AutoCAD, Mechanical Desktop, AutoLISP Programming, Computer Numeric Control, Descriptive Geometry, Mechanical Drafting and Design, (Mechanical Desktop 3.0 & CNC), Electrical/Electronics Drafting, Algebra and Trigonometry, and Professional Project Management. In addition, he was been developing instructional material for the past five years for ITT Educational Services Inc.

ACKNOWLEDGMENTS

The author would like to thank and acknowledge the many professionals who reviewed the manuscript, helping us publish this textbook. The technical edit was performed by John Caplinger, Burrough-Brasuell Corp.

A special acknowledgment is due to the following instructors, who reviewed the chapters in detail:

Richard Gold, Yale School of Drama, New Haven, CT

Bill Kramer, Auto-Code Mechanical, Dublin, OH

The following figures, tables and lists were obtained from Autodesk's Visual LISP Developers Guide:

Page 40, Table 2-2: Escape Codes for GETXXX Functions

Page 56, Table 2-4: Legal Control Codes for PRXXX Functions

Page 157-158, Restrictions and Rules for Symbol Tables (List)

Page 160-161, Restrictions and Rules or Dictonary Objects (List)

Page 163, Table 5-2: Selection Mode Options

Page 190-192, Table 6-1: DIESEL Functions

Page 232, Table 7-1: Action Expression Variables

Page 236, Table 7-2: Symbol Names for Color Attributes

Page 268-269, Table 8-1: Visual LISP Menu Categories Summary

Page 273, Table 8-2: Visual LISP Toolbar Categories Summary

Page 321, Table 8-3 Trace Stack Elements

Page 326, Symbol Flags Options (List)

Page 361, AutoCAD Object Model (Figure)

Page 372-373, Table 9-1: Data Types and Visual LISP Functions Used to Create Them

Page 402-403, Table 10-1: Editor Reactor Identification

Page 405-406, CallBack Functions Guidelines and Restrictions (List)

The author would also like to acknowledge and thank the staff members of Delmar Publishers for their help and support. A thanks also goes to John Shanley of Phoenix Creative Graphics for his help in the production of this book.

DEDICATION

This book is dedicated to my loving and caring wife, Debrah, who supported me through all the long hours. Without her support, and understanding, this project would not have been possible.

Kevin Standiford

CHAPTER 1

Introduction to Programming

OBJECTIVES

Upon completion of this chapter the reader will be able to:

- Describe the difference between Notepad and Wordpad
- Describe the difference between a flowchart and pseudo-code
- Define the eight steps used in problem solving
- Write a basic AutoLISP expression
- Load an AutoLISP expression using the LOAD function and the AutoCAD APPLOAD command
- Define functions and AutoCAD commands using the AutoLISP DEFUN function
- Use comments to describe important operation, steps, and conditions in a user-defined function or sub-function

KEY TERMS AND AUTOLISP PREDEFINED FUNCTIONS

;	Function	Rich Text Format
ANSI	GUI	Software
APPLOAD	Hardware	Source Code
Argument	LOAD	Sub-Function
AutoLISP	Machine Language	Sub-Routine
Binary	Main Function	Syntax
Comment	Non-GUI	Text Document
Common LISP	Notepad	Text Document MS-DOS Format
Compiler	Operating System	Unicode Text Format
DEFUN	Predefined Function	User Defined Function
Expression	Program	Wordpad
Flowchart	Pseudo-Code	

INTRODUCTION TO PROGRAMMING

The computer revolution has had an impact on every facet of today's modern engineering firm, changing everything from the way a company composes its correspondence to the method used to perform the necessary calculations for a particular project. This impact has changed not only the way everyday operations of a company are performed but also the amount and kind of knowledge its employees are expected to possess. For example, twenty-five years ago an entry-level engineer was expected to have a good understanding of the principles of engineering in addition to a working knowledge of manual drafting. Having those manual drafting skills enabled the engineer to communicate readily with the drafting team. Calculations were carried out with tables and/or in many cases, a slide rule. Design concepts were sketched out by hand on either Vellum or Mylar and then passed down to a drafter, who completed the drawing. This process was very time consuming and at times resulted in mistakes from miscommunication between the design team and the engineer. In today's market, companies have been forced to downsize their overhead (staff) in order to stay competitive within a changing global market. This has changed the requirements for employment set forth by today's industry. Employees are now expected to have a larger range of skills that encompass an expanding job description. All companies now expect their entry-level engineers to have basic computer skills with basic office applications (word processors, spreadsheets, and databases), in addition to a fundamental understanding of engineering principles. Preference is now given to persons entering this field with computer programming experience, because they offer greater value per capita to the company. Of particular interest to today's employers are persons who can customize the AutoCAD workstation for engineering design applications using AutoLISP, Visual LISP, Visual Basic, C++, Visual C++, ActiveX, or ObjectARX. Engineers who can program are not limited to the conceptualization and reproduction of a project using a computer. They are also able to create applications (programs) that can alleviate some of the tedious and costly aspects of the design process.

Before learning how to manipulate a computer to perform specialized tasks, it is necessary review a few basic terms and concepts. The components that make up a computer can be divided into two categories: *Hardware* and *Software*, as illustrated in Figure 1–1. *Hardware* is defined as the physical attributes of a computer, for example, the monitor, keyboard, central processing unit (the part of the computer where the logic and arithmetic operations are performed), and mouse. *Software*, on the other hand, is defined as the programs or electronic instructions that are necessary to operate a computer. A *Computer Program* is an application that is designed to carry out a particular task. Examples of computer programs are AutoCAD, Word, Excel, Access, PowerPoint, and Mechanical Desktop. In fact it is the computer programs that have essentially transformed those machines from an intricate maze of resistors, transistors, and wires, into the powerful tools they are today. The way commands are arranged inside a computer program is known as *Syntax*. The actual commands used to construct a computer program constitute the *Programming Language*. Most programming languages are designed to look like naturally spoken languages. This allows the programmer to construct an application using normal thought processes and real world problem-solving techniques. Examples of computer programming languages are C, C+, C++, Pascal, LISP, AutoLISP, Visual LISP, Diesel, Visual Basic, ActiveX, and ObjectARX. When a programmer completes a program, it is converted from the language in which it was created to a format that the computer can understand. This conversion process is carried out by a program called a *Compiler*. The format that the compiler transforms the program into is called *Machine Language*, which is based on the binary system. This system is comprised of 1s and 0s; 1 means on and 0 means off. All computers are controlled by a main program called the *Operating System*. The operating system is nothing more than a set of programs that manages stored information, loads and unloads programs (to and from the computer's memory), reports the results of operations requested by the user, and manages the sequence of actions taken between the computer's hardware and software. Examples of operating systems are Windows 2000, Windows NT 4.0, Windows 98, Windows 95, OS2, Warp, and DOS. Operating systems can also be placed in one of two categories: *GUI* (*G*raphical *U*ser *I*nterface) and *Non-GUI* (*Non-*G*raphical *U*ser *I*nterface). A *Graphical User Interface* operating system allows the user to interact with a computer by selecting icons and pictures that represent programs, commands, data files, and even hardware (see Figure 1–2). A *Non-*G*raphical *U*ser *I*nterface operating system relies on the user to enter commands through the keyboard (see Figure 1–3).

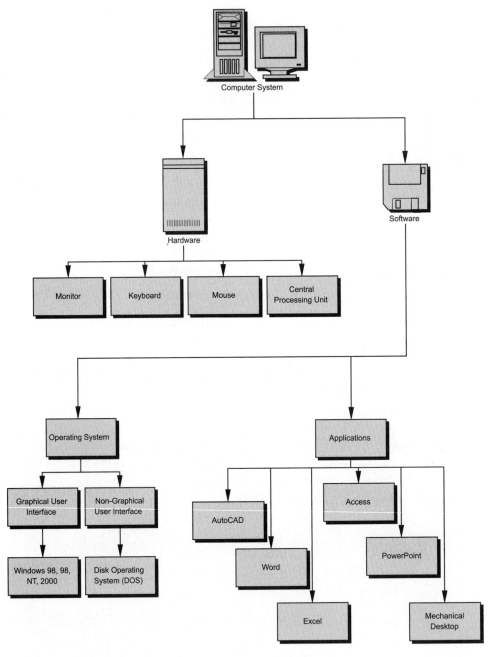

Figure I-I *The components of a computer*

Figure 1–2 *A GUI operating system*

Figure 1–3 *A Non-GUI operating system*

WHAT IS AUTOLISP?

AutoLISP is a programming language developed by Autodesk that is designed to assist the user in increasing productivity by simplifying multiple-step tasks into a routine that requires a single command to execute (in AutoLISP they are often referred to as functions). AutoLISP was initially released in May, 1985, as a separate plug-in package for AutoCAD. It wasn't until January, 1986, that it was integrated into the AutoCAD version 2.18 package, thus making it available to all AutoCAD users. The language is a variation of the common LISP programming language; functions have been added that permit AutoLISP to communicate directly with the AutoCAD drawing environment. LISP, an acronym for List Processing, was developed in the late 1950s by John McCarthy. Its primary use in the early days of development was for the manipulation of data. Today it is used extensively in the fields of research and development as well as being the industry standard for the development of artificial intelligence.

PROGRAMMING TECHNIQUES

Developing a computer program begins with a need. This need can be anything from the automation of a tedious drafting task to the development of a design tool that could aid the engineer with any of the stages of a project. Once the need has been assessed, the programmer can then start the planning phase of the project. This is perhaps the most important phase of any project and should not be taken lightly. The creation of a well-developed and successful program is no accident. Sufficient time should taken in the planning and preparation of the program before a single line of code is actually typed. It is crucial that the programmer get into a habit of planning first and writing the code last. The approach taken in this book is a two-phase process. The first phase is an eight-step procedure that can be implemented to solve any type of technical problem that an engineer, architect, or computer programmer might encounter. Traditionally, this phase has been associated more with the engineering and architectural fields. Because it is included in the planning process covered in this book, however, the programmer becomes better prepared to create programs that are targeted for design applications. This first phase might not seem helpful in the creation of a simple program designed to alleviate a tedious drafting technique, but nevertheless, it is a good habit.

PHASE ONE – EIGHT STEPS TO SOLVING A DESIGN PROBLEM

When solving a technical problem, the engineer does not start by pulling out a calculator and entering numbers; there is a systemic approach that must be taken. This approach starts by defining and then researching the problem. Once sufficient information has been ascertained, the engineer can start determining all possible solutions to the problem. The following steps required for successful problem solving can, in many instances, be executed in order.

1. State the problem

2. List unknown variables

3. List what is given

4. Create diagrams

5. List all formulas

6. List assumptions

7. Perform all necessary calculations

8. Check answer

The problem as shown in Figure 1–4 is used to explain and illustrate each step of this process. Not all steps may apply to all situations, but the overall concept is still the same.

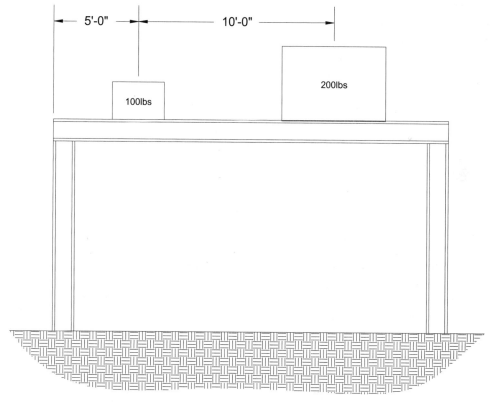

Figure 1–4 *A beam positioned between two upright structural members. For simplicity, the weight of the beam has been omitted.*

Step 1 **State the problem**

In this step, a statement of the problem is created. The statement should be kept as simple and direct as possible. There is no need to list every detail about the problem, just the key points. Additional information can be added later if necessary. A statement for the problem shown in Figure 1–4 might read:

Statement:

Given the load of the two forces acting along the negative Y-axis and their distance from one end of the beam and each other, determine the reactions necessary to keep the beam in equilibrium.

Step 2 **List unknown variables**

Create a list of all the unknown attributes for the stated problem (even though the problem is in its infancy and several minor calculations might have to be executed before all the main components that are to be found can be listed). Be sure to provide ample space for any future elements that might require calculations. In the problem statement above, it is clear that the two reactions (forces) that keep the beam in equilibrium must be calculated. Therefore, the list should look like:

Find:

Reaction #1

Reaction #2

Step 3 **List what is given**

A list of all known attributes associated with the stated problem is created. For example, using Figure 1–4 and the statement from step 1, the following list can be developed:

Given:

Force1 = 100 lbs.

Force2 = 200 lbs.

Distance from Force1 to the End of the Beam = 5'

Distance from Force1 to Force2 = 10'

Distance from Reaction #1 to Reaction #2 = 19'

Note: In this example, the information is already provided. In the case of a computer program, the end user will provide this information, and the list would take on the following appearance:

Given:

Force1 = Prompt user for input.

Force2 = Prompt user for input.

Distance from Force1 to the End of the Beam = Prompt user for input.

Distance from Force1 to Force2 = Prompt user for input.

Distance from Reaction #1 to Reaction #2 = Prompt user for input.

Step 4   Create diagrams

Often the best way to determine exactly what is going on in a problem is to make a simple sketch. The sketch should be void of any unnecessary details that might hinder the interpretation of the actual problem. The sketch should serve as a visual guide or reference for the elements contained within the system and how they interact with one another. Figure 1–5 is a sample sketch showing the main components of the system from Figure 1–4.

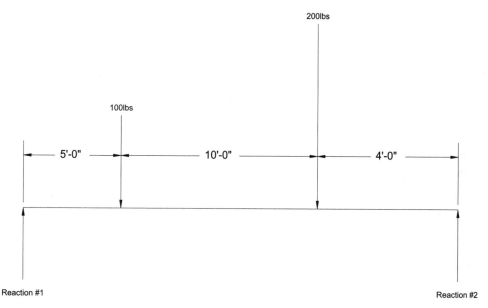

Figure 1–5 *Simple sketch of load problem*

Step 5   List all formulas

In this step, a list is created of all the formulas that will be used to solve the problem, as well as a source reference for each formula. This list will serve as a guide that will facilitate the final answer check. When a technical solution is validated, all stages of the solution are verified, including formulas and their connotation. To solve the problem shown in Figure 1–4, the following formulas will be used:

Formulas:

Formula	Source	Page
$\sum F_y = 0$	Physics: Algebra/Trig 2nd edition, Delmar Publishers, 1997	115
$\sum F_x = 0$	Physics: Algebra/Trig 2nd edition, Delmar Publishers, 1997	115
$\sum M_a = 0$	Physics: Algebra/Trig 2nd edition, Delmar Publishers, 1997	115

Step 6   List assumptions

Often, when designing an application it is necessary to assume some of the details of the project. Suppose that an engineer is calculating the amount of heat that is transferred by conduction through the exterior walls of a building to its surroundings. Although the average outside temperature can be obtained from the ASHRAE *Fundamentals Handbook* for all major cities in the United States, it may be necessary to assume a temperature if the area where the building is located is not listed. In the case of the problem shown in Figure 1–4, an assumption can be made that all forces acting in a downward direction are negative and that all moments are acting in a counterclockwise direction.

Assumptions:

All forces acting in a downward direction are negative.

All moments causing a counterclockwise rotation are negative.

Step 7   Perform all necessary calculations

The necessary calculations are carried out, or in the case of a computer program, the formulas are rearranged so that they can be used later. Solving the problem illustrated in Figure 1–4 will yield the following calculations:

Solving for Reaction #2

$\sum M_a = 0$

$\sum M_a$ = -((Reaction #2)(Distance #3)) + (Force #1)(Distance #1) + (Force #2)(Distance #2)(Reaction #2)(Distance #3) = (Force #2)(Distance #1) + (Force #2)(Distance #2)

Reaction #2 = ((Force #1)(Distance #1) + (Force #2)(Distance #2))/ (Distance #3)

Inserting the Values Listed from the Given Table:

Reaction #2 = ((100 lbs.)(5') + (200 lbs.)(10'))/(19')

Note: Reaction #2 = 131.57 lbs. Answer for Reaction #2

Solving for Reaction #1

$\Sigma F_y = 0$

0 = Reaction #1 + 178.57 lbs. − 100 lbs. − 200 lbs.

100 lbs. + 200 lbs. − 178.57 lbs. = Reaction #1

Note: 168.42 lbs. = Reaction #1 Answer for Reaction #1

In the case of a computer program, the necessary data will not be available to perform the calculations. However, this step is still carried out, resulting in the formulas being rearranged for later use in the actual program. This is illustrated in the following example:

Solving for Reaction #2

$\Sigma M_a = 0$

ΣM_a = -((Reaction #2)(Distance #3)) + (Force #1)(Distance #1) + (Force #2)(Distance #2)(Reaction #2)(Distance #3) = (Force #2)(Distance #1) + (Force #2)(Distance #2)

Reaction #2 = ((Force #1)(Distance #1) + (Force #2)(Distance #2))/ (Distance #3)

Solving for Reaction #1

$\Sigma F_y = 0$

0 = Reaction #1 + Reaction #2 − Force #1 − Force #2

Force #1 + Force#2 − Reaction #2 = Reaction #1

Step 8   **Check answer**

Any time a calculation is made, it must be checked for accuracy. This will ensure that the program is running correctly and that the information the program supplies to the user is valid. To illustrate this point, the answers obtained in the previous section are applied to the formula $\Sigma F_y = 0$.

According to this equation, all forces acting on the beam, once added together, must equal zero for the beam to remain in equilibrium.

Check:

$\Sigma F_y = 0$

-100 lbs. − 200 lbs. + 131.57 lbs. + 168.42 lbs. = 0

Note: If the values are not available during this phase, a formula to check the answers should be derived.

Reaction #1 + Reaction #2 − Force #1 − Force #2 = 0

Putting It All Together

To better illustrate how this would actually appear once on paper and without the explanations of each step, the following example is provided.

Statement:

Given the load of the two forces acting along the negative Y-axis, and their distance from one end of the beam and each other, determine the reactions necessary to keep the beam in equilibrium.

Given:

Force1 = 100 lbs.

Force2 = 200 lbs.

Distance from Force1 = 5'

Distance from Force1 to the end of the beam Force2 = 10'

Distance from Reaction #1 to Reaction #2 = 19'

Find:

Reaction #1

Reaction #2

Diagram:

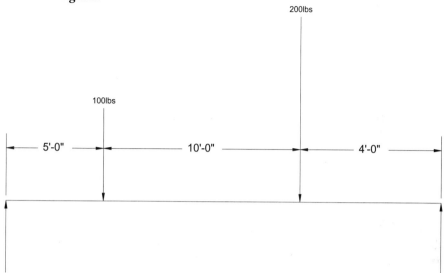

Formulas:

Formula	Source	Page
$\sum F_y = 0$	Physics: Algebra/Trig 2nd edition, Delmar Publishers, 1997	115
$\sum F_x = 0$	Physics: Algebra/Trig 2nd edition, Delmar Publishers, 1997	115
$\sum M_a = 0$	Physics: Algebra/Trig 2nd edition, Delmar Publishers, 1997	115

Assumptions:

All forces acting in a downward direction are negative.

All moments causing a counterclockwise rotation are negative.

Solving for Reaction #2

$\sum M_a = 0$

$\sum M_a$ = -((Reaction #2)(Distance #3)) + (Force #1)(Distance #1) + (Force #2)(Distance #2)(Reaction #2)(Distance #3) = (Force #2)(Distance #1) + (Force #2)(Distance #2)

Reaction #2 = ((Force #1)(Distance #1) + (Force #2)(Distance #2))/(Distance #3)

Inserting the Values Listed in the Given Table:
Reaction #2 = ((100 lbs.)(5') + (200 lbs.)(10'))/(19')

 Note: Reaction #2 = 131.57 lbs. Answer for Reaction #2

Solving for Reaction #1
$\Sigma F_y = 0$

0 = Reaction #1 + 178.57 lbs. − 100 lbs. − 200 lbs.

100 lbs. + 200 lbs. − 178.57 lbs. = Reaction #1

 Note: 168.42 lbs. = Reaction #1 Answer for Reaction #1

Check:
$\Sigma F_y = 0$

-100 lbs. − 200 lbs. + 131.57 lbs. + 168.42 lbs. = 0

PHASE TWO – PLANNING THE SOURCE CODE

After phase one of the planning process has been completed, it is time to move on to the planning stage of the source code (the non-compiled version of a computer program). As with phase one, adequate preparation ensures that the code is written with the greatest possible accuracy and efficiency. It also reduces the amount of time required to complete the program. Start this process by describing the key elements of the main program and their sequence in relation to one another. It is not necessary to completely describe how each aspect will operate; instead use generalized statements or a general overview and build from there. There are two different ways to approach this process: the programmer can either use flowcharts or pseudo-code.

Flowcharts

Flowcharts, the older of the two methods, incorporates the use of graphic symbols to describe a sequence of operations or events. Used frequently to describe a process performed by a computer, flowcharts are also incorporated into almost every branch of engineering, from civil to mechanical. They are used to describe everything from the process involved in the treating of wastewater to the sequence of operations for a microprocessor. A typical example of a flowchart is shown in Figure 1–6.

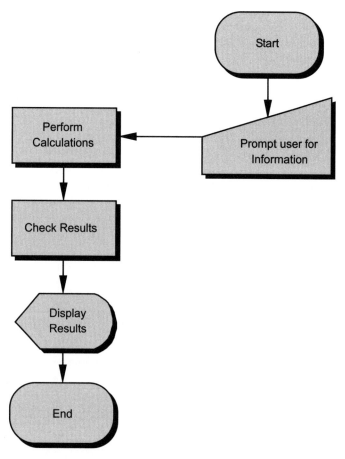

Figure 1–6   *Flowchart of the load problem in Figure 1–5*

In this example, the program is broken down into four main operations and arranged in the logical order that they are encountered. The program starts by prompting the user for the data need to complete the calculations. Next, it performs the calculations using the information that was supplied by the user in the previous step. The program then checks the results returned by the calculations. Finally, the answer is displayed to the user, either on the monitor or with the information sent to a printer (hardcopy). Once this has been completed, the programmer can go back and begin adding more detail by expanding the flowchart. For example, the section where the user is prompted to enter information can be expanded to include exactly what data is needed to perform the calculations (see Figure 1–7).

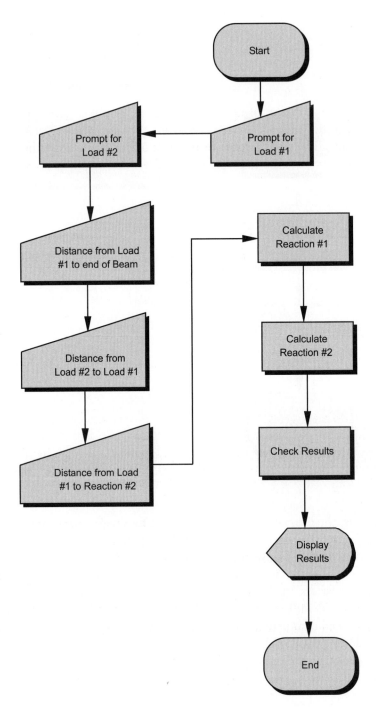

Figure 1–7 *Flowchart expanded to show prompts*

Note: This process is repeated until all aspects of the program have been defined in their proper sequence.

Pseudo-Code

A more natural way of preparing the source code for a computer program is with pseudo-code. In this method, standard English terms are used to describe what the program is doing. The descriptions are arranged in the logical order that they appear. An example of how this method could be used in place of the previous flowchart is shown below.

> Start Program
>
>> Prompt user for Information
>>
>> Perform Calculations
>>
>> Check Answers
>>
>> Display Answers
>
> End Program

 Note: Notice the similarities between the two methods. The main difference is that the flowchart method incorporates the use of graphic symbols.

Once the key elements of the main program have been established, the programmer can start expanding the pseudo-code. Using the previous pseudo-code layout and the example in Figure 1–5, the programmer would start identifying what data is needed to complete the calculations as well as what formulas will be used in the calculations (this information would be taken from step 5 of phase one).

> Start program
>
>> Prompt User to "Enter Load #1 : "
>>
>> Prompt user to "Enter Load #2 :"
>>
>> Prompt User to "Enter Distance from end of Beam to Load #1 :"
>>
>> Prompt User to "Enter Distance from Load #1 to Load #2 :"
>>
>> Prompt User to "Enter Distance from Reaction #1 to Reaction #2 :"
>>
>> Using the Formula $\sum M_a = 0$ and Load #1, Load #2, Distance #1 and Distance #2, Calculate Reaction #2.
>>
>> Using the Answer from the Previous Step and the Formula $\sum F_y = 0$, Calculate Reaction #1.

If the sum of all the Forces in the Negative Y Direction are Equal to the Sum of all the Forces in the Positive Y Direction Then

Display Answers Calculated for Reaction #1 and Reaction #2

End Program

 Note: When the initial source code of a program is planned with either flowcharts or pseudo-code, ample space between entries should be provided for future additions or any oversight that might have occurred from the original process. It is also recommended that the programmer complete this phase of the planning process using a pencil. This will allow the programmer to easily change or correct any portion of the program that may need editing later.

WRITING AN AUTOLISP EXPRESSION

Once the planning phases have been completed, it is time to start writing the actual AutoLISP program. AutoLISP expressions can be brought into the AutoLISP environment from either the keyboard or a file. All AutoLISP expressions follow the same basic format or syntax, which is a (function argument). In this particular arrangement, the function is the actual AutoLISP command used to carry out a particular task. The argument is the information required to proceed with a particular operation. An argument can be in the form of text, numbers or even other expressions. The basic AutoLISP expression is contained within a set of parentheses. These parentheses are used to group together a function and its arguments, a function and another expression, or a function and a group of expressions. (Examples of valid expressions, functions, and arguments are shown in Table 1–1.)

Table 1–1
Examples of Valid AutoLISP Expressions, Functions, and Arguments

Expressions	Description
(SIN 45)	Returns the value of SIN of 45° in Radians
(+ 1 2)	Combines the Integers 1 and 2
(* 1 2 3)	Returns the product of the integers 1, 2 and 3
Functions	
Sin	
+	
*	

Argument(s)
45
1 and 2
1, 2 and 3

Result Returned
0.850904
3
6

Examples of different ways an AutoLISP expression may be arranged are shown below.

```
(Function1 Argument1)
```

This first example shows a basic AutoLISP expression containing a function and an argument. The argument in this example can be either a single argument or a list of arguments.

```
(Function1 (Function2 Argument2))
```

This second example uses an expression as the argument for function1. In this case, function2 evaluates argument2 and passes the result to function1 in the form of an argument.

```
(SIN (+ 1 2))
```

Once the expression (+ 1 2) is evaluated, its result (3) is passed as an argument to function1 producing (SIN 3) which returns 0.14112.

```
(Function1 (Function2 Argument2)(Function3 Argument3))
```

The third example contains two expressions embedded within a single expression. In this example, function2 evaluates argument2 and function3 evaluates argument3. The results returned by both function2 and function3 are then used as arguments for function1.

```
(+ (+ 2 3) (+ 4 5))
```

The result of (+ 2 3) is 5 and the result of (+ 4 5) is 9, producing (+ 5 9) which returns 14.

```
(Function1 (Function2 (Function3 Argument3))(Function4 Argument4))
```

This last example contains two expressions embedded within a single expression, and one of these expressions also contains an embedded expression. In this example, function3 evaluates argument3 and passes the result to function2, where it is used as an argument. Then function 2 and function 4 each evaluate their respective arguments and pass the result to function 1 as arguments.

```
(+ (SQRT (* 7 10))(SQRT 100))
```

The result of (* 7 10) is 70. This is passed to the expression (SQRT 70) resulting in 8.366. This and the result of the expression (SQRT 100) which is 10, are passed to the expression (+ 8.366 10) resulting in 18.366.

When an expression has been entered into AutoCAD (from either the keyboard or a file), it is passed to the AutoLISP interpreter. The interpreter sends the expression to the evaluator, which analyzes the expression and the result is returned. Once all expressions have been evaluated, AutoCAD reestablishes control and the command prompt reappears. For example, entering the expression (+ 1 2) at the AutoCAD command prompt yields:

Command: **(+ 1 2)** (ENTER)
3 *(The result of the expression.)*
Command:

If the expression had contained an embedded expression, as in the case of the second example, then the result displayed by AutoLISP before the command prompt reappeared would be the result produced by the last function (expression) evaluated. For example:

Command: **(SIN (+ 1 2))** (ENTER)
0.14112 *(Returned value from expression.)*
Command:

If a result is not returned and the command prompt is replaced with a *n>* (where *n* is equal to an integer value) then the expression is unable to be evaluated because it has not been closed; the expression is missing *n* number of right parentheses (see Figure 1–8). To close the expression and return a value, simply add the required number of missing parentheses. For example:

Command: **(+ 1 2** (ENTER)
(_>) (ENTER)   *(Indicates that one right parenthesis is missing.)*
3 *(Returned value from expression.)*

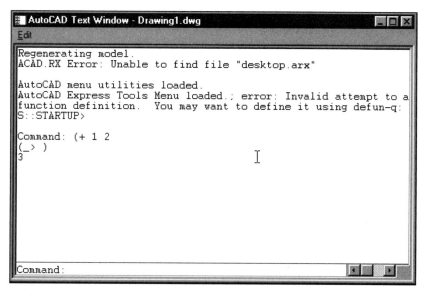

Figure 1–8 *AutoCAD text window*

SPACING AND ALIGNMENT ISSUES

Traditionally, AutoLISP expressions are written with a single space between the function and its arguments. However, if multiple spaces are used, the expression is evaluated as if a single space had been entered. Also, all carriage returns are evaluated as a single space. Therefore, in the example above, the result returned from AutoLISP is still the same whether it is entered from the keyboard as:

Command: **(+ 1 2)** (ENTER)

or

Command: **(+ 1 2** (ENTER)
(_>) (ENTER)

SAVING AN AUTOLISP EXPRESSION

As stated earlier, an AutoLISP expression can be entered from the keyboard or saved in a file along with other expressions and loaded into AutoCAD at any time. When a group of expressions are saved in a file, that file must be an ASCII (American Standard Code for Information Interchange) text file. These files can be created by any word processor, as long as the word processor has the ability to save files in an ASCII format. An ASCII text file, unlike a file created by Microsoft Word, contains no formatting. This makes ASCII text files universal, in the fact that they can be transferred to different operating systems and can be used by almost every program.

Tip: When writing an AutoLISP program, it is recommended that the .LSP extension be assigned to the file. Also, it is important to choose a file name that somewhat describes or has some meaning to the program itself. For example, a program that calculates the amount of heat transferred through a wall due to conduction, a possible name for the program could be *HEAT.LSP*.

WINDOWS-BASED WORD PROCESSING PROGRAMS

Two text-editing applications can be used in creating an AutoLISP program file: Notepad and Wordpad. Because these two programs are capable of saving files in ASCII format and both are included with the Windows 95, 98 and NT operating systems, it will be these programs that will be utilized in the first half of this book. However, feel free to use any word processing application that saves files in the ASCII format.

Notepad

Notepad is a word processing application that saves files up to 64 K in size and only in the ASCII format. Launch it by selecting the **Start** menu followed by **Programs**, then **Accessories**, and then **Notepad** (see Figure 1–9). Once the application has been started and the text has been entered, save the file with the extension .LSP by selecting **File** from the menu bar and then **Save As**. A dialog box will then appear and a file name can be assigned. Even though the application saves files in ASCII format only, if the file type is not changed from the default setting (**Text Documents**) to **All Files**, Notepad will automatically assign an extension of .TXT (see Figure 1–10).

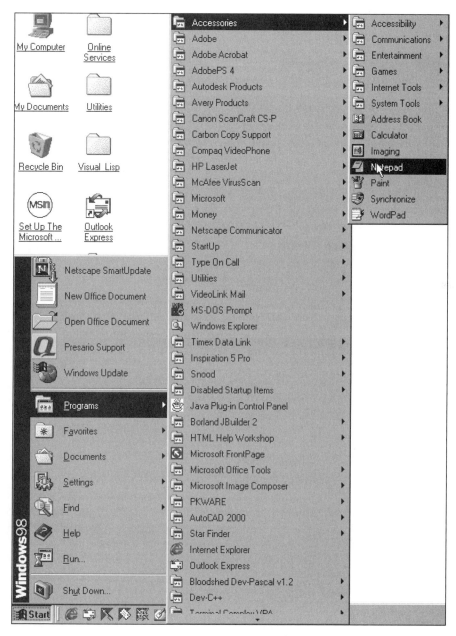

Figure 1–9   *Starting Notepad from the Windows Start Menu*

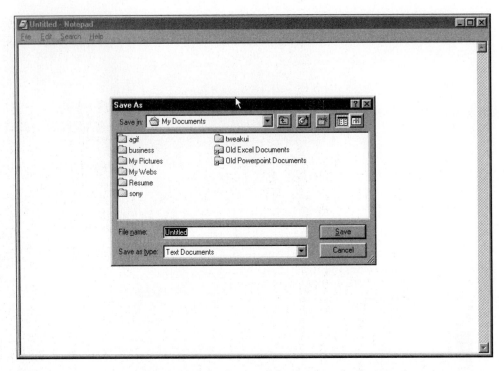

Figure 1–10 *Windows Notepad dialog box used for the Save As Command*

Wordpad

Just as for Notepad, select the **Start** menu followed by **Programs**, then **Accessories**, and then **Wordpad** to launch the application (see Figure 1–11). Wordpad differs from Notepad in the respect that it can handle files larger than 64 K as well as save documents having a variety of fonts and paragraph styles. In addition, the application allows the user to save files in one of five different formats: Word for Windows 6.0, Rich Text Format (RTF), Text Document, Text Document – MS-DOS Format, and Unicode Text Document (see Figure 1–12).

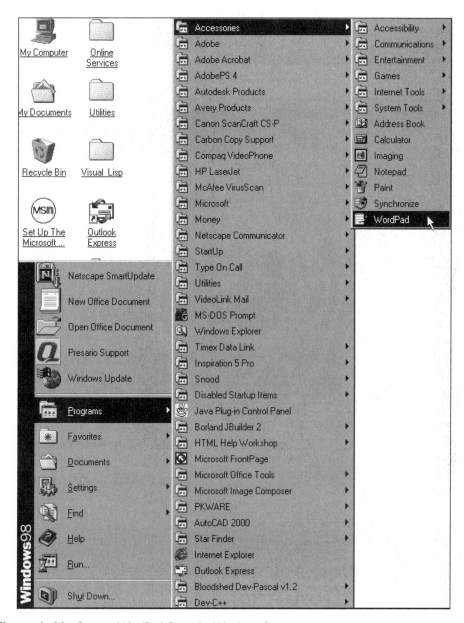

Figure 1–11 *Starting WordPad from the Windows Start menu*

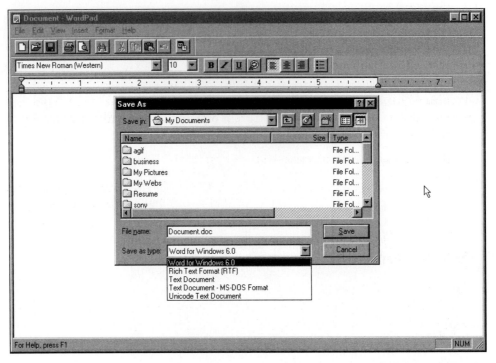

Figure 1–12 *WordPad File Format options*

The **Word for Windows 6.0** option saves the file in a Microsoft Word (Office 95) format, allowing the file to be opened in either Office 95 or Office 97. The **Rich Text Format** is a standard developed by the Microsoft Corporation; the files are saved in the ASCII format with special commands inserted to describe important formatting information (fonts and margins). The **Text Document** option allows files to be saved without formatting using ANSI standards. Like the **Text Document** option, the **Text Document – MS-DOS Format** also saves file without any text formatting; however, this option saves using the ASCII standard and should be used when files are shared with non-Windows applications. Finally, the **Unicode Text Document** option saves files using the Unicode standard. This standard was developed more for the Chinese and Japanese languages. It is a 16-bit standard as opposed to ASCII, which is an 8-bit standard.

DEFINING A FUNCTION

The key to writing an AutoLISP program (also referred to as a user-defined function) is being able to write, combine and group expressions together in an application that serves some useful purpose. Just combining and grouping expressions together is not enough; they must also be controlled so that once a program is loaded into memory it does not automatically execute. If the program is executed automatically upon loading, then each time that program is to be used it must first be loaded. For smaller programs, this is not a problem. However, for larger programs, this could become quite a time-consuming operation. This problem can be solved with the AutoLISP DEFUN function. The DEFUN function is an acronym, like many AutoLISP functions, which stands for "define function." Once a program has been defined, it can be recalled at any time during the current drawing session without its having to be reloaded before each use. The syntax for the function is:

```
(DEFUN symbol-name argument-list expressions)
```

The first component supplied to the DEFUN function is the *symbol name*, the name that will be used to launch the program. Symbol names are not case sensitive; therefore, it has no effect on the program whatsoever if the symbol name is uppercase, lowercase or mixed case. It is a string of characters that can be letters, numbers, or a combination of the two. Unlike other predefined AutoLISP functions that require a string, the name is *not* set in quotation marks; AutoLISP automatically does it. If the programmer does set the name in quotes, the *symbol name* becomes invalid and the program will not execute. Once the symbol or program name has been defined, execute the program by simply entering the name of the program set in parentheses at the AutoCAD command prompt. For example, suppose that a program called "test" has been created and loaded into AutoCAD:

```
(DEFUN test ().
```

To execute this program the user would enter **(test)** at the AutoCAD command prompt.

```
Command: (TEST)  (ENTER)
```

Symbol names are treated as AutoCAD commands if the prefix C: is placed in front of the symbol name. With the C: placed in front of the function name test, the user no longer has to enclose the name in parentheses.

```
(DEFUN C:test ()
```

Once loaded into memory, the user would simply type **test** at the command prompt to start the program.

```
Command: test (ENTER)
```

Tip: When naming a user-defined function, the programmer should select a name that is indicative of the operation for which the function is designed. The symbol name of a user-defined function does not have to match the file name; however, it will help ensure that the user does not have problems remembering the function name and can easily launch the program.

Often a program will consist of a main function or program and several sub-functions, also called sub-routines. The sub-functions or sub-routines often contain expressions that are used more than once by the main program. Therefore, sub-routines can save valuable time in creating the program and reduce the amount of source code necessary. When a sub-routine is defined to be used by the main program or even another sub-routine, the sub-routine's symbol name must not have the C: prefix. In other words, a user-defined function cannot call another user-defined function if the function contains the C: prefix.

The second component of the DEFUN function is the *argument list*. This is a list of either local or global variable names that will be used by the program. The variable names are contained within a single set of parentheses. Whether or not any variable names have been declared, these parentheses are still required. If these parentheses are left empty, it is referred to as an empty set. For example:

```
(defun XXX ( )              ;The user-defined sub-function is
                           ;named XXX with no variables
                           ;declared.
```

or

```
(defun C:XXX ( )            ;The user-defined main function is
                           ;named XXX no variables have
                           ;been declared.
```

or

```
(defun  C:XXX ( Variable names )   ;The user-defined main function is
                                  ;named XXX with variable names
                                  ;declared.
```

 Note: For more information regarding variables, see Chapter 2.

The third component necessary to define a function is *expressions*. The expressions are grouped together using parentheses. For example:

```
(defun XXX ( )                      ;Defines  a  function.
     (function argument argument)    ;Expression.
     (function argument argument)    ;Expression.
     (function argument argument)    ;Expression.
)                                   ;End user defined function.
```

or

```
(defun C:XXX ( )                     ;Defines  a  function.
     (function argument argument)    ;Expression.
     (function argument argument)    ;Expression.
     (function argument argument)    ;Expression.
)                                   ;End user defined function.
```

ADDING COMMENTS TO A FUNCTION

A necessary element in all programming, whether it is AutoLISP, Visual LISP, Visual Basic, or ObjectARX, is documentation. This includes the planning and preparation that goes into the program at the beginning as well as documenting important features, functions, and operations that may be going on inside a program. When documentation is used inside a computer program, it is known as a comment. A comment is the description placed in a program for the sole purpose of aiding the programmer in keeping track of what a program is doing. When the AutoLISP evaluator comes across a comment, it automatically ignores it and the remaining expressions are evaluated. To insert a comment in a program, simply place a semicolon before the comment. For example:

```
(defun c:XXX ( )                            ; Start XXX function.
        (function argument argument)        ; Prompt user for information.
        (function argument argument)        ; Process information.
)                                           ; End function.
```

Comments can be placed anywhere in a program. They can be inserted in the expression or positioned outside. If a comment is located inside an expression, then semicolons set it off. When the evaluator encounters a semicolon, it ignores everything after it until it reaches either the end of the line or another semicolon. For example:

```
(SETQ pt ;used 3 times; (getpoint "\nSelect insertion point : "))
                                   ;Prompts user to select a point.
(SETQ var (+ 1 3))                 ;Adds 1 and 3 setting their value to var.
```

In the example above, a comment is inserted in the expression reminding the programmer that the variable pt is used three different times. Also, in this example, comments have been inserted between expressions as well as at the end of an expression.

LOADING AN AUTOLISP PROGRAM

Once the AutoLISP program has been written and saved as an ASCII text file, it must be loaded into AutoCAD's memory so that it can be executed at any time. This can be accomplished with the AutoLISP LOAD function or the AutoCAD APPLOAD command. The advantages in using either of the methods depends upon the circumstances that dictate why the programmer or user is loading the program in the first place. If the programmer is loading a program from a menu option or from within another program, then the AutoLISP LOAD function must be used. If the programmer or user is loading the program from the AutoCAD command prompt, it is easier to use the APPLOAD command.

AUTOLISP LOAD FUNCTION

The AutoLISP LOAD function is used to load a program into memory. When a program is loaded, each expression is evaluated and checked for common errors (missing quotation marks, parentheses, periods, etc.) before the program can be executed. If the evaluator detects a common error, the program is not loaded and AutoLISP returns an error message to the AutoCAD command line. If the program does not contain any common errors, then the result of the last expression is returned to the AutoCAD command line. The syntax for the LOAD function is as follows:

```
(load "filename" [onfailure])
```

The *filename* argument is a string representing the name and location of the AutoLISP file to be loaded. It is not necessary to include the .LSP extension when specifying this name. AutoLISP automatically defaults to this extension if one is not supplied. If the AutoLISP program that is to be loaded has an extension other than .LSP, then it becomes necessary to specify that extension. The *filename* argument is enclosed in quotations unless the name is set to a variable, in which case the quotation marks are omitted.

The second argument, *onfailure*, can be a valid function or a string (contained in quotations). If the argument is a valid function, then it is evaluated. The purpose of this argument is to give AutoLISP an alternate action if the LOAD function fails. For example, suppose that the file *test.lsp* is not defined. If the programmer tries to load the file, the following error will occur:

Command: **(load "test")** (ENTER)
Can't open "test.lsp" for input *(Load has failed causing AutoLISP to error out. If the expression had been contained inside a program, then the function would have stopped due to an error.)*
Cancel
Command:

If the *onfailure* argument is a defined function or a sub-routine, then when the programmer tries to load a file that does not exist. the *onfailure* function is evaluated and the error is prevented. For example:

```
(defun onfailure ()
(princ "\nFile not found")
(princ)
)
```

To illustrate how the onfailure argument is handled by AutoLISP, entering the following expression at the AutoCAD command prompt returns the message "File not found".

Command: **(load "test" onfailure)** (ENTER)
File not found *(Load failed but no error has occurred. Again, if this expression had been contained inside a program, then the function would not have stopped due to an error.)*
Command:

If the *onfailure* argument is a string, then the following action will occur when the programmer tries to load a file that does not exist:

Command: **(load "test" "file not found")** (ENTER)

"file not found" *(Load failed but no error has occurred. Again, if this expression had been contained inside a program, then the function would not have stopped due to an error.)*

Command:

When AutoLISP is instructed to load a file and the path is not supplied, AutoCAD searches the AutoCAD library path by default. The AutoCAD library path is a list of directories that contain files necessary for the execution of the AutoCAD program. Any directory can be added to this list that might contain files to be used by AutoCAD or AutoLISP. For example, suppose that a directory containing only AutoLISP program files is created. To add a directory to the library path, the user would select the Files tab after issuing the AutoCAD PREFERENCES command (see Figure 1–13).

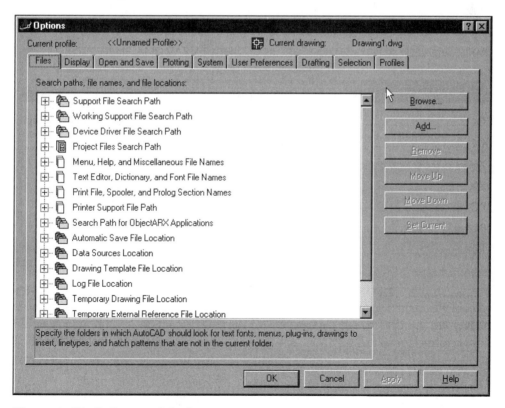

Figure 1–13 *Preferences dialog box*

Typically, when AutoCAD searches for a file, it searches in the following locations in the following order:

1. The working directory

2. The directory that contains the current drawing file

3. Directories specified by the Device Driver File Search Path

4. The directory that contains the *ACAD.EXE* file

To specify the location of a file when using the LOAD function, it is only necessary to supply the drive letter when the file is located on a drive other than the current drive; otherwise always start at the root partition and work up. To indicate a directory prefix, either a single forward slash "/" or two backslashes "\\" can be used. For example, to load the file TEST.LSP from the sub-directory /Programs/LISP, the following syntax would be used:

```
(load "/PROGRAMS/LISP/TEST")
```
or

```
(load "\\PROGRAM\\LISP\\TEST")
```

AUTOCAD APPLOAD COMMAND

The second method for loading an AutoLISP program file is to use the AutoCAD APPLOAD command. This command provides a dynamic dialog box in which the file can be visually selected and loaded. The APPLOAD command is not limited to AutoLISP files; AutoCAD Development Systems (ADS) and ObjectARX (ARX) program files can also be loaded with this command (see Figures 1–14 and 1–15).

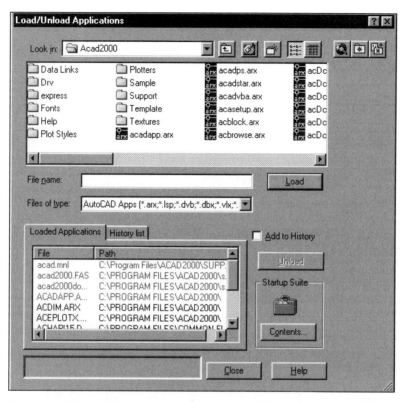

Figure 1–14 *APPLOAD dialog box*

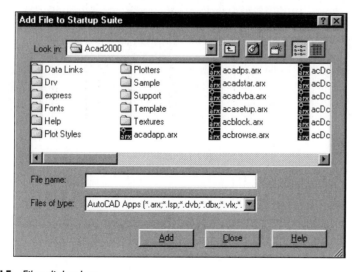

Figure 1–15 *Files dialog box*

SUMMARY

Computer systems can be broken down into two categories: hardware and software. Hardware is defined as the physical aspects of the computer, while software is the electronic instruction necessary to operate a computer. Software can be broken down even further into applications and operating systems. Computer applications, or programs as they are sometimes called, are instructions that are arranged in a definite order that tell the computer what to do. The way that commands in a computer program are arranged is called the syntax, while the actual commands used make up the language in which the program is created. Once a program has been written, it is converted (compiled) from original language to a format the computer can understand. The conversion process is carried out by a computer program called a compiler. The format to which the compiler converts the program is called machine language (binary 1 and 0). An operating system is defined as the software that controls the operation of the computer. Operating systems can also be broken down into two categories: GUI (Graphical User Interface) and Non-GUI. A graphical user interface uses icons and pictures to represent programs, commands, data files, and even hardware.

Before a computer program is created, a planning process must be undertaken to ensure the quality and accuracy of the program. Since most computer programs are tailored to solve a specific problem, an eight-step systematic approach to problem solving can be implemented. After this has been completed, the programmer is ready to start planning the source code. This is accomplished by the laying out of the program in either flowchart or pseudo-code format. The flowchart format uses both graphical symbols and text to represent the logic of a computer program, whereas the pseudo-code uses only text.

Once the planning process of a program has been completed, the actual code can be written. In AutoLISP, a line of code is referred to as an expression. The syntax for a basic AutoLISP expression is (function argument). The function is typically a predefined AutoLISP command (function). Expressions can be entered into AutoCAD from either the command prompt or a file. To load an AutoLISP program, the user can use either the LOAD function or the APPLOAD command. The LOAD function is used most often in either menu files or other AutoLISP expressions, while the APPLOAD command is used to load a program from the command prompt.

REVIEW QUESTIONS

1. Define the following terms:

   Syntax

   Source Code

   Flowchart

   Pseudo-Code

   GUI

   Programming Language

   Machine Language

   Hardware

   Software

   Binary

2. Explain the importance of planning a program before the actual source code is written.

3. List the eight steps used in phase one of the planning process.

4. Explain the difference between a flowchart and pseudo-code. Give an example of each.

5. Describe the difference between the two methods used to load an AutoLISP program and when they would be used.

6. What is the purpose of the compiler?

7. What is the basic format of an AutoLISP expression?

8. What is the difference between Rich Text Format and ASCII?

9. List the two categories of computer operating systems and give examples of each.

10. What is the difference between programs and operating systems?

11. What is the purpose of an operating system?

12. What is the purpose of comments, and how are they added to computer programs?

CHAPTER 2

Interfacing with the User, Saving Output, Math Functions

OBJECTIVES

Upon completion of this chapter the reader will be able to:

- Identify and describe the ten data types present in AutoLISP
- Understand the difference between global and local variables
- Set a variable using the SETQ function
- Recall a variable from the AutoCAD command prompt
- Use the GETSTRING, GETREAL, GETINT, GETDIST, GETANGLE, GETORIENT, GETCORNER, and GETPOINT functions to prompt the user for information
- Use the GETVAR function to retrieve AutoCAD system information
- Use the GETENV function to retrieve AutoCAD environment information
- Write the result of a program to the screen using the AutoLISP functions PRIN, PRINC, PRINT, WRITE-LINE, and WRITE-CHAR
- Write the result from a program to a file using the AutoLISP functions PRIN1, PRINC, PRINT, WRITE-LINE, and WRITE-CHAR
- Use the OPEN function to read, write, and append to a file
- Use the CLOSE function to close an open file
- Use the four basic math function to solve math problems
- Describe the difference between the less than function and greater than function
- Use the functions LOG, EXP, SQRT and EXPT in complex mathematical calculations
- Use the trigonometric functions SIN, COS, and ATAN in a complex mathematical calculation

KEY TERMS AND AUTOLISP PREDEFINED FUNCTIONS

-	GETANGLE	Local Variable
!	GETCORNER	OPEN
+	GETDIST	Prefix Notation
ATAN	GETENV	Real
CLOSE	GETINT	Selection Sets
Control Character	GETORIENT	SETQ
Data Type	GETPOINT	Side Effects
Entity Name	GETREAL	SIN
Escape Code	GETSTRING	Strings
EXP	GETXXX Functions	Subrs
EXPT	Gobal Variable	Symbols
External Subrs	Infix Notation	Variables
File Descriptors	Integers	VLA
Garbage Collection	List	

AUTOLISP DATA TYPES

As mentioned in Chapter 1, the AutoLISP evaluator analyzes each expression contained within a program and returns the result of the last expression evaluated to the AutoCAD command prompt. The way the expressions are processed is dependent upon the order in which they are arranged and their data type. Currently there are ten data types associated with AutoLISP: Subrs and External Subrs, Integers, Real, Strings, Lists, Selection Sets, Entity Names, File Descriptors, VLA, and Symbols and Variables. Each data type is explained in the sections that follow, and additional information on this subject can be found in Chapter 7 of the *AutoCAD 2000 Customization Guide*.

SUBRS AND EXTERNAL SUBRS

Most of the predefined AutoLISP functions that are used to create a program are either built-in or external sub-routines. A sub-routine is a function that is defined within another function. Its purpose is to conserve keystrokes and reduce the amount of code required for a particular program by grouping frequently used operations in a function that may be executed any number of times within a program simply by the calling of that function. In AutoLISP the predefined functions OPEN, SETQ, +, -, /, *, etc. are classified as built-in sub-routines. These sub-routines are assigned the data type *subrs*. A list of built-in sub-routines can be found in Chapter 13 of the *AutoLISP Function Catalog*. When a function is defined outside AutoLISP, it is known as an external sub-routine. These are functions created by either ADSRX or ARX applications. External sub-routines are given the classification of *exrxsubr* data types.

INTEGERS

An integer is defined as any whole number, including zero that is positive or negative, factorable or non-factorable. Examples of integers are 0, 1, 4, −5, and −9. In mathematics, integers range from negative to positive infinity; however, in AutoLISP integers range from −2,147,483,648 to +2,147,483,647. This is because AutoLISP integers are restricted to 32 bits and as long as the number stays in AutoLISP, it will remain a 32-bit number. If an integer is passed from AutoLISP to AutoCAD, it is reduced from a 32-bit number to a 16-bit number, making the range for possible integers from −32768 to +32767. When an integer is entered in AutoLISP, it is known as a constant. Examples of legal AutoLISP integers are −100, −3, 0, −4, and −5. When an integer value exceeds the maximum allowable value, AutoLISP converts the value from an integer to a real number. For example, the integer 2,312,457,898,754 would be converted to the real number 2.31246e+012 (e + 012 is equal to × 10^{12}). When this occurs, it is known as an integer overflow.

 Note: The maximum value of a 32-bit integer is found by raising the number 2 to the 32nd power (2^{32}). This produces a value of 4,294,967,296; however, because AutoLISP uses signed numbers, the maximum value of an integer now becomes 2^{31}, which produces the value ±2,147,483,647. This difference in value is a result of the signs, which are assigned to the values. The same principle also applies to the 16-bit format used when an integer is passed to AutoCAD.

REAL NUMBERS

A real number is any number including zero that is positive or negative, rational (a number that can be expressed as an integer or a ratio of two integers, ½, 2, −5) or irrational (square root of 2). Examples of real numbers are −4, −4.25, −3, 0, and 2 6.5. AutoLISP defines a real number as any number that contains a decimal point. Even if the number is a whole number, it must contain a decimal point or AutoLISP considers it an integer. Also, numbers that are greater than −1 and less than +1 must start with a zero, or AutoLISP will return an error. Table 2–1 shows examples of AutoLISP integers and real numbers.

Table 2–1   Examples of Integers and Real Numbers in AutoLISP

Real Numbers	Description
4.0	Valid
–4.0	Valid
.25	Invalid; does not have a zero in front of decimal point.
25.25	Valid
–.1	Invalid; does not have a zero in front of decimal point.
Integers	
5	Valid
–4	Valid

When a real number is entered into AutoLISP, it is stored in double precision floating-point format (a computer's approximation of a real number) allowing for decimal accuracy of fourteen significant digits. Real numbers can also be expressed in terms of scientific notation (using either **e** or **E** notation). For example, the value 0.0000005 when expressed in scientific notation becomes 5.0 e-7. Just like integers, when a real number is entered into an AutoLISP expression, that value is known as a constant.

STRINGS

In AutoLISP a group of characters surrounded by quotations is known as a string. Like integers, strings have a limitation imposed upon their length: they can not exceed 132 characters. The characters that make up a string can be alphabetic, numeric or a combination of the two. Strings can also contain special characters called *Control Characters*, also known as *Escape Codes*. Escape codes enable the programmer to add special characters and/or formatting instructions to a text string. An escape code is inserted in a string with a backslash (\) preceding it. For example, to display a string of text on a new line the escape code *n* is used (\n).

Command: **(princ "Test\nTest1")** (ENTER)
Test *(Result returned up to the escape code.)*
Test1 *(Result returned after the escape code.)*
Command:

Other escape codes that may be incorporated into a string are listed in Table 2–2.

Table 2–2   Escape Codes for GETXXX Functions

Code	Description	Example
\\	\ Character	"\\"
\"	" Character	"\""

Code	Description	Example
\e	Escape Character	"\e"
\n	Newline Character	"\n"
\r	Return Character	"\r"
\t	Tab Character	"\t"
\nnn	Character whose octal code is nnn	"\nnn"
\U+xxxx	Unicode Character Sequence	
\M+xxxx	Multi-bite Character Sequence	

LISTS

One of the more important data types encountered in AutoLISP is lists. Similar to other applications, AutoLISP uses lists for storing related information. The data contained in a list is separated by spaces and enclosed in parenthesis. For example, the coordinate of a three-dimensional point displayed in a list would appear as (3.96987 2.22201 0.0). A list can also contain other lists. For example:

```
((-1 . <Entity name: 3160500>) (0 . "LINE") (5 . "20") (100 .
    "AcDbEntity") (67 . 0) (8 . "0")
(100 . "AcDbLine") (10 3.96987 2.22201 0.0) (11 8.3149 5.14366 0.0) (210
    0.0 0.0 1.0))
```

SELECTION SETS

Just as in AutoCAD, a selection set in AutoLISP is nothing more than a group of entities (internally AutoCAD refer to objects as entities). Once the selection set has been built, entities can be modified and/or deleted singularly or as a whole. Also, as in AutoCAD, entities in an AutoLISP selection set can be added and/or subtracted. For example, the following AutoLISP expression creates a selection set consisting of all the lines in a drawing:

Command: **(SSGET "X" (LIST (CONS 0 "LINE")))** (ENTER)
<Selection set: 1>

ENTITY NAMES

When an object is created in AutoCAD, it is given a unique name known as a *handle* that is used when referencing information concerning that object. The handle that is assigned to an entity remains constant throughout the drawing's life cycle. However, when information concerning an entity is needed for an AutoLISP program, it is usually obtained by reference to the entity's name. Entity names are similar to AutoCAD handles in one way: they are both alphanumeric names that are unique to each entity in a drawing. Unlike handles that retain their value throughout the life cycle of the drawing file, entity names are applicable only to a particular object in the current drawing session. Each time a drawing is reopened, the entity name for a particular object will be different. Once the entity name has been obtained, AutoLISP is free to manipulate its data in a variety of ways. The following example illustrates how the entity name of the line shown in Figure 2–1 will appear once obtained through the AutoLISP functions (CAR (ENTSEL)).

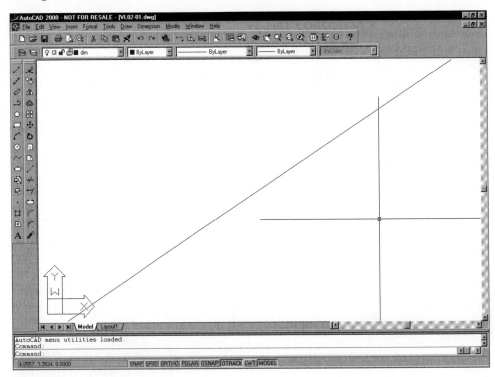

Figure 2–1 *Line in an AutoCAD drawing*

Command: **(CAR (ENTSEL))** (ENTER)
<Entity name: 3160500>

Note: When an entity's definition data is obtained, then the entity name will always appear as the first sub-list in the overall list. For example, ((-1 . <Entity name: 3160500>) (0 . "LINE") (5 . "20") (100 . "AcDbEntity") (67 . 0) (8 . "0")(100 . "AcDbLine") (10 3.96987 2.22201 0.0) (11 8.3149 5.14366 0.0) (210 0.0 0.0 1.0))

FILE DESCRIPTORS

Often in computer programming it is necessary to either read and/or write information to an external file. When an external file is to be manipulated, AutoLISP assigns that file a unique name known as a *File Descriptor*. File descriptors, just like entity names, are alphanumeric in nature and are retained only in the current drawing session or as long as the external file is open for input/output. Any time that an attempt is made to read, write, or append to a file, the file descriptor must be identified in that expression. The following example illustrates how a file descriptor would appear with the AutoLISP OPEN function.

> Command: **(SETQ name (OPEN "test" "w"))** (ENTER)
> <File: #26804c8>

VLA

This is a new data type within AutoLISP that resulted from the introduction of Visual LISP. This data type allows objects in a drawing to be represented as Visual LISP ActiveX objects. This data type will be covered in more detail later.

SYMBOLS AND VARIABLES

The primary function of any computer program is to take information that has been supplied by the user, process that information, and return a result. For this to be accomplished, the program must have the capability of temporarily storing information that was entered from the user or calculated by a program. One way programs temporarily store information is through the use of variables. A variable can be thought of as a placeholder, or a container for data storage. In programming, this information or data can often be used later in some particular operation or procedure. Once a variable has been set, then it may be recalled at any time until the variable has been removed from memory. In AutoLISP, the term *symbol* is often used to describe the data that is assigned to a variable. Hence the term *symbol-name* can be interchanged with the term *variable name* (The term *symbol-name* is often used in the *AutoLISP Programming Reference*). Variable names can be alphanumeric, notation or a combination of the two. Variable names are not case sensitive and cannot be all numeric characters. They cannot contain any of the following special codes: single quote, double quote, parenthesis, periods, or semicolons. Table 2–3 shows examples or legal and illegal variable names.

Table 2–3   Legal and Illegal Variable Names

Valid Variable Name	Invalid Variable Names
point1	point;1
POINT2	point.2
first1	first'1
first345	first"345
entityname34	entityname(34)
Variable_name	Variable;name

LOCAL AND GLOBAL VARIABLES

As stated earlier, a variable can be thought of as a container where information ranging in size from one character to an entire list is held. In AutoLISP variables are divided into two categories: *Local* and *Global*. A local variable retains its value only as long as the program is running. Once the program has finished, the value of the variable is returned to nil (nothing). This is advantageous because the memory space that was allocated for the variable is now freed. This process is known as a *Garbage Collection*. Global variables, on the other hand, retain their value even after the program has finished running. The memory space reserved by this variable remains reserved until either the variable is reset to nil or the user performs a garbage collection. For manually performing a garbage collection, the AutoLISP function (GC) is used. Memory is not the only issue concerning local and global variables. Since global variables retain their value long after the program has completed running, this sometimes results in interference between programs known as a *Side Effect*. The chance of this occurring greatly increases with each program that is loaded into a computer. However, side effects can be used to achieve a desired result, especially when information is to be passed from one program to another. In addition, the use of global variables can be quite helpful when a programmer is trying to debug an AutoLISP program. If a program contains variables that are widely scattered, it is often possible to determine where a program is running into trouble by checking the values of each variable. Even though global variables can be useful for debugging an AutoLISP program, once a program has been completely debugged, all global variables should be converted to local.

DECLARING LOCAL VARIABLES

All variables in AutoLISP are considered to be global variables until they have been declared otherwise. Recall that when a program is created through the DEFUN function, the programmer has the option of supplying an *argument list*. It is the *argument list* portion of this function where local variables are distinguished. Since this *argument list* can contain both arguments and variables, a forward slash is placed before the first local variable name, to separate it from the arguments. For example, to declare the variable pt1 as a local variable, the following syntax would be used:

```
(DEFUN program (/ pt1)          ;Pt1 is declared as a local
                                ;variable.
    (function argument)         ;Expression.
    (function argument)         ;Expression.
)                               ;End program.
```

If a value is passed from one program to another or from one sub-routine to another, the variable is then declared as an argument. For example:

```
(DEFUN program (pt1 / pt2)      ;The variable pt1 is supplied as an
                                ;argument, while the variable pt2 is
                                ;declared to be local.
```

```
(SETQ pt2
    (GETREAL "\nEnter value of second number : ")
    (+ pt1 pt2))
)                                    ;End of program.
```

To illustrate how values are passed into the program featured above, the following expression would be entered from the AutoCAD command prompt:

Command: **(program 3)** (ENTER)
Enter value of second number : **3**
6.0

Note: The argument that is supplied along with the function name is passed to the variable pt1.

The *argument list* of the DEFUN function can contain all arguments, no arguments, all local variables, no local variables or any combination. If no arguments or variables are used, the opening and closing parentheses must still be supplied. For example:

```
(DEFUN program (pt1 pt2 pt3) ..........)
```

In this example the user-defined function has three arguments: pt1, pt2, and pt3. No local variables are defined. For this reason all variables are considered global.

```
(DEFUN program (/ pt1 pt2 pt3) ..........)
```

In this example the user-defined function has defined Pt1, pt2, and pt3 as local variables. No arguments are supplied to this function.

```
(DEFUN program ( ) ..........)
```

This example shows as empty set. No local variables or arguments are defined. All variables will be global.

```
(DEFUN program (pt1 /pt2 pt3) ..........)
```

In this example one argument is supplied and two local variables have been defined. Pt1 is an argument; pt2 and pt3 are local variables.

DECLARING A VARIABLE IN AUTOLISP

Whether a variable is local or global, the method used to assign the values to those variables is the same. In AutoLISP this can be accomplished through the SETQ (SET Quote) function. For example:

```
(SETQ variable-name1 value1 [variable-name2 value2 ....])
```

According to the syntax of this function, a variable name and value must be supplied; otherwise an error is returned. The syntax also indicates that the programmer has the option of supplying more than one variable name and value (this is indicated by the brackets). The brackets are not used in the actual program; they are merely inserted into the syntax to indicate that this is an optional feature. If the programmer is supplying more than one variable name and value, then the value that is associated with a particular variable name must follow that variable name. Finally, the value that is assigned to a particular variable name can be any of the ten different data types. For example:

```
(SETQ r 5.0)
```

In this example the real number 5.0 is set to the variable r.

```
(SETQ r 5.0
      a 6)
```

In this example the real number 5.0 is set to the variable r and the integer 6 is set to the variable a.

```
(SETQ r 5.0
      a 6
      d 5.6
      tex "text")
```

In this example the real number 5.0 is set to the variable r, the integer 6 is set to the variable a, the real number 5.6 is set to the variable d, and the text string "text" is set to the variable tex.

RECALLING THE VALUE OF A GLOBAL VARIABLE AT THE AUTOCAD COMMAND PROMPT

Once a global variable has been set, its value can be recalled at any time from the AutoCAD command prompt. As stated earlier, this ability to examine the value of a variable can make the task of debugging a program much easier. To display the value of a variable at the AutoCAD command prompt, the programmer would enter an exclamation point followed by the variable name. For example:

```
Command: !d  (ENTER)
5.6
Command: !r  (ENTER)
5.0
Command: !a  (ENTER)
6
Command: !tex  (ENTER)
"TEST"
```

Aside from debugging a program, the ability to recall the value of a variable at the AutoCAD command prompt has other possibilities. It provides a means of passing the value of a variable to an AutoCAD command. For example, the value of the variable d could be used as an offset distance for the AutoCAD OFFSET command. To use the value assigned to a variable with an AutoCAD command, an exclamation point before the variable name is inserted at the prompt where the data is to be used. For example:

> Command: **OFFSET** (ENTER)
> Offset distance or Through <1.0000>: **!d** (ENTER)
> Select object to offset:
> Side to offset?
> Select object to offset: (ENTER)

PROMPTING THE USER FOR INPUT

Before a program can be considered versatile, it must have the capability of requesting information from the user. This information can be anything from the coordinates of a point to the thermal properties of copper. In AutoLISP, data is entered into the program from a file, selected from the AutoCAD graphic screen, or typed in from the keyboard. With the exception of reading the information from a file, the majority of the information that is provided to an AutoLISP program is done through the GETXXX functions. These functions allow for information to be entered in a program in the form of strings, real numbers, integers, coordinates, file names, angles, distances, AutoCAD system variables, and system environment variables.

REQUESTING STRING INFORMATION

One of the most common and basic types of information often needed for a program is text or string information. This information may be needed to describe a material or provide the name of a file to be created. To prompt the user to enter string information, the GETSTRING function must be used. This function allows the programmer to construct a prompt and pause for user input. All information, regardless of whether or not it is alphabetic or numeric, is returned in the form of a string. The syntax for this command is (GETSTRING [cr] [message]). The *cr* argument of this function allows the program to accept spaces. If this argument is supplied and its value is not equal to nil, then the user is allowed to use the SPACEBAR to input spaces. If this argument is supplied and its value is equal to nil, then the SPACEBAR is treated as a carriage return. This function does have one limitation: the length of the input string cannot exceed 132 characters. If the string does exceed this limit, then only the first 132 characters are accepted. If the input string contains a backslash (\) then the backslash is converted to a double backslash (\\). This gives the programmer the option of using the string input as part of a file name. Following are several examples of using the GETSTRING function.

```
(GETSTRING "Enter material name : ")
```

In this example the user is prompted to enter the name of a material. The user is not allowed to enter spaces.

```
(GETSTRING T "Enter material name : ")
```

In this example, the user is again prompted to enter the name of a material. The capital T is a predefined AutoLISP variable that has a value is equal to the constant T (true) and can be used anywhere a non nil value is required. As a result, the user can now input spaces using the SPACEBAR.

```
(GETSTRING T "\nEnter material name : ")
```

In this example, the user is again prompted to enter the name of a material. The escape code with lowercase n instructs the program to enter the prompt on a new line.

REQUESTING NUMERIC INFORMATION

Another common type of information that is often requested in a program is numeric data. Numeric data can be real numbers, integers, distances, angles, etc. In AutoLISP, there are five functions that are used to prompt the user to enter numeric data: GETREAL, GETINT, GETDIST, GETANGLE, GETORIENT.

Using the GETREAL Function

To input a real number into an AutoLISP program, the GETREAL (GETREAL [message]) function is used. Much like the GETSTRING function, this function allows the programmer to construct a prompt and pause for user input. It differs from the GETSTRING function by allowing the input of only numeric data with no spaces. The SPACEBAR is used strictly as a carriage return. If an integer is entered, it is automatically converted to a real number. Examples of this function are as follows:

```
(GETREAL "Enter outdoor air temperature : ")
```

In this example, the user is prompted to enter the outdoor air temperature. The prompt is not started on a new line.

```
(GETREAL "\nEnter outdoor air temperature: ")
```

In this example, the user is also prompted to enter the outdoor air temperature, but this time the prompt is started on a new line.

Note: Starting a prompt on a new line makes it easier for the user to read and understand messages that are displayed. This will reduce any confusion that might arise from several prompts appearing on the same line and make the program more user friendly.

Using the GETINT Function

The GETINT (GETINT [message]) function works almost exactly like the GETREAL function, the only difference being that instead of returning a real number, the function returns an integer. This function does not allow for the use of the SPACEBAR except for the termination of the function. Unlike the GETREAL function, which converts integers that are entered to real numbers, the GETINT function will not accept a real number. If the user enters a real number, the function displays the message "Requires an integer value". This is followed by the function repeating the request.

```
(GETINT "Enter indoor air temperature : ")        ;Prompts for indoor air
                                                  ;temperature.
```

Using the GETDIST Function

Real numbers can also be entered in a program with the GETDIST function. This function, like the other GETXXX functions, allows the programmer to construct a prompt and pause for user input. Unlike with the GETREAL function, the user has the option of either entering the requested information from the keyboard or selecting two points from the graphics screen. The syntax for this function is (GETDIST [pt] [message]). The *pt* argument can contain a two-dimensional or three-dimensional point; it is strictly optional. This argument, if supplied, allows the user to specify a distance by selecting a single point. The distance is then calculated from the point that was supplied by the programmer (*pt*) and the point selected by the user. Whether or not this argument is supplied, the user still has the option of entering a point or distance from the keyboard. If the user specifies the distance by selecting points, AutoCAD responds by drawing a rubber-band line between the first point selected and the current position of the crosshairs (see Figure 2–2). This rubber-band serves as an aid that helps the user visualize the distance specified. If a two-dimensional point is selected, then the distance returned is calculated from the X and Y components of the point selected. If a three-dimensional point is selected, the distance returned by the function is based on the three coordinates, but by using the INITGET function, the programmer can instruct the GETDIST function to ignore the Z components.

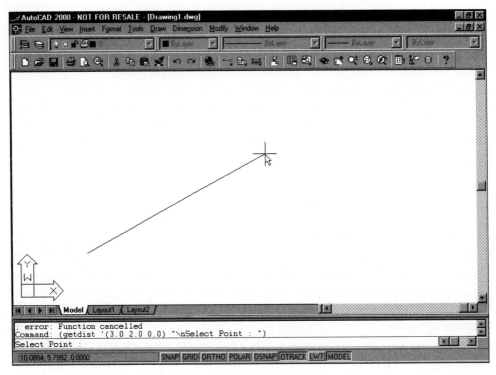

Figure 2–2 *Rubber-band effect on the* GETDIST *function*

 Note: The distance between two points can be determined through the formula $\sqrt{(X_2 - X_1)^2 + (Y_2 - Y_1)^2 + (Z_2 - Z_1)^2}$.

```
(SETQ dis (GETDIST "\nEnter shaft diameter : "))
```
In this example the user is prompted to enter the diameter of the shaft. The diameter can be entered from the keyboard or the two points can be selected (graphically representing the shaft diameter).

```
(SETQ pt '(0 0 0))
(SETQ dis (GETDIST pt "\nEnter shaft diameter : "))
```
In this example the variable pt is set equal to the point 0,0,0. This point is then used in conjunction with the GETDIST function and the user is prompted to enter the diameter of the shaft.

REQUESTING ANGULAR INFORMATION
Engineering is the art of solving technical problems using the theories and concepts associated with the fields of mathematics and science. This requires that the engineer

have not only a good understanding of mathematics and physical science but also an excellent understanding of three-dimensional spatial problems and how to solve them using either descriptive geometry and/or trigonometry. It is crucial when a program is written for solving engineering problems that the user be able to input angular information. In AutoLISP this can be accomplished using either the GETREAL or the GETINT function. However, if the programmer wants to provide the user with a means of specifying angular information by selecting points from the AutoCAD graphics screen, then either the GETANGLE or GETORIENT function should be used. These functions are similar to the GETDIST function in syntax only; the results returned by these functions are the angles measured in radians.

Using the GETANGLE Function

Like the GETDIST function, this function allows the programmer to construct a prompt and pause for user input. This function also allows the user to enter the requested angular data from either the keyboard or the graphics screen. The syntax for this function is (GETANGLE [pt] [message]). Just as in the GETDIST function, the *pt* argument is optional and, if supplied, allows the user to enter angular data by simply selecting a single point. The angular data that is returned uses the current setting of the AutoCAD ANGBASE system variable with respect to the current XY plane, current elevation and current UCS. The angle is always measured in a counterclockwise direction regardless of the value of the AutoCAD ANGBASE system variable.

Using the GETORIENT Function

The other function used to get angular data is the GETORIENT function. This function is similar to the GETANGLE function in syntax and the value returned, but it differs in the fact that the value is not affected by the AutoCAD system variables ANGDIR and ANGBASE. Depending upon the application for which the program is written, the programmer would select which function to use based on whether the returned value can be the relative or absolute angle.

REQUESTING COORDINATES AND OTHER MISCELLANEOUS INFORMATION

As mentioned earlier string, integer, real numbers and angular information are not the only types of information the programmer can prompt the user for or even obtain from AutoCAD. Information concerning points, AutoCAD system settings as well as other information can be obtained through GETXXX functions. In this section, only five of these functions are covered. It is recommended that the programmer refer to the *AutoCAD 2000 Customization Guide* to reference the remaining GETXXX functions.

Obtaining Coordinates

Because AutoCAD is a graphically based software program and because graphics play an important role in the engineering and architectural fields, the ability to retrieve and process information concerning the location of entities in three-dimensional space is not just necessary, but it is crucial in most applications. AutoLISP has two GETXXX functions that are at the programmer's disposal for obtaining the coordinates of a point. They are the GETPOINT and GETCORNER functions.

Using the GETPOINT Function

To retrieve information concerning a point in AutoCAD, the AutoLISP function GETPOINT is used. This function allows the programmer to construct a prompt and pause for user input. The user can either enter the coordinates from the keyboard or select a point from the graphics screen. The coordinates that are returned are in respect to the current UCS. The syntax for the GETPOINT function is (GETPOINT [pt] [message]). The *pt* argument can be a two-dimensional or three-dimensional point, and if furnished, AutoCAD will construct a rubber-band line from the point specified to the current crosshairs position (see Figure 2–3). The value returned by the *pt* argument is the three-dimensional coordinates of the point selected with respect to the current UCS.

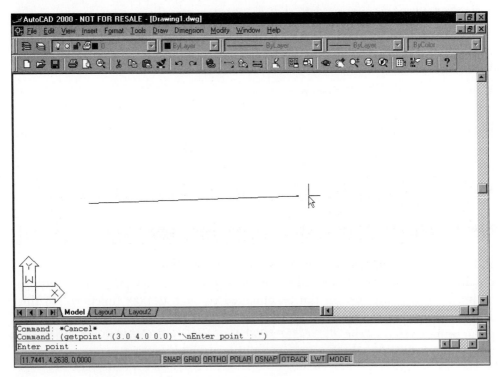

Figure 2–3 *Rubber-band effect on the* GETPOINT *function*

Using the GETCORNER Function

The other function designed to retrieve point data is the GETCORNER function. Similar to the GETPOINT function, this function also returns the value of the point selected with respect to the current UCS. This function does have one limitation: it uses the current elevation as the Z component for the point selected. The syntax for this function is identical to that of the GETPOINT function, but this function does require that the *pt* argument be supplied. Otherwise an error is returned. Also, the GETCORNER function differs from the GETPOINT function because it constructs a rectangle from the point supplied to the current crosshairs position (see Figure 2–4).

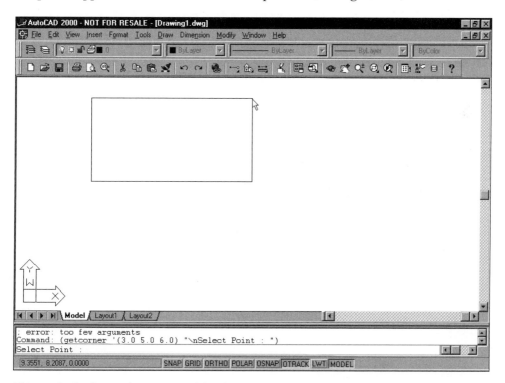

Figure 2–4 *Rectangle constructed by the* GETCORNER *function*

Obtaining System Information

Because AutoCAD is a graphical application, it is often necessary for AutoLISP to interact with AutoCAD either in obtaining information to be used in a program or in processing the data graphically once the computations have been completed. This information is often in the form of AutoCAD system, and/or AutoCAD environment information. For example, suppose that a program is needed that will calculate the amount of material to be removed from a cross section. A layer is created where a hatch pattern designates the area to be removed, all without permanently changing the current layer. In this particular example, the programmer would need to access AutoCAD system variables to determine what the current layer is so that, before the program has finished executing, it will reset the current layer. Currently there are two GETXXX functions that allow the programmer to access both the environment and system settings. They are GETVAR and GETENV.

Using the GETVAR Function

The GETVAR function allows the programmer to obtain AutoCAD system information (current layer, current text style, height, font, etc.). The syntax for this function is (GETVAR Variable-Name). The *variable-name* argument is the name of the system variable that AutoCAD uses to store information. The variable name is a string and must be contained in quotation marks. If an invalid variable name is entered, the function returns nil; otherwise the current value of that variable is returned. For example:

Valid System Variable

```
Command: (GETVAR "clayer")   (ENTER)
"0"
Command:
```

Invalid System Variable

```
Command: (GETVAR "layer")   (ENTER)
nil
Command:
```

 Note: A list of valid AutoCAD system variables can be found in the *AutoCAD 2000 Command Reference.*

Using the GETENV Function

The GETENV function is similar to the GETVAR function in the fact that critical AutoCAD information is returned. However, this function returns the AutoCAD environment setting. The syntax for this function is (GETENV 'Variable-Name"). For example:

Valid Environment Variable

```
Command: (GETENV "ACAD") (ENTER)
"C:\\Program Files\\AutoCAD R14\\SUPPORT;C:\\Program Files\\AutoCAD
R14\\FONTS;C:\\Program Files\\AutoCAD "
```

Invalid Environment Variable

Command: **(GETENV "system")** (ENTER)

nil

UTILIZING THE RESULTS

Obtaining information from the user and performing the calculations is only part of the requirements for a computer program; the program must be able to make that information available to the user. If a program is able to receive and process information but has no means of returning the results to the end user, then the user has not benefited from the use of that program. In AutoLISP there are three main ways that information processed from a program can be passed on to the user. The results can be written to the screen, to a file, or used by the AutoCAD graphics package.

WRITING THE RESULTS TO THE SCREEN

Currently AutoCAD has five functions that allow the user to write messages to the command line portion of the screen. This allows for the results of a program or other important information concerning the processes involved in making calculations to be displayed. These five functions are PRIN1, PRINC, PRINT, WRITE-LINE and WRITE-CHAR.

Using the PRXXX Functions

The most frequently used functions for writing information to the AutoCAD command line are the PRIN1, PRINC, and PRINT functions. Even though these functions are identical in their syntax and their functionality, it is the result that is returned once their expressions have been evaluated that differentiates them. All three functions can be used to write messages to the AutoCAD command prompt as well as to an open file. The PRIN1 function writes the message exactly as it appears (control codes and all). The PRINT function automatically returns a new line before the message is printed and a space after, while the PRINC function prints a message without printing the escape codes or a space thereafter. The syntax for these functions is (PRXXX [expression/string [file descriptor]]).

The *expression/string* argument associated with this function can be either a string value or another AutoLISP expression. If an AutoLISP expression is used, the value returned by the PRXXX function will equal the value of the expression supplied. For example, using the SSGET function, all the entities that are contained in a drawing can be set to a selection set. The value that is returned from this function is the name of the selection set created.

Command: **(SSGET "X")** (ENTER)

<Selection set: 1>

If one of the PRXXX functions is incorporated into the expression, the outcome is still the same. A selection set is created containing all the entities of a drawing, and the name of that selection set is passed to the PRINC function.

Command: **(PRINC (SSGET "X"))** (ENTER)
<Selection set: I><Selection set: I>

As previously stated, AutoLISP expressions are not the only thing that can be supplied as an argument to the PRXXX functions; these functions also allow for string input. Just as with the GETXXX functions, the PRXXX functions allow a string to contain escape codes. A list of legal control codes that can be used with these functions are shown in Table 2–4.

Table 2–4   Legal Control Codes for PRXXX Functions

Code	Description	Example
\\	\ Character	"\\"
\"	" Character	"\""
\e	Escape Character	"\e"
\n	Newline Character	"\n"
\r	Return Character	"\r"
\t	Tab Character	"\t"
\nnn	Character whose octal code is nnn	"\nnn"

The second argument that can be supplied to this function is the *file descriptor*. This is a name assigned by AutoLISP to an open file. This is an optional argument; if it is supplied, then the message is written to the specified file (provided that the value of the argument is not nil, in which case an error is returned). If this argument is not supplied, the message is written to the AutoCAD command line.

Using the WRITE-XXX Functions

The next group of functions that can be used to write messages to the command prompt area are the WRITE-LINE and WRITE-CHAR functions. These functions are similar to the PRXXX functions in syntax and operation. Both groups of functions are able to write to the screen as well as an open file. Both functions have the optional file descriptor argument, but only the PRXXX functions allow for an expression that does not return either a string or an integer to be used as the first argument supplied. The WRITE-XXX functions themselves differ in the data type that can be supplied to them. The WRITE-LINE function accepts only a string argument while the WRITE-CHAR function accepts only numeric arguments. That is because the WRITE-CHAR function is primarily designed to write a character to the screen or an open file. The numeric value supplied to this function is the ASCII decimal code for that character. The syntax for this group of functions is (WRITE-XXX argument [file descriptor]).

WRITING THE RESULTS TO A FILE

Before information can be written to a file using one of the five functions previously addressed, the file this information is to be written to must exist and be open. This is accomplished through the OPEN function (OPEN filename mode). This function must be executed before an attempt is made to write to a file. The *filename* argument is the name of the file that is to be either open or created. The *mode* argument determines whether the file that is opened will be read from or written to. This argument has three possible settings: "r" (read), "w" (write), and "a" (append). If the "r" option is supplied, then the file is opened as read only and no data can be written to it. If the "w" option is supplied, then the file is opened in the write mode and information can only be written to this file. The "w" option has one more characteristic: if the file already exists, then it is automatically overwritten; if the file does not exist, then it is created. Option "a" also writes information to a file, but if the file already exists, then the information is appended to the file (starting on a new line after the last character). The value returned by this function is the alphanumeric file descriptor, which is assigned by AutoLISP. Once the file descriptor has been defined, its value must be saved to a variable and that variable used with PRXXX and/or WRITE-XXX functions. Examples of how this function and its returned values are used are shown below.

```
(SETQ FIL (OPEN "test_file" "w"))      ;Opens a file for
                                       ;input.
(PRINC "\nThis line will be written to the file" fil)
                                       ;Writes data to file.
(PRIN1 "\nThis line will be written to the file" fil)
                                       ;Writes data to file.
(PRINT "\nThis line will be written to the file" fil)
                                       ;Writes data to file.
(WRITE-LINE "\nThis line will be written the file" fil)
                                       ;Writes data to file.
(WRITE-CHAR 44 fil)                    ;Writes data to file.
```

Once all the information has been written to or read from a file, then that file must be closed. This is accomplished through the CLOSE function (CLOSE file_description). The *file description* argument supplied to this function is the file descriptor that was established by the OPEN function. Adding the CLOSE function to the previous example yields:

```
(SETQ FIL (OPEN "test_file" "w"))      ;Opens a file for
                                       ;input.
(PRINC "\nThis line will be written to the file" fil)
                                       ;Writes data to file.
```

```
(PRIN1 "\nThis line will be written to the file" fil)
                                        ;Writes data to file.
(PRINT "\nThis line will be written to the file" fil)
                                        ;Writes data to file.
(WRITE-LINE "\nThis line will be written the file" fil)
                                        ;Writes data to file.
(WRITE-CHAR 44 fil)                     ;Writes data to file.
(CLOSE fil)                             ;Closes file.
```

READING INFORMATION FROM A FILE

As mentioned, information can be read into a program from an external file once that file has been opened. In AutoLISP this is accomplished using either the READ-LINE (read-line [file-desc]) or READ-CHAR (read-char [file-desc]) function. The READ-LINE function when supplied with a *file-desc* returns a string value representing the information contained within the file. The function reads the information until it encounters either an end-of-line or end-of-file marker. If a *file-desc* is not furnished, then the keyboard provides the input. On the other hand, the READ-CHAR function, when supplied with a *file-desc*, returns the ASCII code representing the character read from a file. If a *file-desc* is not furnished, then the input is supplied from the keyboard. For example:

```
(SETQ fil (OPEN "input_file.asc" "r")   ;Opens the file
                                        ;input_file.asc in the
                                        ;read mode and sets
                                        ;its file descriptor
                                        ;to the variable fil.

        input_line (READ-LINE fil)      ;Reads the first line
                                        ;contained in the file
                                        ;input_fil.asc and
                                        ;returns it as a
                                        ;string.  The result is
                                        ;then set to the
                                        ;variable input_line.
)
(SETQ input_line (READ-LINE))           ;Accepts input from the
                                        ;keyboard since a file
                                        ;descriptor was not
                                        ;furnished.

(SETQ input_line (READ-CHAR fil))       ;Reads the first
                                        ;character from the
                                        ;file input_line and
                                        ;returns the results
                                        ;as an ASCII CODE.
```

```
(SETQ input_line (READ-CAR))        ;Accepts input from the
                                    ;keyboard since a file
                                    ;descriptor was not
                                    ;furnished.
```

PERFORMING MATHEMATICAL OPERATIONS

Mathematics has played a critical role in shaping the world in which we live; without it none of the technical advances that are enjoyed today would be possible. This is why any programming language that is selected to solve engineering applications must have a well-defined arsenal of basic as well as advanced functions at its disposal. AutoLISP does indeed have such an array of predefined functions that are designed to carry out mathematical computations. The following sections discuss some of the functions defined by AutoLISP. These functions are categorized by the results returned by that function. The categories used in this book are basic mathematical operations, equalities and inequalities, exponential and logarithm, and trigonometry. All math functions defined in AutoLISP use *Prefix Notation*. Prefix notation is a method of ordering functions and arguments where the argument precedes the function (operation). In standard notation (also known as infix notation) the function precedes the argument. For example, in standard notation, the equation A + B would be written as A+B. In prefix notation the same formula is now written as + A B.

BASIC MATHEMATICAL OPERATIONS

Addition +

The first of the four basic mathematical functions that the programmer must master is addition (+ [number number]). This function allows the programmer to add either a set of numbers or an entire series. If two or more integers are combined, then the result will also be an integer. If two or more real numbers are combined, then the result will be a real number. However, if a real number and an integer are combined, then the result is a real number. Examples of the addition function are as follows.

```
(+)
```

In this example no arguments were supplied and AutoLISP will assign a zero for each missing argument, thus returning value of zero. If the example had been (+ 2), then the result returned would have been 2.

```
(+ 2 2)
```

In this example the two integers 2 and 2 are combined. The value that is returned will be the integer 4.

```
(+ 2 3 1)
```

In this example the series of integers 2, 3, and 1 are combined to return an integer value of 6.

```
(+ 2 2.0)
```

This example combines the integer 2 and the real number of 2.0, returning the real number 4.0.

```
(+ 2.0 3.0)
```

This example combines two real numbers 2.0 and 3.0 to return the real number 5.0.

Subtraction -

The next of the four basic math operations is subtraction (- [number number]). In reality, subtraction is nothing more than the addition of negative numbers. Phrasing it another way, subtraction is the difference between two or more numbers. Just as in the natural world, AutoLISP allows the programmer to find the difference between either two numbers or a set of numbers. If two integers are subtracted, then the result is an integer. If two real numbers are subtracted, then their difference will be a real number. Finally, if a real number and an integer are subtracted, then the result will be a real number. Examples of the subtraction function are as follows.

```
(-)
```

In this example no arguments were supplied and AutoLISP will assign a zero for each missing argument, thus returning a result of zero. If the example were (- 2), then the result returned would have been −2.

```
(- 2 1)
```

In this example the difference of the integers 2 and 1 is determined (argument1 is subtracted from 2). The value that is returned by this expression is the integer 1.

```
(- 2 3 1)
```

In this example the difference of the series of integers 2, 3, and 1 is determined (argument 3 is subtracted from 2 yielding −1; −1 is then subtracted from −1). The value that is returned by this expression is −2. This would be read as +2 minus 3 minus 1 (2–3–1).

```
(- 2 3.0)
```

In this example the difference of the integer 2 and 3.0 is determined. The value that is returned by this expression is the real number −1.0.

```
(- 2.0 3.0)
```

In this example the difference of the real numbers 2.0 and 3.0 is determined. The value that is returned by this expression is the real number −1.0.

Multiplication*

The third basic math operation is the multiplication function (* [number number]).
Much like the other two functions, this function allows the programmer to multiply
either two numbers or a series of numbers. If no arguments are supplied, then the
function returns zero. If only one argument is supplied, then the product that is
returned is the same as the number that was supplied; in other words, it is the same
as multiplying that number by 1. If two or more integers are multiplied together,
then the value returned is an integer. If two or more real numbers are multiplied
together, then their product is a real number. Finally, if a real number and an
integer are multiplied together, then their product is a real number. Examples of the
multiplication function are as follows.

```
(*)
```

In this example no arguments were supplied and AutoLISP will return a product of
zero. If the example were (* 2), then the result returned would have been 2.

```
(* 2 1)
```

In this example the product of the integers 2 and 1 is determined. The value that is
returned is the integer 2.

```
(* 2 3 1)
```

In this example the product of the series of integers 2, 3, and 1 is determined. This
would be read as 2 * 3 * 1. The value that is returned is the integer 6.

```
(* 2 3.0)
```

This example multiplies the integer 2 to the real number of 3.0; this returns the real
number 6.0.

```
(* 2.0 3.0)
```

This example multiplies two real numbers, 2.0 and 3.0. This also returns a real
number of 6.0.

Division/

The last of the four basic math operations is division. This function allows for the
division of either two numbers or a series of numbers. If more than two numbers are
supplied, then the quotient is found by dividing the first number by the second, after
which their quotient is divided by the third number, and so on. Again, just like the
other basic math functions, if the arguments supplied to the function are integers,
then the quotient that is returned is an integer. If the two integers do not divide
evenly, then the remainder is discarded. If the arguments supplied are real numbers,
then the quotient returned is also a real number. If the two real numbers do not

divide evenly, then the remainder is returned in the form of a decimal. Finally, if the arguments supplied to the function are mixed, then the quotient returned is a real number. Much like the multiplication function, if no arguments are supplied, then the function returns a value of zero. If only one argument is supplied, then the quotient that is returned is the same value of the argument supplied; in other words the argument is divided by 1. The syntax for this function is (/ [number number]). Examples of the division function are as follows.

> (/)

In this example no arguments were supplied and AutoLISP will return a quotient of zero. If the example were (/ 2), then the result returned would have been 2.

> (/ 42 2)

In this example the integer 42 is divided by the integer 2, yielding the integer quotient 21.

> (/ 2 3 1)

In the example the integer 2 is divided by the integer 3, whose quotient is then divided by the integer 1. Since the first two arguments supplied to this function are integers and since 3 will not divide into 2 evenly, the quotient returned will be zero. The result is then divided by the integer 1. Whenever zero is divided by another number greater than zero, the result will always be equal to zero.

> (/ 2 3.0)

In this example the integer 2 is divided by the real number 3.0, returning the quotient 0.666667.

> (/ 2.0 3.0)

In this example the real number 2.0 is divided by the real number 3.0, returning the quotient 0.666667.

EQUALITIES AND INEQUALITIES

The ability to make comparisons is crucial to all computer programming languages. If this ability is not present, then the functionality of any computer program is greatly diminished. It is absolutely essential that the programmer become well versed with these functions, since they will be used later as part of the decision-making process.

Equal to =

This function is used almost exclusively in AutoLISP programs to make comparisons between two arguments or a group of arguments. It can be used to compare any of the ten data types previously covered. The syntax for this function is (= number [number]…). Upon completion of this function evaluation, the result that is returned is either T (true) or nil (false). If only one argument is furnished, the function returns the value T. If more than one argument is supplied, then the function compares all arguments and returns the value of T only if all arguments are equal. If no arguments are supplied, then the function returns an error; at least one argument must be present or an error will occur. If an integer and a real number are entered as the arguments then, as long as the values are numerically equal, the function will return the value T. Examples of this function are as follows.

    (= 4)

In this example only one argument is supplied; therefore the value that is returned for the equal to function is T.

    (= "Yes" "yes")

In this example two strings are compared. The first string contains an uppercase Y while the second string does not contain any uppercase letters. The value returned by the function is nil.

    (= "yes" "yes")

In this example two string are also compared, but this time the cases are identical so the value returned by the function is T.

    (= 4 4)

In this example two integers that are numerically equal are compared, and the result returned by the function is T.

    (= 2.0 4 2)

In this example three arguments are furnished. Even though the first and the last arguments are numerically equal, the second argument is not equal to either the first or the last argument; the value returned by this function is nil.

Not Equal To /=

The not equal to function is similar to the equal to function in some ways. It can be used to compare any of the ten data types present in AutoLISP. It returns either a T or nil once the expression has been evaluated, the difference between the two functions being that all arguments presented must be the same for the equal function to return T. No arguments presented must be equal for the unequal function to return a T. If any two or more arguments in the list presented are equal, the function will return nil. The syntax for this function is (/= number [number]…).

Less Than <

Just as important as the equal to and not equal to functions is the less than function. This function is used to determine whether or not one argument is less than another argument. The function can evaluate two arguments or a series of arguments. If two arguments are presented, then the function returns a T value if the first argument is less than the second. If more than two arguments are supplied, then the function starts with the first argument and compares it to the second argument; the second argument is compared to the third, and so on. This process is repeated until all arguments have been compared. All arguments presented must be less than the successive argument for this function to return a T. If only one argument is furnished, then the function returns a value of T. If no arguments are present, the function returns an error. The syntax for this function is (< number [number]…). Examples of the less than function are as follows:

    (< 1 3)

In this example the first argument is less than the second argument, and the value returned by the function is T.

    (< 1 3 5)

In this example three arguments are supplied; the function compares the first argument to the second argument and the second argument to the third argument. Since each successive argument is greater that the last, the function returns the value T.

    (< 4 2)

In this example two integers are compared. Since the first argument is greater than the second argument, the function returns the value nil.

    (< 2 4 7 1)

In this example a series of numbers is compared; this time the function returns a value of nil because the integer 7 is greater than the integer 4.

Less Than or Equal To <=

This function is almost like a combination of the = and < functions. The arguments being compared can now be either equal to each other or the first less than the next. All arguments presented must be less than or equal to their successive argument for this function to return T. The syntax for this function is (<= number [number]…).

Greater Than >

Similar to the less than function, this function evaluates arguments and returns a value of T if the second argument is greater than the first. It can be used to evaluate two arguments or a series of arguments. The syntax for this function is (> number [number]…).

Greater Than or Equal To >=

This function is almost like a combination of the = and > functions. The arguments being compared can now be either equal to each other or the first greater than the next. All arguments presented must be greater than or equal to their successive argument for this function to return a T. The syntax for this function is (>= number [number]…).

PERFORMING COMPLEX MATHEMATICAL OPERATIONS

Unfortunately, not all equations can be solved using basic math functions; therefore, the programmer must also be fluent in all higher mathematical operations that can be performed using AutoLISP functions. The higher math functions that AutoLISP currently offers can be broken up into two categories: exponential functions and trigonometry functions.

EXPONENTIAL AND LOGARITHM FUNCTIONS

These are the functions that deal with exponents, roots, and logs. The four most common of these functions are LOG, EXP, SQRT, and EXPT. The LOG function returns the natural log of the numeric argument supplied. The EXP function returns the constant e (the base of the natural logarithm) to a power. The EXPT function returns the numeric argument raised to a power, and the SQRT function returns the square root of the numeric argument supplied. The syntax for these functions are:

```
(LOG number)
(EXP number)
(EXPT number power)
(SQRT number)
```

TRIGONOMETRY FUNCTIONS

In trigonometry there are six basic functions (sine, cosine, tangent, secant, cosecant, cotangent) that can be used to solve problems. The vast majority of problems dealing with angles can be solved using either these functions or a derivative of these functions. Currently AutoLISP offers the direct use of three of these functions: sine, cosine and arc tangent (inverse of tangent). The syntax for these functions are:

```
(SIN ang)
(COS ang)
(ATAN num1 num2)
```

 Note: sin θ = opposite / hypotenuse, cosine θ = adjacent / hypotenuse, Tan θ = sin θ / cos θ, cotan θ = cos θ/sin θ. Sine + Cosine is a dimensionless ratio. In other words the result has no units. See Figure 2–5.

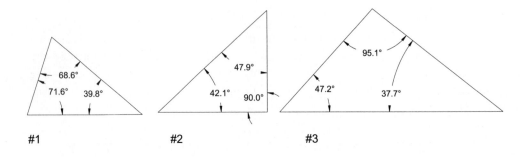

#1   #2   #3

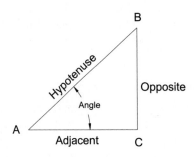

$$Sine = \frac{Opposite}{Hypotenuse}$$

$$Cosine = \frac{Adjacent}{Hypotenuse}$$

$$Tan = \frac{Opposite}{Adjacent}$$

Figure 2–5 *Trigonometric functions*

The SIN function returns in radians the sine of the angle specified. For example:

    Command: **(SIN 55)** (ENTER)
    -0.999755 *(Sine of the angle returned in radians.)*
    Command: **(SIN 90)** (ENTER)
    0.893997 *(Sine of the angle returned in radians.)*
    Command: **(SIN 180)** (ENTER)
    -0.801153 *(Sine of the angle returned in radians.)*

The COS function returns in radians the cosine of the angle specified. For example:

Command: **(COS 55)** (ENTER)
0.0221268 *(Cosine of the angle returned in radians.)*
Command: **(COS 90)** (ENTER)
-0.448074 *(Cosine of the angle returned in radians.)*
Command: **(COS 180)** (ENTER)
-0.59846 *(Cosine of the angle returned in radians.)*

The ATAN function does a little more than just return the arc tangent of a specified angle. If only one argument is furnished, then the function returns the arc tangent of that number in radians. If two numbers are supplied, then the function finds the arc tangent of the first number divided by the second number. Again, this is returned in radians. It should be noted that if the second number supplied is zero, the function returns a value of ±1.570796. The determining factor as to whether or not the value is positive or negative is the sign of the first number supplied. For example:

Command: **(ATAN 4 5)** (ENTER)
0.674741 *(Tangent of the angle returned in radians.)*
Command: **(ATAN 45 23)** (ENTER)
1.0983 *(Tangent of the angle returned in radians.)*
Command: **(ATAN 2 0)** (ENTER)
1.5708 *(Tangent of the angle returned in radians.)*
Command: **(ATAN -4 0)** (ENTER)
-1.5708 *(Tangent of the angle returned in radians.)*

 Note: To return the sine, cosine or tangent of an angle in degrees, the angle supplied to the corresponding trig function must be multiplied by (3.14 / 180) or 0.017453292. For example, (COS (* 55 0.017453292)) yields 0.573576436.

APPLICATION – CALCULATING BEND ALLOWANCES

Now that the groundwork has been established for planning and writing an AutoLISP program, it is time to apply these concepts to a program that will calculate the bend allowance necessary to develop flat patterns of various rectangular shapes. These shapes will all be constructed of the same material (soft steel) but will vary in thickness. The following phases illustrate how this program could be written.

Phase 1   Planning the Program

Statement of Problem

Construct a program that will calculate the bend allowance necessary to construct a flat pattern of various rectangular containers.

Given:

Width of sheet-metal used = varied

Type of metal used = soft Steel

Find:

Bend allowance	(calculate)
Material Thickness	(prompt user)
Radius of Bend	(prompt user)
Angle of Bend	(prompt user)

Formulas:

Allowance =[(0.64 × Thickness) + (0.5 × 3.14 × Radius of Bend)] × (angle°/90°)

Taken from Chapter 6 of *Descriptive Geometry: An Integrated Approach Using AutoCAD*, Delmar Publishers

Diagram:

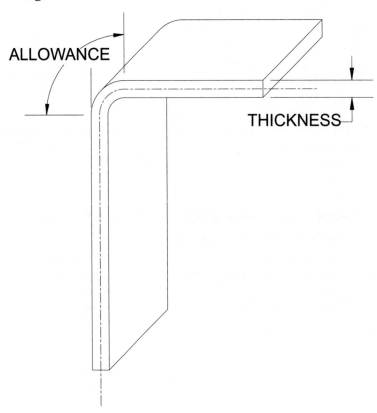

Assumptions:

None

Calculations:

Allowance = (* (+ (* 0.64 thickness) (* (* 0.5 3.14) radius_of_bend)) (/ angle 90))

Check Answer:

Allowance = [(0.64 × Thickness) + (0.5 × 3.14 × Radius of Bend)] × (angle°/90°)

$$\frac{(90°)(Allowance)}{Angle} = 0.64 Thickness + 1.57 Radius\ of\ Bend$$

$$\frac{(90°)(Allowance)}{Angle} - 0.64 thickness = 1.57\ Radius\ of\ Bend$$

$$[\frac{(90°)(Allowance)}{Angle} - 0.64 thickness] / 1.57 = Radius\ of\ Bend$$

 Note: This step will not be inserted into the program at this point in time.

Phase 2   Writing the Pseudo-Code

Start program

    Prompt User to "Enter Thickness of Metal: ".

    Prompt user to "Enter Radius of Bend :".

    Prompt User to "Enter Angle of Bend :".

    Using the Formula (* (+ (* 0.64 thickness) (* (* 0.5 3.14) radius_of_bend)) (/ angle 90)) calculate bend allowance.

    Display Bend Allowance

End Program

Once this phase has been completed, the code is created using a word processor and saved as an ASCII text file. See Figure 2–6.

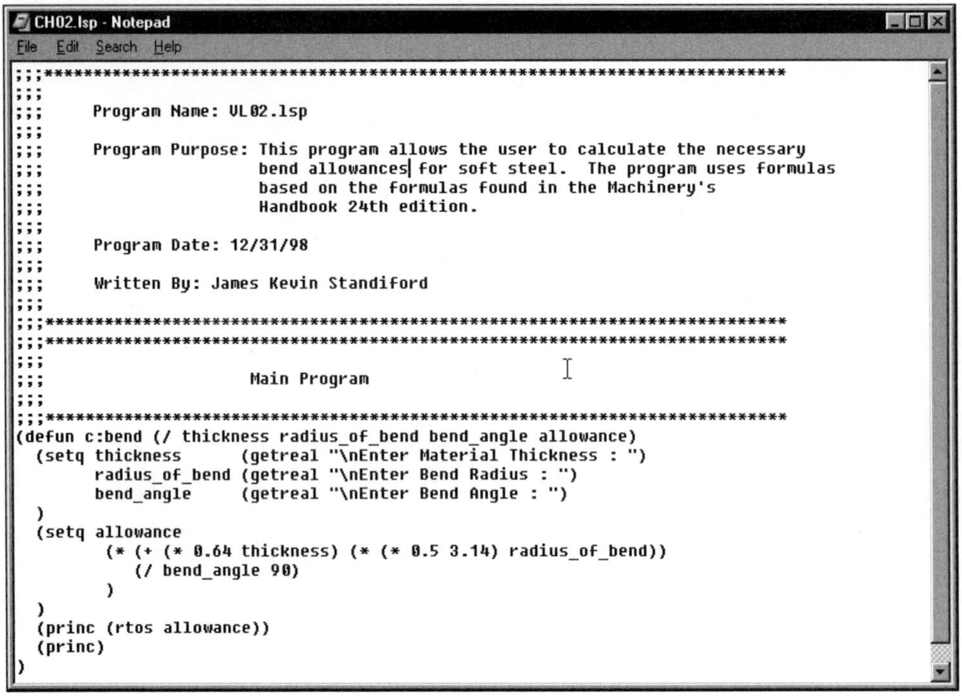

Figure 2–6 *Chapter 2 program in WordPad*

The completed program should look like:

```
;;;*********************************************************************
;;;
;;;     Program Name: VLO2.lsp
;;;
;;;     Program Purpose: This program allows the user to calculate the
;;;                      necessary bend allowances for soft steel.  The
;;;                      program uses formulas based on the formulas
;;;                      found in the Machinery's Handbook 24th edition.
;;;
;;;     Program Date: 12/31/98
;;;
;;;     Written By: James Kevin Standiford
;;;
;;;*********************************************************************
;;;*********************************************************************
;;;
```

```
;;;              Main Program
;;;
;;;******************************************************************
(defun  c:bend  (/  thickness  radius_of_bend  bend_angle  allowance)
    (setq        thickness        (getreal "\nEnter Material Thickness : ")
        radius_of_bend (getreal "\nEnter Bend Radius : ")
        bend_angle      (getreal "\nEnter Bend Angle : ")
    )
    (setq                              allowance
        (* (+ (* 0.64 thickness) (* (* 0.5 3.14) radius_of_bend))
            (/ bend_angle 90)
        )
    )
    (princ (rtos allowance))
    (princ)
)
(princ)
```

SUMMARY

Currently there are ten data types that the AutoLISP evaluator will recognize when processing an expression. These data types are Subrs and External Subrs, Integers, Real (Real numbers), Strings, Lists, Selection Sets, Entity Names, File Descriptors, VLA, and Symbols and Variables. The data type variables can be further broken down into the sub-classifications Global and Local. A global variable is one that retains its value after the program has finished executing, whereas a local variable value returns to nil once the program has finished executing. Both local and global variables in AutoLISP are declared through the SETQ function. However, local variables are distinguished by being listed in the argument-list option of the DEFUN function.

The prompting and receiving of information is a critical aspect for any computer program. In AutoLISP one way of accomplishing this is with the GETXXX functions. These functions enable the programmer to retrieve a variety of information, ranging from AutoCAD system information to real numbers. It is essential that the programmer master the use of these functions.

An important aspect of any computer program is its ability to interact with its human counterpart. A program must have the capability of providing the user with its results. Two methods of doing this are by writing the information to the command prompt area of AutoCAD and to a file. Either one of these methods can be accomplished through the PRXXX functions or the WRITE-XXX functions. The PRXXX functions have one advantage over the WRITE-XXX functions: they allow the programmer to display either a string or the result returned from another AutoLISP expression. Before information can be written to a file, that file must be open and ready to receive information. The AutoLISP function that is designed to prepare files for the exchange of information is the OPEN function. This function has three options associated with it: a file can be read only, write only or appended. Once a file has been opened and the information either retrieved from it or written to it, that file must be closed. This can be accomplished through the CLOSE function.

Because engineering is deeply rooted in mathematics and physics, it is imperative that any computer programming language used to solve these types of problems have an arsenal of predefined math functions. In AutoLISP there are several functions designed for number processing. These functions range from basic math functions (addition, subtraction, etc.) to trigonometry functions (sine, cosine and arc tangent).

REVIEW QUESTIONS

1. Describe the difference between an AutoLISP integer and an AutoLISP real number.

2. How does the GETREAL function differ from the GETSTRING function in the way they handle real numbers?

3. How do entity names differ from handles? What is the purpose of both?

4. Describe the difference between local and global variables. When are local variables preferred? When are global variables preferred?

5. How are local variables declared?

6. Compare the AutoLISP data types **file descriptors** and **entity names**.

7. What AutoLISP function is used to open a file?

8. Describe in detail the difference between the AutoLISP functions PRIN1, PRINC, and PRINT. How do these functions differ from WRITE-LINE?

9. When is the WRITE-LINE function most often used?

10. Given the following equations, write an AutoLISP expression for each.

$X^2 + 2X - 4$

$X^4 + 4X^3 + 7X^2 + 1$

$R_1 + R_2 / (R_1)(R_2)$

$(2 + 5) * (2 * (4 - 5) / 6)$

L / L_o

CHAPTER 3

List Processing, Manipulating and Converting Strings, Making Decisions

OBJECTIVES

Upon completion of this chapter the reader will be able to:

- Define the term element
- Create a list using the AutoLISP LIST function
- Combine two or more lists into a single list using the APPEND function
- Extract elements from a list using the CAR, CDR, CADR and CADDR functions
- Determine the length of a list using the LENGTH function
- Convert numeric data to string data using the RTOS and ITOA functions
- Convert string data to numeric data using the ATOF and ATOI functons
- Merge two or more text strings together using the STRCAT function
- Truncate a text string using the SUBSTR function
- Use the IF and COND functions to conditionally evaluate expressions
- Use the AND and OR functions in conjuction with the IF and COND functions to evaluate two or more test expressions
- Use the PROGN function along with the IF function to evaluate more than one THEN and ELSE expression

KEY TERMS AND AUTOLISP PREDEFINED FUNCTIONS

AND	COND	NTH
ANGTOS	Element	OR
ANGTOF	IF	PRINC
APPEND	ITOA	PRINT
ATOF	LENGTH	PROGN
ATOM	LIST	RTOS
CADDR	LISTP	STRCAT
CADR	Logic And Truth Table	SUBSTR
CAR	Logic OR Truth Table	TYPE
CDR	Nesting Functions	

INTRODUCTION TO LISTS AND LIST PROCESSING

One of the many attributes that makes Common LISP an excellent candidate for artificial intelligence and the perfect model for AutoLISP is its ability to store large amounts of data in the form of a list. A list is nothing more than a collection of data that is either related or non-related to a particular entity or event. For example, a list describing the characteristics of a line could include the Entity Name, Entity Type, Entity Handle, Entity Color, Layer, Starting Point, and Ending Point. A list can also consist of a set of instructions or commands describing a particular sequence of events. For example, a list containing the steps for determining the area of a circle given only its diameter would be:

1. Calculate the radius of the circle by dividing its diameter by 2.

2. Square the radius and multiplying the result by 3.14, (producing the area of the circle).

In most cases, lists provide an efficient as well as practical way of storing information over the traditional method of using variables (variables are able to store only a single item, known as an atom). This is especially true if the information used by a program varies from execution to execution. For example, to write a program that would find the equivalent resistance of a series circuit and calculate the voltage drop for each resistor at first seems like an easy task to accomplish. First, add up the resistance of all the resistors in the circuit to find the equivalent resistance. Next, divide the total voltage supplied to the circuit by the equivalent resistance, producing the current supplied to the circuit. Finally, find the voltage drop for each resistor by multiplying the current by the resistance of the individual resistor. However, there is one small problem: the number of resistors could vary from circuit to circuit making it extremely difficult, if not impossible, to create a generic program using variables.

An alternative is to construct a list containing values assigned to each resistor and then set that list equal to a single variable. Not only does this solve the problem of the variant amount of information entered into the program, but it also reduces the number of variables used in the program and the amount of code required to create the program. A list can also contain other lists (these are often referred to as sub-lists). Each entry in a list is referred to as an *Element*. For example, the list (apple orange peach) contains the elements apple, orange and peach. In AutoLISP a list will always start with an opening parenthesis and end with a closing parenthesis.

LISTS IN AUTOCAD

When information is requested from AutoCAD, that information is returned in the form of a list. This can be illustrated with the GETPOINT function. Recall from Chapter 2 that the GETPOINT function prompts the user to either select or enter a point. The result returned by this function is in the form of a list (7.4366 6.16185 3.80986) regardless of whether the user entered the information from the keyboard or selected a point from the graphics screen.

CREATING A LIST

In AutoLISP a list can be generated either by the LIST function (LIST [expr...]) or by placing the elements of the list inside opening and closing parentheses, preceded by a single quote '(4 5 6). The LIST function, when supplied with one or more arguments (expressions and/or non-expressions) returns a list. For example, to create a list containing coordinates 5, 4, 3 the following syntax would be used:

```
(SETQ point (LIST 5 4 3))              ;The expression returns (5 4 3).
```

Once a list containing coordinate information has been created and its value set to a variable, then it may be used with any of the AutoCAD commands where point information is required. For example, to construct a line starting at the coordinates supplied in the previous example, the variable name preceded by an exclamation point would by entered at the "Specify first point" LINE prompt. This would extract the information held by the variable and return it to the LINE command.

```
Command: LINE (ENTER)
Specify first point: !point (ENTER)
(5 4 3)
Specify next point or [Undo]:
```

Other examples of lists created through the LIST function include:

```
(SETQ list_example (LIST "A" "B" "C"))
```

In this example a list is created containing the strings A, B, and C and set to the variable list_example.

```
(SETQ list_example (LIST 0.5 10 "ORANGE" "GRAPE"))
```

In this example a list is created containing the real number 0.5, the integer 10, and the strings ORANGE and GRAPE.

```
(SETQ list_example (LIST '(SETQ d 5) '(SETQ f 7)))
```

In this example a list is constructed containing AutoLISP expressions. In order for a list to contain actual expressions and not their evaluated results, a single quotation mark must precede the list. The single quotation mark tells AutoLISP to return the expression without evaluating it.

COMBINING TWO OR MORE LISTS INTO A SINGLE LIST

Although the LIST function can be used to combine two of more lists into a single list, this function has one major disadvantage in using it in this capacity. Each list that is supplied to the function becomes an element of the newly created list. For example, using the LIST function to combine the lists (5.5 4.3 2.3) and (44.7 39.6 12.4) would yield the list ((5.5 4.3 2.3) (44.7 39.6 12.4)). This is illustrated in the following example:

```
(SETQ list1 (LIST 5.5 4.3 2.3))        ;Creates a new list setting it to
                                       ;the variable list1.
(5.5 4.3 2.3)                          ;Result returned from the previous
                                       ;expression.
(SETQ list2 (LIST 44.7 39.6 12.4))     ;Creates a new list setting it to
                                       ;the variable list2.
(44.7 39.6 12.4)                       ;Result returned from the previous
                                       ;expression.
(SETQ list3 (LIST list1 list2))        ;Combines list1 and list2 into a
                                       ;single list setting it to the
                                       ;variable list3.
((5.5 4.3 2.3) (44.7 39.6 12.4))       ;Result returned from the previous
                                       ;expression.
```

As illustrated in the example above, the lists themselves become entities that are considered as single elements in the newly formed list. In some cases this result may be desirable. In those instances where this result is not acceptable, another method must be employed. In cases where the elements of the individual list are needed to create a new list, then the APPEND function must be used. The APPEND function (APPEND list ...) extracts the elements from each individual list supplied and returns a single list. For example, if the lists from the previous example were supplied as arguments to the APPEND function, the result (5.5 4.3 2.3 44.7 39.6 12.4) would be returned. This is illustrated in the following example:

```
(SETQ list1 (LIST 5.5 4.3 2.3))        ;Creates a new list setting it to
                                       ;the variable list1.
```

```
(5.5  4.3  2.3)                        ;Result  returned  from  the  previous
                                       ;expression.
(SETQ  list2  (LIST  44.7  39.6  12.4))  ;Creates  a  new  list  setting  it  to
                                       ;the  variable  list2.
(44.7  39.6  12.4)                     ;Result  returned  from  the  previous
                                       ;expression.
(SETQ  list3  (APPEND  list1  list2))  ;Combines  both  list1  and  list2  into
                                       ;a  single  list  setting  it  to  the
                                       ;variable  list3.
(5.5  4.3  2.3  44.7  39.6  12.4)      ;Result  returned  from  the  previous
                                       ;expression.
```

DETERMINING IF AN ITEM IS A LIST

From time to time it becomes necessary to determine if the item that is currently set to a variable or the information that is passed to a program is in the form of a list. AutoLISP currently supplies two functions for this purpose, the LISTP function and the TYPE function. While the syntax for both functions is identical, ((LISTP item) and (TYPE item)), the result returned by each these functions is different. The LISTP function when supplied with an item returns either T (if the item is a list) or nil (if the item is not a list). The TYPE function returns the actual data type. Therefore, this function has a wider range of applications that it can be used for; it is not limited to identifying lists. In the following example, both the LISTP and the TYPE functions are used to check a variety of data types.

```
(LISTP  3.4)                           ;Tests  the  item  3.4  to  determine  if
                                       ;the  item  is  a  list.
nil                                    ;The  item  is  a  real  number  and  the  value
                                       ;returned  by  this  function  is  nil.
(LISTP  3)                             ;Tests  the  item  3  to  determine  if
                                       ;the  item  is  a  list.
nil                                    ;The  item  is  an  integer  and  the  value
                                       ;returned  by  this  function  is  nil.
(LISTP  "Text  String")                ;Tests  the  item  "Text  String"  to
                                       ;determine  if  the  item  is  a  list.
nil                                    ;The  item  is  a  string  and  the  value
                                       ;returned  by  this  function  is  nil.
(LISTP  (LIST  "apple"  "orange"  "grape"))  ;Test  the  result  supplied  by  the  LIST
                                       ;function.  Because  the  LIST  function  is
                                       ;used  to  create  a  list,  the  result
                                       ;returned  by  this  function  is  T.
(TYPE  3.4)                            ;Tests  the  item  3.4  to  determine  its
                                       ;data  type.
REAL                                   ;The  result  returned  by  the  previous
                                       ;test  indicates  that  the  item  is  a
                                       ;real  number.
```

(TYPE 3)	;Tests the item 3 to determine its ;data type.
INT	;The result returned by the previous ;test indicates that the item is an ;integer.
(TYPE "Text String")	;Tests the item "Text String" to ;determine its data type.
STR	;The result returned by the previous ;test indicates that the item is a ;string.
(TYPE (LIST "apple" "orange" "grape"))	;Tests the result supplied by the ;LIST function.
LIST	;Because the LIST function is used to ;create a list, the result returned ;by this function is a list.

RETRIEVING INFORMATION FROM A LIST

As mentioned earlier, the individual entries in a list are known as elements. An element can be a string, a real number, an integer, another list, an expression (single quoted), or any combination. So how are the elements of a list extracted? By using one or a combination of two or more of the five basic list retrieval functions: CAR, CDR, CADR, CADDR, and NTH. Although, the syntax for the CXXXX functions is the same, (CXXXX list), their outputs are different. The CAR function returns the first element of a list, while the CDR function creates a list starting with the second element. For example, supplying the list (4.002 3.45 5.7) to both the CAR and CDR functions returns the following results:

(SETQ list_example (LIST 4.002 3.45 5.7))	;Creates a list and sets it to the ;variable list_example.
(CAR list_example)	;Returns the first element of the ;list.
4.002	;Returned value.
(CDR list_example)	;Creates a new list starting with the ;second element.
(3.45 5.7)	;New list created from the CDR function.

The CADR function combines the capabilities of the CAR and CDR functions. It returns only the second element of a supplied list. To illustrate this, the following example uses the list created in the previous example to compare the results returned by the CADR function and an expression containing both the CAR and CDR functions.

(CAR (CDR list_example))	;The CDR function creates a new list ;starting with the second element and ;returning that list to the CAR function. ;The CAR function then returns the

```
                                      ;first element of the list supplied
                                      ;by the CDR function.
3.45                                  ;Returned value.
(CADR list_example)                   ;The CADR function returns only the
                                      ;second element of the supplied list.
3.45                                  ;Returned value.
```

 Tip: Although both the CADR and the combination CAR/CDR functions return the same value, the CADR function provides a shorter route to obtaining the same results. If only the second element of a list is required, then the CADR function not only saves valuable time and the possibility of making a mistake, it could also reduce the overall size of the program, resulting in a program that loads and runs faster.

The last of the CXXXX functions is the CADDR function. This function returns the third element of a list. It could be compared to using the CDR function twice, followed by the CAR function, or the CDR function followed by the CADR function. Again, the advantage of using this function in place of any of the combinations reduces the possibility of making a mistake as well as shortens the length of the program. This is illustrated in the following example, where both the CADDR function and the two combinations are supplied with the list from the previous two examples:

```
(CADDR list_example)                  ;The CADDR function returns only the
                                      ;third element of a list.
5.7                                   ;Returned value.
(CADR (CDR list_example))             ;The CDR function creates a new list
                                      ;starting with the second element
                                      ;and returns that list to the CADR
                                      ;function. The CADR function then
                                      ;returns the second element of the
                                      ;list supplied by the CDR function.
5.7                                   ;Returned value.
(CAR (CDR (CDR list_example)))        ;The first CDR function creates a
                                      ;new list starting with the second
                                      ;element and returns that list to
                                      ;the next CDR function in the expression.
                                      ;This CDR function then creates
                                      ;another list starting from the
                                      ;second element supplied to it from
                                      ;the previous CDR function. The list
                                      ;is then passed on to the CAR function
                                      ;where the first element is returned.
5.7                                   ;Returned value.
```

An alternative to using the CXXXX functions is the NTH function (NTH number list). This function, when supplied with a list, returns the element specified by the programmer. The programmer specifies which element is to be returned by entering the integer value representing the position of the element in the supplied list. The elements are numbered starting with zero and continuing in the positive direction. For example, using the list provided in the previous example, the third element is extracted as follows.

```
(NTH 2 list_example)        ;The NTH function returns an elements
                            ;specified by the programmer. In this
                            ;example the programmer has specified
                            ;the third element. The first element
                            ;is considered to be located at the
                            ;zero position while the third
                            ;element is considered to be at the
                            ;second position.
        5.7                 ;Returned value.
```

DETERMINING THE LENGTH OF A LIST

One of the disadvantages of using the NTH function is that the programmer must know the exact position of the element before that element can be extracted. If the last element of a list is required for a particular calculation, then the programmer would need to know its exact position or at least the length of the list before the information could be extracted. This presents a problem because lists can vary in length and the exact length of a list may not be known. The programmer must have a means of obtaining the length of a list in order to complete this task. AutoLISP does supply a function for determining the length of a list; it is the LENGTH function (LENGTH list). This function returns a list's length in the form of an integer. For example, if the list used in the previous examples were supplied to the LENGTH function, the integer 3 would be returned, representing the length of the supplied list. The result determines only the length of a supplied list; it does not reveal the position of any of the elements. To extract the last element from a list, the integer value returned would have to be reduced by one. The following example illustrates how the integer value returned from the LENGTH function combined with the NTH function can be used to extract the last element in a list.

```
(SETQ list_example (LIST  5.66  7.34  2.1  3.4  5.8  9.99))
                                ;Creates a list, setting it equal
                                ;to the variable list_example.
(SETQ list_length (LENGTH list_example))
                                ;Returns the length of the list previously
                                ;created and sets that value equal
                                ;to the variable list_length.
(SETQ list_length (- list_length 1))  ;Reduces the integer representing the
                                ;length of the list by one and sets that
                                ;value to the variable list_length.
```

```
(SETQ list_element (NTH list_length list_example))
                                   ;Extracts the last element of the
                                   ;list supplied.
```

If expressions are combined, the same result can be achieved using fewer lines of code. Although this method reduces the amount of typing and the possibility of making a mistake, it is somewhat difficult for beginning programmers to interpret.

```
(SETQ list_example (LIST  5.66  7.34  2.1  3.4  5.8  9.99))
                                   ;Creates a list, setting it equal to
                                   ;the variable list_example.

(SETQ list_element (NTH (- (LENGTH list_example) 1) list_example))
                                   ;This expression is evaluated by first
                                   ;determining the number of elements
                                   ;assigned to the variable list_example.
                                   ;The result is then reduced by one and
                                   ;passed to the NTH function as the
                                   ;number argument for that function.
                                   ;The NTH function then extracts the last
                                   ;element from the variable list_example.
                                   ;The result is passed to the SETQ function
                                   ;where it is set to the variable
                                   ;list_element.
```

PROCESSING THE ELEMENTS OF A LIST

Once the elements of a list have been extracted, their values can then be used to carry out necessary data manipulations to achieve desired results. In Chapter 2, basic techniques were discussed where information can be supplied to a program, from either the keyboard or file or by selection from the graphics screen, and then processed, with the result either displayed or stored in a file. Until now, the primary emphasis has been the manipulation of numeric data. However, numeric data is not the only data type that may require manipulation; string data can also undergo processing.

MANIPULATING AND CONVERTING STRINGS

Altering string data can be as simple as changing the string's format, combining two or more strings, or converting numeric data to a single string. Numeric data must be converted from its native data type to a string data type before a merger can take place. A program that calculates the equivalent resistance, current, and voltage drop of each resistor in a series circuit is not only expected to properly function and return an accurate result, it is also expected to present the result in a form that is easily understood. To accomplish this, the numeric data returned from the calculations must be enhanced with text descriptors, completely describing the program's output. Otherwise the user is bombarded with a series of numbers that might be extremely difficult, if not impossible, to interpret. In cases like this, the information must undergo a refinement process before it is presented to the user.

CONVERTING NUMERIC DATA TO STRING DATA

Before any numeric data (real or integers) can be combined with a string, it must undergo a conversion process to transform it into a string data type. For integers or real numbers, the process is accomplished by either the RTOS or ITOA function. If the numeric data is an angle, then a different function must be used.

To convert a real number to a string, the RTOS function is used. The syntax for this function is (RTOS number [mode [precision]]). The *mode* argument of this function determines whether the value is returned in scientific, decimal, engineering, architectural, or fractional form. The *precision* argument sets the number of decimal places returned by this function. Both the *mode* and *precision* argument values correspond to the AutoCAD LUNITS and LUPREC system variables. If no arguments are supplied, then the function uses the current settings of both the LUNITS and LUPREC system variables. Setting the LUNITS system variable to an integer between 1 and 5 causes AutoCAD to display linear units in scientific, decimal, engineering, architectural, or fractional notation. For example, to convert the real number 4.2568 to a string using the architectural format, the following syntax, would be used:

```
(RTOS 4.2568 4)                    ;This returns a value of "4 1/4\"".
```

 Note: The \" at the end of the returned value is the escape code for the inch symbol.

To convert an integer to a string, the ITOA (ITOA int) function is used. This function is similar to the RTOS function in that the value returned is a string, but it differs from the RTOS function in the fact that it does not allow the programmer to specify the mode or precision displayed. If the numeric data must be returned in a particular form, then the RTOS function must be used. For example, to convert the integer 5 to a string, the following syntax would be used:

```
(ITOA 5)                           ;This returns a value of "5".
```

 Note: If a real number is passed to this function, the function then returns the error message *** ERROR: bad argument type: fixnump.

CONVERTING STRING DATA TO NUMERIC DATA

If numeric data can be converted to string, then it stands to reason that it can be converted back again. In AutoLISP this can be accomplished by either the ATOF or ATOI function. The ATOF (ATOF string) function converts string values back to their native format, real numbers. For example, to convert the string "4.5" back to a real number the following would be used:

```
(ATOF "4.5")                          ;This returns a value of 4.5.
```

 Note: If a string is passed to this function that does not contain a numeric value, then the function returns a value of 0.0. If a string is passed where the first character is not a number, then the function again returns the value of 0.0. Finally, if an alphanumeric string is passed to the function and the first character is a number, then the function returns a value equal to the real number complement of the numeric portion of the string.

If a string can be converted to a real number, then it can also be converted to an integer. This is accomplished by the ATOI (ATOI string) function. To convert the string "5" to an integer the following expression would b e used:

```
(ATOI "5")                            ;This returns a value of 5.
```

 Note: If a string is passed to this function that does not contain a numeric value, then the function returns a value of 0. If a string is passed where the first character is not a number, then the function again returns the value of 0. If a real number is passed to the function, then the decimal portion of the number is truncated and only an integer is returned. Finally, if an alphanumeric string is passed to the function and the first character is a number, then the function returns a value equal to the integer complement of the numeric portion of the string.

MERGING TEXT STRINGS

Once a number has been converted from its numeric data type to a string, it can then be combined with other string values. For example, suppose that a program is needed to calculate the voltage drop across a group of resistors arranged in series. If the program calculates and returns the value for one resistor, then the result returned by the program can be easily interrupted. However, if the program calculates the value of several resistors, the program must also provide a description of the results so that the user can associate the voltage drop with its corresponding resistors. In other words, the results should be combined with a description that provides the user with a detailed analysis of the results. To merge two or more text strings, the AutoLISP function STRCAT (strcat [string [string]...]) is employed. The *string* arguments associated with this function can either be strings or expressions that when evaluated result in a string. This is illustrated in the following program, where the user is prompted to enter the voltage for the circuit and the resistance for two resistors in series. The program then uses information to calculate the circuit's equivalent resistance, overall current, and the voltage drop across each resistor. Finally, the results are displayed to the screen.

```
;;;******************************************************************
;;;     Program Name: Resistance.lsp
;;;     Program Purpose: Calculate the voltage drop across two resistors
;;;                         in series, as well as the current for the circuit.
;;;     Date: 1/13/99
;;;     Programmed By: James Kevin Standiford
;;;
;;;******************************************************************
;;;
;;;     Main Program
;;;
;;;******************************************************************
(DEFUN C:resistance (/      voltage       resistance_1
                resistance_2    equivalent    value_1
                value_2    current
                )
;;;
;;; Obtain Information and Perform Calculations
;;;
  (SETQ
    voltage          (GETREAL "\nEnter voltage of circuit : ")
    resistance_1    (GETREAL "\nEnter resistance of resistor #1 : ")
    resistance_2    (GETREAL "\nEnter resistance of resistor #2 : ")
    equivalent        (+ resistance_1 resistance_2)
    current            (/ voltage equivalent)
    value_1            (* current resistance_1)
    value_2            (* current resistance_2)
  )
;;;
;;; Print Information to Both a File and Screen
;;;
  (PRINC
    (STRCAT "\nThe overall current for this circuit is "
                    (RTOS current) "amps"
            )
  )
  (PRINC
```

```
      (STRCAT "\nThe equivalent resistance for this circuit is "
                   (RTOS equivalent) "ohms"
           )
    )
    (PRINC
      (STRCAT "\nThe voltage drop across resistor #1 is "
                   (RTOS value_1) "volts"
           )
    )
    (PRINC
      (STRCAT "\nThe voltage drop across resistor #2 is "
                   (RTOS value_2) "volts"
           )
    )
    (PRINC)
  )
(princ)
(princ "\nEnter Resistance to start program ")
(princ)
```

TRUNCATING A TEXT STRING

From time to time it becomes necessary to truncate string data, (also referred to as a *Sub-String*). For example, this would be necessary if it is decided that the results from the previous example should be saved to a file that has the same name as the drawing file from which the program is executed but contains a different extension. In this particular case, the file name could be extracted from the drawing using the AutoLISP function GETVAR and the AutoCAD system variable DWGNAME. This poses a problem because the DWGNAME system variable also stores the file extension. Therefore, to save the information to a file with a different extension the string must be truncated. To truncate a string, the AutoLISP SUBSTR (sub-string) function must be used. This function returns a sub-string and allows the programmer to specify the starting and ending points of the string. The syntax for this function is (SUBSTR string start [length]). The *start* argument of this function determines the first character of the string to be returned. This is not an optional argument and must be supplied in the form of a positive non-zero integer (a positive integer greater than zero). If the argument is not supplied, then an error is returned and the program stops execution. If the integer supplied is greater in value than the string length, then the function returns "". The *length* argument determines the length of the sub-string to be returned beginning with the character specified by start argument. This is an optional argument; if it is not supplied then the function returns the remaining portion of the string. This argument is also supplied as a positive integer, but zero may be used (in which case the function would return ""). If an integer is specified that is greater than the remaining portion of the string, then the function will again behave as if the argument has been omitted. To add this capability to the program in the previous example, the following syntax is used:

```
(SETQ name (GETVAR "dwgname"))
(SETQ name_truncated (SUBSTR name 1 8))
(SETQ file_name (STRCAT name_truncated ".RLT"))
```

or

```
(SETQ file_name (STRCAT (SUBSTR (GETVAR "dwgname") 1 8) ".RLT"))
```

 Note: Using fewer variables and combining lines of code in a program will create smaller files as well as faster, cleaner programs. Therefore, the second example would be the better of the two choices. However, for beginning programmers, the first example may be easier to interpret.

DETERMINING THE LENGTH OF A STRING

In the previous example, the variable *name_truncated* is set to a string that is a total of eight characters in length. If the drawing is assigned a name that is less than eight characters in length, then the previous expression would return a value that still incorporates part if not all of the original extension. To compensate for this, the length argument supplied to the SUBSTR function must be flexible. In other words, the length of the string minus four digits (allowing for a three-digit extension plus the period.) would determine its value. To determine the length of a string, the AutoLISP function STRLEN (STRLEN string) must be used. This function returns an integer value representing the overall string length. Adding this function to the previous example allows the program to determine the overall length of the file name, thereby enabling the program to adjust for file names of various length. For example:

```
(SETQ name (GETVAR "dwgname"))
(SETQ length_string (STRLEN name))
(SETQ name_truncated (SUBSTR name 1 (- length_string 4)))
(SETQ file_name (STRCAT name_truncated ".RLT"))
```

or

```
(SETQ
   file_name (STRCAT (SUBSTR (GETVAR "dwgname")
                        1
                        (- (STRLEN (GETVAR "dwgname")) 4)
                     )
                     ".RLT"
           )
)
```

After this is inserted into the resistance program from the previous example, the code now appears as:

```
;;;*********************************************************************
;;; Program Name: Resistance.lsp
;;; Program Purpose: Calculate the voltage drop across two resistors
;;;                  in a series, as well as the current for the cirrcuit.
;;; Date: 1/13/99
;;; Programmed By: James Kevin Standiford
;;;
;;;*********************************************************************
;;;
;;; Main Program
;;;
;;;*********************************************************************
(DEFUN C:resistance (/              voltage          resistance_1
            resistance_2    equivalent     value_1
            value_2         current        name
            length_string name_truncated file_name
           )
;;;
;;; Obtain Information and Perform Calculations
;;;
  (SETQ
    file_name (STRCAT (SUBSTR (GETVAR "dwgname")
                         1
                         (- (STRLEN (GETVAR "dwgname")) 4)
                      )
                      ".RLT"
              )
    file       (open file_name "a")
    voltage    (GETREAL "\nEnter voltage of circuit : ")
    resistance_1 (GETREAL "\nEnter resistance of resistor #1 : ")
    resistance_2 (GETREAL "\nEnter resistance of resistor #2 : ")
    equivalent      (+ resistance_1 resistance_2)
    current    (/ voltage equivalent)
    value_1    (* current resistance_1)
    value_2    (* current resistance_2)
  )
;;;
;;; Print Information to Both a File and Screen
;;;
  (PRINC
    (PRINC (STRCAT "\nThe current drawn by this circuit is "
```

```
                          (RTOS current)
          )
          file
     )
   )
   (PRINC
     (PRINC (STRCAT "\nThe equivalent resistance for this circuit is "
                 (RTOS equivalent)
          )
          file
     )
   )
   (PRINC
     (PRINC (STRCAT "\nThe voltage drop across resistor #1 is "
                 (RTOS value_1)
          )
          file
     )
   )
   (PRINC
     (PRINC (STRCAT "\nThe voltage drop across resistor #2 is "
                 (RTOS value_2)
          )
          file
     )
   )
   (close file)
   (PRINC)
  )
  (princ)
  (princ "\nEnter Resistance to start program ")
  (princ)
```

MISCELLANEOUS STRING MANIPULATIONS

The major string manipulation functions that a programmer is mostly likely to use have been addressed in the previous section. However, there are occasions that may require changes to a string's case or conversion of a string representing an angle to its radian equivalent or vice versa. AutoLISP has three string manipulation functions for these situations: ANGTOF, ANGTOS, and STRCASE. Both the ANGTOF and ANGTOS functions are designed to work with angular information. The ANGTOF function (ANGTOF string [units]) converts a string representing an angle to a corresponding real number representing the angle in radians. The *string* argument of this function is the string value of the angle to be converted to radians. The *units* argument of this function is an integer used to specify what unit the string is currently in (Degrees - 0,

Degrees/Minutes/Seconds - 1, Grads - 2, Radians - 3, or Surveyor's Units - 4). This argument corresponds with the AutoCAD system variable AUNITS, and if a value is not specified, the settings of the AUNITS system variable are used. An example of this function is provided below; a string representing an angle in gradients is converted to its equivalent real number in radians.

```
(ANGTOF "200g" 4)                    ;This returns a value of 3.14159.
```

The counterpart to ANGTOF function is the ANGTOS function (ANGTOS angle [units [precision]]). This function when supplied with a real number (specified in radians) converts it to a string in either Degrees - 0, Degrees/Minutes/Seconds - 1, Gradients - 2, Radians - 3 or Surveyor's Units - 4. The returned value is determined by the *units* argument. If this argument is not supplied, then the function uses the current value assigned to the AutoCAD AUNITS system variable. Along with selecting the units of the returned value, the precision with which the conversion is carried out can also be specified. If this argument is not furnished, then the function uses the current value assigned to the AUPREC system variable. An example of this function is provided below, where a string representing an angle in radians is converted to a string representing the angle in decimal degrees with a precision of two decimal places.

```
(ANGTOS 0.78656 0 2)                 ;This returns a value of "45.07".
```

Finally, the last of the string manipulation functions is the STRCASE function (STRCASE string [which]). This function, when supplied with a string, returns that string in either all uppercase or lowercase letters. The decisive factor as to which case the string is converted to is the *which* argument. If the *which* argument is a variable whose value is equal to nil, then the string is returned in all uppercase letters. If the *which* argument is not equal to nil, then the function returns a string with all lowercase letters. For example:

```
(SETQ sample "This is a test")       ;This returns "This is a test".
(STRCASE sample T)                   ;This returns "this is a test".
(STRCASE sample nil)                 ;This returns "THIS IS A TEST".
```

MAKING DECISIONS

Frequently in computer programs, when information is obtained, a decision must be made by the program as to what to do with that information either in the processing or displaying phase of the program. For example, when using the AutoCAD CIRCLE command, the user is prompted to either construct a circle by using the 3Point, 2Point, Tangent Tangent Radius options or select the center point of the circle. If the user selects a point for the center of the circle, then the command prompts the user to specify either the Radius or the Diameter of the circle. The radius option of this process is the default; therefore to specify the diameter of the circle the user would first enter **D** and (ENTER) in place of specifying the radius. If the user chooses the diameter option, the command then prompts the user to enter a diameter for the circle. In every step of this process a decision by the CIRCLE command has been made on exactly how to proceed based on the information provided by the user. This process is summed up in the flowchart presented in Figure 3–1.

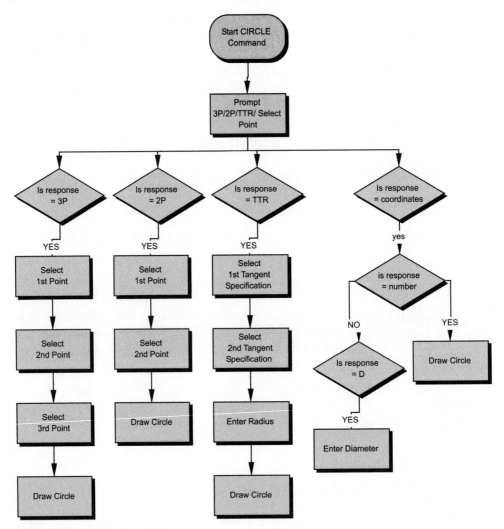

Figure 3–1 *Flowchart of the AutoCAD* CIRCLE *command*

One way a program can make decisions is with the use of IF statements. IF statements (also known as conditional evaluators) test an argument supplied against a specified condition. If the argument meets the criteria, then the program is directed to react in one direction. If the arguments fail to meet the criteria, then the program will either skip that set of instructions or execute a totally different set of instructions. Either way the program has made a decision based upon a set of criteria and supplied information.

USING THE AUTOLISP IF FUNCTION

All computer programming languages provide a means of testing the condition of an argument and then performing a sequence of steps based on the result of that test. In most computer programming languages this is done with an IF THEN ELSE statement. AutoLISP has its own version of the IF THEN ELSE statement called the IF (IF testexpr thenexpr [elseexpr]) function. The IF function, when supplied with a test expression, evaluates that expression and executes either the *thenexpr* or the *elseexpr* argument. The *testexpr* argument is the expression that is to be tested and can be any of the AutoLISP data types previously discussed. For example, to determine if the data supplied by the user is equal to 4, then either of the following test expressions can be used:

```
(equal input 4)
(= input 4)
```

In both expressions if the variable input is equal to 4, then the expression returns the symbol T (true); otherwise the symbol nil (false) is returned. Once the *testexpr* has been evaluated, the result is then passed to the IF function. If the symbol T is passed, then the *thenexpr* argument is evaluated, but if the symbol nil is passed, then the *elseexpr* argument is evaluated. This concept is illustrated in the following example, where the user is prompted to enter a number. If the number supplied is less than or equal to 10, the number is divided by 1.5 and the quotient is set equal to the variable less. However, if the number is greater than 10, the number is divided by 4.5 and the quotient is set equal to the variable greater.

```
(SETQ answer (GETREAL "\nEnter a number : "))
(IF (<= answer 10)
  (SETQ less (/ answer 1.5))
  (SETQ greater (/ answer 4.5))
)
```

 Note: The elseexpr is optional, and if not supplied, then the function simply executes the next expression.

Evaluating More than One Thenexpr or Elseexpr Argument

Often when a computer program is developed it becomes necessary to supply more than one *thenexpr* argument to an IF function. For example, suppose that in the previous section the quotient produced by the expression (/ answer 1.5) is to be multiplied by a second number supplied by the user. The user is prompted only if the *testexpr* argument is evaluated as true and only after the expression (/ answer 1.5) has been evaluated. To accomplish this, multiple IF expressions can be written or a single expression where the PROGN function is used. The PROGN function (PROGN [expr]...) allows for multiple expressions to be assigned as a single argument. The function evaluates each expression supplied sequentially and returns the result of the last expression evaluated. When it is used with the IF function, the need for multiple IF expressions is reduced. For example:

```
(SETQ answer (GETREAL "\nEnter a number : "))
(IF (<= answer 10)
  (PROGN
   (SETQ less (/ answer 1.5)
      ans (GETREAL "\nEnter a second number : ")
      less (* less ans)
   )
   (PRINC (STRCAT "\nThe answer is " (RTOS less)))
  )
  (SETQ greater (/ answer 4.5))
 )
```

The same technique can be used if more that one *elseexpr* argument is to be supplied to an IF expression. For example, suppose that the quotient produced by the expression (/ answer 4.5) is to be raised to a power specified by the user. Again, the user is prompted to enter the exponent only if the *testexpr* evaluates nil and only after the (/ answer 4.5) expression has been evaluated.

```
(SETQ answer (GETREAL "\nEnter a number : "))
(IF (<= answer 10)
(PROGN                              ;Start of first PROGN
                                    ;function.
   (SETQ less (/ answer 1.5)
         ans (GETREAL "\nEnter a second number : ")
         less (* less ans))
   (PRINC (STRCAT "\nThe answer is " (RTOS less)))
 )                                  ;End of first PROGN ;function.
(progn                              ;Start of second PROGN ;function.
   (SETQ greater (/ answer 4.5)
         ans    (GETREAL "\nEnter exponent : ")
         greater (EXPT greater ans))
   (PRINC (STRCAT "\nThe answer is " (RTOS greater)))
 )                                  ;End of second PROGN
                                    ;function.
 )                                  ;End of IF functon.
```

Nesting Functions

Often a decision cannot be made based on a single IF function but instead requires a series of decision expressions where the results from the previous test expression determine the next phase of the decision process. In other words, the final decision is made with a single IF expression that contains one or more embedded IF functions. This process of embedding IF expressions within one another is referred to as *Nesting*. In AutoLISP there is no limit to the number of expressions that can be nested inside another, but extreme caution should be taken to ensure that each expression is properly closed; otherwise the task of debugging could become overwhelming. Figure 3–2 provides an example of a nested IF expression.

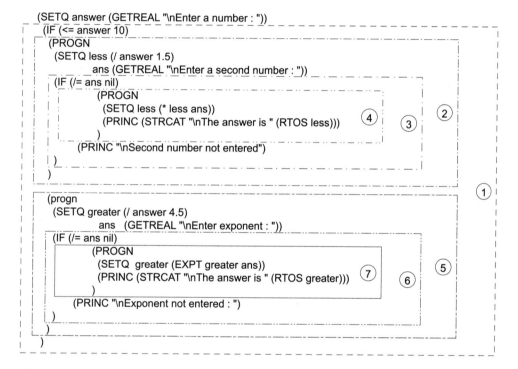

1. Main IF expression
2. Thenexpr argument of main IF expression
3. Nested IF expression
4. Thenexpr argument of nested IF expression {section three}
5. Elseexpr argument of main IF expression
6. Nested IF expression
7. Thenexpr argument of nested IF expression {section six}

Figure 3–2 *Nested IF expressions*

In the example above, the section of the program labeled 1 is the main IF expression. This expression tests the value of the variable answer to determine if the value is either less than or equal to 10. If the value is less than or equal to 10, the program advances to section 2. This section is the *thenexpr* argument of the main IF expression. In this section, the value of the variable answer is divided by 1.5 and the result is set to the variable less. Next the user is prompted to enter a number. Once this has been completed, the program then passes that result to the nested IF expression labeled 3. If the user entered a value other than nil, then the program moves on to section 4, where the value of the variable less is multiplied by the input supplied by the user from section 3. If the user fails to supply a value, then the program skips section 4 and goes straight to the expression (PRINC "\nSecond number not entered).

If the value of the variable answer is greater than 10, then the program executes the *elseexpr* of the main IF function labeled section 5. This section then divides the input supplied by the user and sets the result equal to the variable greater. Once this has been completed, the user is prompted to enter an exponent. The information is then passed to section 6, where it is tested to determine if the user entered a number or simply pressed ENTER. If the user pressed ENTER, then the program proceeds to the *elseexpr* argument of section 7. If the user did enter a number, then the program runs section 7.

Using More than One Test Expression Argument

In the previous section, the concept of nesting IF expressions was addressed as a means of making a decision based upon the results of a previous decision. However, there are times when a decision must be made where multiple test expressions must be evaluated simultaneously. In these situations a nested IF expression may not return the desired results. For example, suppose that an application is needed to select a bearing from a vendor's catalog based upon the bearing's bore size and its ± tolerances. If the program is patterned using the chart shown in Figure 3–3, then in order for the bearing 308K to be selected, the bore size must be equal to 1.5748 inches with a tolerance of +0.0000 and − 0.005 inches. To accomplish this, the program would need to compare the tolerances and bore sizes entered by the user to those stored in either a database or within the program itself. In AutoLISP, multiple test expressions can be evaluated through either the AND function or the OR function.

Dimensions—Tolerances

Bearing Number	Bore B				Outside Diameter D				Width W		Fillet Radius (1)		Wt.		Static Load Rating C_o		Extended Dynamic Load Rating C_E	
			tolerance +.0000" +.000 mm to minus				tolerance +.0000" +.000 mm to minus											
	mm	in.	in.	mm	in.	mm	in.	mm	in.	mm	in.	mm	lbs.	kg	lbs.	N	lbs.	N
300K	10	.3937	.0003	.008	1.3780	35	.0005	.013	.433	11	.024	.6	.12	.054	850	3750	2000	9000
301K	12	.4724	.0003	.008	1.4567	37	.0005	.013	.472	12	.039	1.0	.14	.064	850	3750	2080	9150
302K	15	.5906	.0003	.008	1.6535	42	.0005	.013	.512	13	.039	1.0	.18	.082	1270	5600	2900	13200
303K	17	.6693	.0003	.008	1.8504	47	.0005	.013	.551	14	.039	1.0	.24	.109	1460	6550	3350	15000
304K	20	.7874	.0004	.010	2.0472	52	.0005	.013	.591	15	.039	1.0	.31	.141	1760	7800	4000	17600
305K	25	.9843	.0004	.010	2.4409	62	.0005	.013	.669	17	.039	1.0	.52	.236	2750	12200	5850	26000
306K	30	1.1811	.0004	.010	2.8346	72	.0005	.013	.748	19	.039	1.0	.78	.354	3550	15600	7500	33500
307K	35	1.3780	.0005	.013	3.1496	80	.0005	.013	.827	21	.059	1.5	1.04	.472	4500	20000	9150	40500
308K	40	1.5748	.0005	.013	3.5433	90	.0006	.015	.906	23	.059	1.5	1.42	.644	5600	24500	11000	49000
309K	45	1.7717	.0005	.013	3.9370	100	.0006	.015	.984	25	.059	1.5	1.90	.862	6700	30000	13200	58500
310K	50	1.9685	.0005	.013	4.3307	110	.0006	.015	1.063	27	.079	2.0	2.48	1.125	8000	35500	15300	68000
311K	55	2.1654	.0006	.015	4.7244	120	.0006	.015	1.142	29	.079	2.0	3.14	1.424	9500	41500	18000	80000
312K	60	2.3622	.0006	.015	5.1181	130	.0008	.020	1.220	31	.079	2.0	3.89	1.765	10800	48000	20400	90000

Width W tolerance: +.000", –.005" +.00 mm, –.13 mm

Chart courtesy of The Torrington Company

(1) Maximum shaft or housing fillet radius which bearing corners will clear.

Figure 3–3   *Bearing selection chart*

Using the AutoLISP AND Function

The AutoLISP AND function is used to logically compare two or more test expressions. If any of the test expressions evaluates as nil, then the AND function returns a nil value. All test expressions supplied to the AND function must return a T value. This is illustrated in Table 3–1.

Table 3–1   Logic AND Chart

Test Expr #1	Test Expr #1	Test Expr #1	Result Returned by AND function
NIL	NIL	NIL	NIL
T	NIL	NIL	NIL
NIL	T	NIL	NIL
NIL	NIL	T	NIL
T	T	NIL	NIL
T	NIL	T	NIL
NIL	T	T	NIL
T	T	T	T

The table indicates that when all three test expressions are evaluated to nil, then the result returned by the AND function is also nil. This is the case in every situation except the last, where all test expression are evaluated to T; in this case the result returned by the AND function is also T. The syntax for the AND function is (AND [expr ...]). The expression arguments of this function are the test expressions to be evaluated. When this function is integrated into an IF expression, the possibilities for evaluating multiple test expressions are almost unlimited. For example, the bearing program mentioned earlier could be written as follows:

```
(SETQ bore_size      (GETREAL "\nEnter bore size : ")
    plus_tolerance (GETREAL "\nEnter + tolerance : ")
    minus_tolerance (GETREAL "\nEnter - tolerance : ")
)
(IF (AND (= bore_size 1.5748)
    (= plus_tolerance 0.0)
    (= minus_tolerance 0.0005)
  )
  (PRINC "\nUse bearing number 308K : ")
)
(IF (AND (= bore_size 1.3780)
    (= plus_tolerance 0.0)
    (= minus_tolerance 0.0005)
  )
```

```
   (PRINC "\nUse bearing number 307K : ")
  )
  (IF (AND (= bore_size 1.1811)
           (= plus_tolerance 0.0)
           (= minus_tolerance 0.0004)
      )
      (PRINC "\nUse bearing number 306K : ")
  )
```

In this example the bore size must be equal to 1.5748 and have a tolerance of +0.0 and − 0.0005 inches before bearing 308K is recommended.

Using the AutoLISP OR Function

The AutoLISP OR function is used to logically compare two or more test expressions. If one or all of these expressions evaluate to T, then the function returns a value of T. Only when all test expressions evaluate to nil does the function return nil. This is illustrated in Table 3–2.

Table 3–2 Logic OR Chart

Test Expr #1	Test Expr #1	Test Expr #1	Result Returned by OR function
NIL	NIL	NIL	NIL
T	NIL	NIL	T
NIL	T	NIL	T
NIL	NIL	T	T
T	T	NIL	T
T	NIL	T	T
NIL	T	T	T
T	T	T	T

The syntax for the OR function is (or [expr...]). The expression arguments of this function are the test expressions to be evaluated. When this function is integrated into an IF expression, the possibilities for evaluating multiple test expressions are greatly enhanced. For example, to determine if a number is greater than 10 or less than 5, the following IF statement would be used:

```
(SETQ number (GETREAL "\nEnter number : "))
(IF (OR (> number 10.0)(< number 5.0))
    (PROGN
             ....
    )
)
```

In this example the number entered by the user must either be less than 5 or greater than 10 before the OR function returns a value of T.

EVALUATING MULTIPLE TEST EXPRESSIONS USING THE COND FUNCTION

The IF function (also referred to as a conditional function) is not the only one that can be used for making decisions; there is another one, the COND function. The COND (COND [(test result ...) ...]) function is the primary conditional function provided for AutoLISP programming. Although, this function is similar to the IF function because both functions are used for decision making, the COND function differs from the IF function by allowing the use of multiple test expressions, with each having its own THEN ELSE arguments. This function is extremely useful for programs where the user is provided with multiple options for a single prompt (as in the CIRCLE command). For example:

```
(DEFUN C:example ()
 (SETQ example (GETSTRING "\nEnter a letter : "))
 (COND
  ((= example "R") (first))
  ((= example "r") (second))
  ((= example "u") (third))
  ((= example "U") (fourth))
  ((= example "H") (fifth))
  ((= example "h") (sixth))
 )
)
(DEFUN first ()
 (PRINC "\nFirst")
)
(DEFUN second ()
 (PRINC "\nSecond")
)
```

In this example, the user is prompted to enter a letter. Once the user has responded, the value entered is set to the variable example. Next, the COND function evaluates the first expression (test expression) in each list (starting in the order in which they are supplied) until a value other than nil is returned. At that point the COND function evaluates the remaining expressions that follow the test argument that returned a non nil value.

Evaluating More Than One Test Expression at a Time

The COND function has the same limitation as the IF function. The first expression in each list is regarded as the test expression, with all other expressions following considered to be the THEN expression(s). If more than one test expression is to be evaluated in the same list, then the AND and/or OR functions must be employed. This is illustrated in the following example, where the portion of the bearing program previously featured is rewritten to use the COND function:

```
(SETQ bore_size      (GETREAL "\nEnter bore size : ")
      plus_tolerance (GETREAL "\nEnter + tolerance : ")
      minus_tolerance (GETREAL "\nEnter - tolerance : ")
)
(COND
  ((AND      (= bore_size 1.5748)
             (= plus_tolerance 0.0)
             (= minus_tolerance 0.0005)
   )                                            ;Close of AND function.
   (PRINC "\nUse bearing number 308K : ") ;Then expression.
  )                                             ;Close of first list.
  ((AND      (= bore_size 1.3780)
             (= plus_tolerance 0.0)
             (= minus_tolerance 0.0005)
   )                                            ;Close of AND function.
   (PRINC "\nUse bearing number 307K : ") ;Then expression.
  )                                             ;Close of second list.
  ((AND      (= bore_size 1.1811)
             (= plus_tolerance 0.0)
             (= minus_tolerance 0.0004)
   )                                            ;Close of AND function.
   (PRINC "\nUse bearing number 306K : ") ;Then expression.
  )                                             ;Close of third list.
)
   ;Close of COND function.
```

APPLICATION – CONSTRUCTING A SIMPLE HARMONIC MOTION CAM DISPLACEMENT DIAGRAM

To illustrate how the concepts in this chapter can be combined to create some very useful programs, the following application creates a cam displacement diagram and cam profile based on the information provided by the user. The program itself illustrates several examples of list manipulation, decision making, basic math, complex math, interfacing with the user and interfacing with AutoCAD. The theories on which the calculations are based can be found in almost any mechanical drafting and design handbook as well as engineering machine design textbooks.

The program starts by prompting the user to enter the base circle's diameter and the follower height. If the user fails to enter a value, the program automatically displays a message that insufficient data has been obtained. Otherwise, the program prompts the user to select the location of the displacement diagram. Once this has been done, the program calculates the locations of the vertical and horizontal lines that serve as the grid for the graph, as well as the points where the cam displacement graph intersects the grid lines. After the points have been calculated and the grid lines drawn, the program then constructs the graph using the AutoCAD SPLINE command. Finally, the program prompts the user to select the location of the CAM profile, and through basic math and list manipulation techniques the points defining the profile are then calculated and used with the AutoCAD SPLINE function to create the cam profile. The output from this program is shown in Figure 3–4.

CAM Displacement Diagram

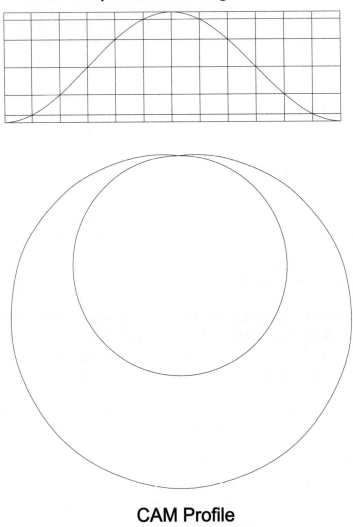

CAM Profile

Figure 3–4 *Cam displacement diagram and profile created by the sample program*

```
;;;********************************************************************
;;;
;;; Program Name: CAM.lsp
;;; Program Purpose: Create a Simple harmonic motion cam displacement
;;;                       diagram and cam profile
;;; Program Written By: James Kevin Standiford
;;; Program Date : 01-31-99
;;;
;;;********************************************************************
;;;
;;; Main Program
;;;
;;;********************************************************************
(defun c:cam (/      cir  lo_l_c inc      up_r_c pt1a   pt2a   pt1b
       pt2b   pt1c   pt2c   pt1d      pt2d   pt1e   pt2e   pt1f
       pt2f   pt1g   pt2g   pt1h      pt2h   pt1i   pt2i   pt1j
       pt2j   pt1k   pt2k   pt1l      pt2l
       )
  (setvar "cmdecho" 0)
  (setq base (getreal "\nEnter base circle diameter : "))
  (setq fo_heigh (getreal "\nEnter follower height : "))
;;********************************************************
;;
;; Conditional functions <, >, <=, >=, =, /= can be found in chapter
;; 2.
;;
;;********************************************************
  (if (and (/= base nil) (/= fo_heigh nil))
   (progn
    (setq
    cir  (* base pi)
    lo_l_c (getpoint "\nSelect lower left hand corner of graph : ")
;;
;;
;; Determines the outer edges of the grid, and calculates the vertical
;; lines of the grid. The variable lo_l_c is the lower left-hand corner
;; of the graph. The variable up_r_c is the upper right hand corner of
;; the graph. The variables pt1? represent the lower portion of the
;; vertical grid line. The variables pt2? represent the upper portion
;; of the vertical grid line.
;;
;;
    inc  (/ cir 12)
    up_r_c lo_l_c
```

```
up_r_c (subst (+ cir (car lo_l_c))
          (car lo_l_c)
          (subst (+ fo_heigh (cadr lo_l_c))
               (car (cdr lo_l_c))
               lo_l_c
          )
      )
ptla   (list (+ inc (car lo_l_c)) (cadr lo_l_c))
pt2a   (list (car ptla) (cadr up_r_c))
pt1b   (list (+ (* 2 inc) (car lo_l_c))
          (car (cdr lo_l_c))
      )
pt2b   (list (car pt1b) (cadr up_r_c))
pt1c   (list (+ (* 3 inc) (car lo_l_c))
          (car (cdr lo_l_c))
      )
pt2c   (list (car pt1c) (cadr up_r_c))
pt1d   (list (+ (* 4 inc) (car lo_l_c))
          (car (cdr lo_l_c))
      )
pt2d   (list (car pt1d) (cadr up_r_c))
pt1e   (list (+ (* 5 inc) (car lo_l_c))
          (car (cdr lo_l_c))
      )
pt2e   (list (car pt1e) (cadr up_r_c))
pt1f   (list (+ inc (car lo_l_c))
          (car (cdr lo_l_c))
      )
pt2f   (list (car pt1f) (cadr up_r_c))
pt1g   (list (+ (* 6 inc) (car lo_l_c))
          (car (cdr lo_l_c))
      )
pt2g   (list (car pt1g) (cadr up_r_c))
pt1h   (list (+ (* 7 inc) (car lo_l_c))
          (car (cdr lo_l_c))
      )
pt2h   (list (car pt1h) (cadr up_r_c))
pt1i   (list (+ (* 8 inc) (car lo_l_c))
          (car (cdr lo_l_c))
      )
pt2i   (list (car pt1i) (cadr up_r_c))
pt1j   (list (+ (* 9 inc) (car lo_l_c))
          (car (cdr lo_l_c))
      )
```

```
    pt2j   (list (car pt1j) (cadr up_r_c))
    pt1k   (list (+ (* 10 inc) (car lo_l_c))
               (car (cdr lo_l_c))
        )
    pt2k   (list (car pt1k) (cadr up_r_c))
    pt1l   (list (+ (* 11 inc) (car lo_l_c))
               (car (cdr lo_l_c))
        )
    pt2l   (list (car pt1l) (cadr up_r_c))
    )
;;
;;
;;
;; Constructs a rectangle that represents the outline of the grid and
;; all vertical grid lines.
;;
;;
    (command "line"      lo_l_c    (subst (car up_r_c) (car lo_l_c) lo_l_c)
                         up_r_c    (subst (cadr up_r_c) (cadr lo_l_c) lo_l_c)
        "c"
        "line"    pt1a    pt2a    ""
        "line"    pt1b    pt2b    ""
        "line"    pt1c    pt2c    ""
        "line"    pt1d    pt2d    ""
        "line"    pt1e    pt2e    ""
        "line"    pt1g    pt2g    ""
        "line"    pt1h    pt2h    ""
        "line"    pt1i    pt2i    ""
        "line"    pt1j    pt2j    ""
        "line"    pt1k    pt2k    ""
        "line"    pt1l    pt2l    ""
    )
;;
;;
;; Calculates the horizontal grid lines and the intersections
;; of the cam displacement graph with the grid, thus establishing the
;; location of the displacement graph.
;;
;;
    (setq first_y_di
                (abs (- (+ (* (cos (/ (* 30 pi) 180)) (/ fo_heigh 2))
                       (/ fo_heigh 2)
                   )
                   fo_heigh
                )
```

```
                          )
         second_y_di
                          (abs (- (+ (* (cos (/ (* 60 pi) 180)) (/ fo_heigh 2))
                                    (/ fo_heigh 2)
                                    )
                                    fo_heigh
                             )
                          )
         third_y_di      (/ fo_heigh 2)
         fourth_y_di

                          (abs (- (+ (* (cos (/ (* 120 pi) 180)) (/ fo_heigh 2))
                                    (/ fo_heigh 2)
                                    )
                                    fo_heigh
                             )
                          )
         fifth_y_di

                          (abs (- (+ (* (cos (/ (* 150 pi) 180)) (/ fo_heigh 2))
                                    (/ fo_heigh 2)
                                    )
                                    fo_heigh
                             )
                          )
         pt3a (list (car lo_l_c)
                       (setq fya (+ first_y_di (cadr lo_l_c)))
                    )
         pt4a (list (car up_r_c) fya)
         pt3b (list (car lo_l_c)
                       (setq fyb (+ second_y_di (cadr lo_l_c)))
                    )
         pt4b (list (car up_r_c) fyb)
         pt3c (list (car lo_l_c)
                       (setq fyc (+ third_y_di (cadr lo_l_c)))
                    )
         pt4c (list (car up_r_c) fyc)
         pt3d (list (car lo_l_c)
                       (setq fyd (+ fourth_y_di (cadr lo_l_c)))
                    )
         pt4d (list (car up_r_c) fyd)
         pt3e (list (car lo_l_c)
                       (setq fye (+ fifth_y_di (cadr lo_l_c)))
                    )
         pt4e (list (car up_r_c) fye)
     )
```

```
;;
;;
;; Draws the horizontal lines and displacement diagram.
;;
;;
    (command "line"    pt3a    pt4a    ""
         "line"    pt3b    pt4b    ""
         "line"    pt3c    pt4c    ""
         "line"    pt3d    pt4d    ""
         "line"    pt3e    pt4e    ""
         "spline"
         lo_l_c
         (list (car pt1a) (cadr pt3a))
         (list (car pt1b) (cadr pt3b))
         (list (car pt1c) (cadr pt3c))
         (list (car pt1d) (cadr pt3d))
         (list (car pt1e) (cadr pt3e))
         (list (car pt1g) (cadr up_r_c))
         (list (car pt1h) (cadr pt3e))
         (list (car pt1i) (cadr pt3d))
         (list (car pt1j) (cadr pt3c))
         (list (car pt1k) (cadr pt3b))
         (list (car pt1l) (cadr pt3a))
         (list (car up_r_c) (cadr lo_l_c))
         ""    ""    ""
    )
;;
;;
;; Prompts the user to select the location of the CAM profile.
;; Calculates the point defining the CAM profile. Draws the CAM
;; Profile.
;;
;;

    (setq basepoint (getpoint "\nSelect location for cam profile : "))
    (command
    "spline"
    (list (car basepoint) (+ (cadr basepoint) base))
    (list (- (car basepoint) (* (+ first_y_di base 0.05) 0.5))
       (+ (cadr basepoint)
            (* (+ first_y_di base 0.05) 0.866025403)
       )
    )
    (list (- (car basepoint)
            (* (+ second_y_di base 0.05) 0.866025403)
```

```
        )
     (+ (cadr basepoint) (* (+ second_y_di base 0.05) 0.5))
  )
  (list (- (car basepoint) (* (+ third_y_di base 0.05) 1))
     (+ (cadr basepoint) (* (+ third_y_di base 0.05) 0))
  )
  (list (- (car basepoint)
        (* (+ fourth_y_di base 0.05) 0.866025403)
     )
     (- (cadr basepoint) (* (+ fourth_y_di base 0.05) 0.5))
  )
  (list (- (car basepoint) (* (+ fifth_y_di base 0.05) 0.5))
     (- (cadr basepoint)
        (* (+ fifth_y_di base 0.05) 0.866025403)
     )
  )
  (list (car basepoint)
     (- (cadr basepoint) (+ fo_heigh base 0.05))
  )
  (list (+ (car basepoint) (* (+ fifth_y_di base 0.05) 0.5))
     (- (cadr basepoint)
        (* (+ fifth_y_di base 0.05) 0.866025403)
     )
  )
  (list (+ (car basepoint)
        (* (+ fourth_y_di base 0.05) 0.866025403)
     )
     (- (cadr basepoint) (* (+ fourth_y_di base 0.05) 0.5))
  )
  (list (+ (car basepoint) (* (+ third_y_di base 0.05) 1))
     (+ (cadr basepoint) (* (+ third_y_di base 0.05) 0))
  )
  (list (+ (car basepoint)
        (* (+ second_y_di base 0.05) 0.866025403)
     )
     (+ (cadr basepoint) (* (+ second_y_di base 0.05) 0.5))
  )
  (list (+ (car basepoint) (* (+ first_y_di base 0.05) 0.5))
     (+ (cadr basepoint)
        (* (+ first_y_di base 0.05) 0.866025403)
     )
  )
```

```
    (list (car basepoint) (+ (cadr basepoint) base))
      ""      ""      ""
    "circle"
    basepoint
    base
   )
  )
  (princ "\nInsufficient Data Press enter to try again : ")
 )
)
(princ "\nTo execute enter cam at the command prompt ")
(princ)
```

SUMMARY

Unlike other programming languages, LISP uses the concept of a list as its primary means of storing large amount of information in a single variable. This same ability is integrated into AutoLISP, making it a powerful tool for the customization of the AutoCAD workstation. In AutoLISP a list can be created using the LIST function. When this function is evaluated, it first constructs a list and then populates that list with any argument (also know as elements) that has been supplied to that function. The LIST function can also be used to group multiple lists together into a single list, with each list supplied being an element of the newly created list. However, when two or more lists are to be combined, then the AutoLISP APPEND function must be used. Once a list as been constructed its elements may be extracted for processing by the application through one of the CXXXX functions. These functions allow the programmer to extract either the first, second, or third elements from a list. An alternative to the CXXXX functions is the NTH function. This function allows the programmer to extract any element from a list by simply specifying the list and the position of the element to be extracted.

Numeric information is not the only data type that can undergo processing; string data may also be manipulated. Strings can be merged, truncated, formatted, and/or converted to different data types. To merge two or more strings into a single string, the AutoLISP STRCAT function must be used. The STRCAT function is limited to merging only string data; therefore if a numeric data type is to be merged with a string, then the number must first be converted from a numeric data type to a string data type. In AutoLISP the conversion from a numeric data type to a string data type is accomplished through either the RTOS or ITOA function. Once a number has been converted to a string it may be converted back to a number through either the ATOF or ATOI function. String data that has been input in a program may be used in its entirety or may be shortened. The processing of shortening a string from its original length is referred to as truncating. To truncate a string in AutoLISP the SUBSTR function must be employed.

At some point in almost every computer program, a decision must be made by the program based upon information that has supplied by the user. In AutoLISP decisions are made through either the COND or IF function. The COND function is the primary conditional function used in AutoLISP and it has the ability of evaluating multiple lists with each containing a test expression and THEN expression(s). The IF function, on the other hand, is limited to a single test expression, which is followed by a THEN and an ELSE expression. To evaluate multiple THEN or ELSE expressions in an IF function, the PROGN function must be used. Multiple test expressions can be supplied to either the IF or COND function as a single test expression through either the AND or OR function.

REVIEW QUESTIONS AND EXERCISES

1. Define the following terms:
   Element
   Truncate
   Atom
   Nested Function

2. List the main advantages of using a list to store information as opposed to using variables.

3. What AutoLISP function is used to create a list? Give an example of how that function would be used.

4. What AutoLISP function is used to merge multiple lists into a single list? Give an example of how that function would be used.

5. Explain the difference between the CAR, CDR, CADR and CADDR functions and give an example of each.

6. How are multiple strings merged into a single string in AutoLISP?

7. True or False: Numeric data (in its native format) can be merged with string data. (If false. explain why.)

8. True or False: AutoLISP does not supply any functions for truncating string data.

9. In AutoLISP decisions can be made by an application using which AutoLISP functions? What are the differences between these functions?

10. Which AutoLISP function can be used to evaluate multiple expressions sequentially and returns the value of the last expression?

Given the data types listed below, write an expression or group of expressions that will merge the data types into a single string.

11. "This is an example" "of an AutoLISP" "string"

12. "The amount of heat loss is" 456.87 "btu/hr"

13. "The square root of" 2 "is" 1.4142

14. "The program has completed" 57 "cycles"

15. "The product of the two numbers is" (* numberOne numberTwo)

What is the value returned by the following expressions?

16. (substr "This is an example of a truncated string" 1 3)

17. (substr "The answer is 57" 15 2)

18. (substr "The amount of heat entering the room is 579.90 btu/hr" 12 5)

19. (substr (strcat "AutoLISP is a powerful language that" " can be useful") 25 9)

20. (substr (strcat "The voltage drop for resistor two is " (rtos (* 0.25 2000)) " volts") 21)

Write an IF expression for each of the following:

21. If the answer is less than 10, set variable to 11.

22. If the answer is less than 10 and more than 5, set variableOne to 7 and variableTwo to 20.

23. If the answer is less than 6 and more than 2, set VariableOne to 15 and variableTwo to 50, else set variableOne to 1 and variableTwo to 2.

24. If the answer is less than 6 or more than 2, set VariableOne to 15 and variableTwo to 50, else set variableOne to 1 and variableTwo to 2.

What is the value returned by each expression?

25. (setq exampleList (list "this is an" "example list" "1" 3 4 5 't))

26. (car exampleList)

27. (cadr exampleList)

28. (cdr exampleList)

29. (caddr exampleList)

30. (nth exampleList 4)

31. (nth 4 exampleList)

32. (nth 7 enampleList)

33. (nth 7 exampleList)

34. (nth 6 exampleList)

35. (nth 1 ExampleList)

Given the following lists, write an expression or group of expressions that will achieve the indicated results.

Lists

(Setq one (list 2 3 4 5)

   two (list 'A 'B 'C 'D)

   three (list "This" "Is" "A" "List")

)

Result

36. (A B C D)

37. (2 3 4 5 A B C D)

38. ((2 3 4 5) (A B C D))

39. ((2 3 4 5) (A B C D) "This" "Is" "A" "List")

40. ((2 3 4 5) (A B C D) ("This" "Is" "A" "List") "This" "Is" "A" "List" 2 3 4 5)

41. 11

42. ("This" "Is" "A" "List")

43. ((((2 3 4 5) (A B C D) ("This" "Is" "A" "List")) ("This" "Is" "A" "List") (2 3 4 5 A B C D) (A B C D)("This" "Is" "A" "List") "This" "Is" "A" "List" A B C D) ("This" "Is" "A" "List") (2 3 4 5))

Advanced List Processing and Loops

OBJECTIVES

Upon completion of this chapter the reader will be able to:

- Describe the difference between association list and regular list
- Use the ENTSEL function to extract the entity name of an object contained within an AutoCAD drawing
- Use the CAR function in conjuction with the ENTGET function to return an association list of an AutoCAD entity
- Create an association list using the QUOTE function
- Describe the difference between association list and dotted pairs
- Construct a dotted pair using the QUOTE function
- Using the CONS function to construct a dotted pair
- Extract information from a dotted pair using the ASSOC function
- Substitute an element in a list, association list, or dotted pair using the SUBST function
- Use the ENTLAST, ENTNEXT, ENTSEL, and NENTSEL functions to aquire an entity name
- Describe the difference between the NENTSEL and ENTSEL functions
- Use the ENTDEL function to delete or restore an entity from the AutoCAD graphics screen
- Use the AutoLISP ENTSEL, CAR, and ENTGET functions to extract an association list for a particular AutoCAD entity
- Modify an AutoCAD entity's association list once that list has been obtained
- Use the ENTMOD function to update an entity's definition in the AutoCAD database

- Use the ENTUPD function to update an entity shown on the AutoCAD graphics screen
- Describe the difference between the ENTMAKE and ENTMAKEX functions
- Use the ENTMAKE and ENTMAKEX functions to construct AutoCAD graphic and non-graphic entities
- Use the ENTMAKEX function to create an AutoCAD entity that is not assigned an owner
- Describe the difference between graphic and non-graphic entities
- Define the term *loops* and describe what they are used for
- Construct a loop using the WHILE and REPEAT functions
- Describe the difference between the WHILE function and the REPEAT function

KEY WORDS AND AUTOLISP FUNCTIONS

ASSOC	ENTMAKEX	Loop
Association List	ENTMOD	NENTSEL
CONS	ENTNEXT	Non-graphic Entities
Dotted Pair	ENTSEL	QUOTE
ENTDEL	ENTUPD	REPEAT
ENTGET	Extended Entity Data	SUBST
ENTMAKE	Graphic entities	WHILE

INTRODUCTION TO ADVANCED LIST PROCESSING

Chapter 3 introduced the basic concepts of lists and list processing. However, the subject extends far beyond the depth of the material covered in that chapter. This becomes apparent when one examines the way AutoCAD returns information to AutoLISP concerning an entity. AutoCAD retains information regarding an entity's layer, color, thickness, font, style, linetype, coordinates, handle, etc. in the drawing file where the entity belongs. When AutoLISP requests information about an AutoCAD entity, that information is returned in the form of a list known as an association list. Association lists are the primary method used by AutoCAD to maintain an entity's definition data.

ASSOCIATION LIST

Unlike the list introduced in Chapter 3, an association list contains elements that are somewhat linked by an index, making it easier to control the information it contains. Imagine all the information that is retained by AutoCAD about a particular entity displayed in a standard list. At first glance this might seem to be an extremely useful asset because all the information concerning an entity can be located within a single source. However, after further examination it becomes apparent that a serious problem exists. How does the programmer distinguish between the different elements if the order of the list is not known, or even changed from entity to entity? If a list is to be constructed where key information about an entity is to be contained, then there must be a way that information can easily be sorted and the proper values returned. This is the main advantage of an association list over a regular list.

In an association list the elements are linked to key values. For example, an association list for a string of text shown in Figure 4–1 would appear as follows:

```
((-1 . <Entity name: 162c960>) (0 . "TEXT") (330 . <Entity name:
   162c8f8>) (5 . "2C") (100 . AcDbEntity") (67 . 0) (301 . "Model") (8 .
   "0") (100 . "AcDbText") (10 8.39276 2.09439 0.0) (40 . 0.2) (1 .
   "test") (50 . 0.0) (41 . 1.0) (51 . 0.0) (7 . "Standard") (71 . 0) (72
   . 0) (11 0.0 0.0 0.0) (210 0.0 0.0 1.0) (100 . "AcDbText") (73 . 0))
```

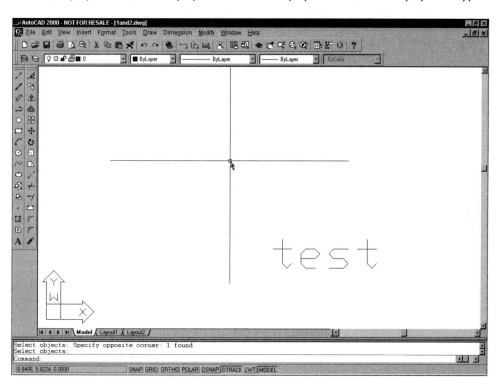

Figure 4–1 *A string of text in AutoCAD*

 Note: In an association list each group of elements is contained within a sub-list.

In this example each element is bound to the list with a group code. Each group code is used to designate a particular piece of information associated with the text, and therefore each group code will be unique to the information that it represents. For example, the group code used to indicate the layer that an entity currently resides on is integer 8, while the group code used to represent an entity's text height is the integer 40. Each entity type has its own predefined group code. However, group codes can overlap with group codes used with other entities. For example, if an association list is displayed for the polyline featured in Figure 4–2, the programmer would notice that the group code used to indicate the layer the polyline resides on is the same as the group code used for the text in the previous example. In addition, both entities share the group codes 0, which indicates entity type, and 301, used to designate the mode where the entity currently resides (model space or paper space).

```
((-1 . <Entity name: 162c958>) (0 . "LWPOLYLINE") (330 . <Entity name:
   162c8f8>) (5 . "2B") (100 . "AcDbEntity") (67 . 0) (301 . "Model") (8
   . "0") (100 . "AcDbPolyline") (90 . 2) (70 . 0) (43 . 0.0) (38 . 0.0)
   (39 . 0.0) (10 4.20059 2.19533) (40 . 0.0) (41 . 0.0) (42 . 0.0) (10
   10.3626 5.71121) (40 . 0.0) (41 . 0.0) (42 . 0.0) (210 0.0 0.0 1.0))
```

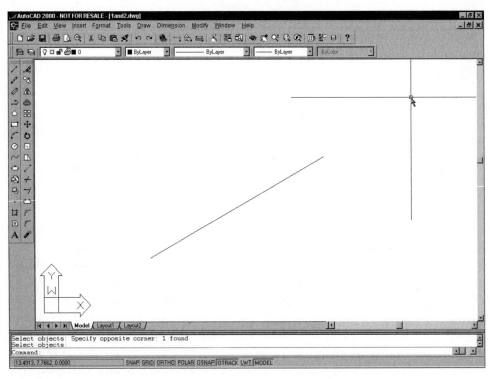

Figure 4–2 *A polyline in AutoCAD*

The group codes used in an association list returned by AutoCAD are in the DXF (Drawing eXchange File) format. However, the DXF codes used in an association list are different from those used to generate a DXF drawing file. A list of the DXF codes used in an AutoLISP association list can be obtained in the DXF Group Codes section of the *AutoCAD 2000 Customization Guide*.

Creating an Association List

The simplest way to construct an association list is with the QUOTE function (QUOTE expr). This function returns the expression passed to it without evaluating it. For example, to set the symbol Y to the variable *test* the expression (SETQ test (quote y)) would be used. In this example, AutoLISP simply passes anything after the QUOTE function to the variable *test*, thus resulting in a returned value of the symbol Y. If the QUOTE function had not been used in the previous expression, AutoLISP would evaluate the symbol Y as a variable and set its value to the variable *test*. Two examples are provided below to illustrate how the QUOTE function can be used to construct an association list. The first example constructs an association list where each sub-list contains three elements. The first element in each sub-list is the group code for that sub-list. The second example builds an association list where each sub-list contains two elements. Again, the first element is used as a group code for that sub-list.

```
(QUOTE ((1 F1 "one")(2 F2 "two")(3 F3 "three")))
(QUOTE ((1 "one")(2 "two")(3 "three")))
```

Resulting in:

```
((1 F1 "one") (2 F2 "two") (3 F3 "three"))
((1 "one") (2 "two") (3 "three"))
```

Note: The QUOTE function can be substituted with a single quote ('). For example, '((I fI "one")(2 F2 "two")(3 F3 "three")) would yield the same results as using the QUOTE function.

Dotted Pairs

As mentioned earlier, when AutoLISP makes a request concerning an AutoCAD entity, that information is returned in the form of an association list. This was illustrated at the beginning of the chapter where an association list of an AutoCAD text string was presented. If a comparison is made of the association list returned by AutoCAD and the association lists created in the previous example, several differences emerge. The sub-lists of the association lists created in the previous example contained three elements each that are separated with spaces; while the sub-lists of the association list for the AutoCAD text string contained only two elements each separated by a period. The sub-lists of the association list returned by AutoCAD are a special type of sub-list known as a *Dotted Pair*. Dotted pairs have two basic requirements: first, they must contain two elements; second, a period (or a dot as it is often called) must separate the elements. A dotted pair can be constructed in AutoLISP using the QUOTE function, but they are typically created using the CONS (CONS new-first-element list-or-atom) function. This function allows the programmer to supply two arguments, which are then used as the elements for the dotted pair. These arguments can be any of the valid AutoLISP data types covered in Chapter 2. It should be noted that since a dotted pair can contain only two elements separated by a period, the first element supplied to the CONS function will be the group code and should be a value that is unique to the data stored as the second element. Examples of this function are as follows:

```
(SETQ number_of_teeth (CONS 1 20))
(SETQ ratio (CONS "gear_ratio" (/ gear_a_rmp gear_b_rpm)))
(SETQ shaft_dia (CONS 2 2.5))
(SETQ shaft_dia (CONS 2 (GETREAL "\nEnter shaft diameter : ")))
(SETQ shaft_dia (CONS shaft_d (GETREAL "\nEnter shaft diameter : ")))
```

Note: Most list-handling functions cannot accept dotted pairs as their arguments. Therefore, care should be taken to ensure that a function used for list handling can accept dotted pairs as arguments.

Creating an Association List Containing Dotted Pairs

When the CONS function is used in conjunction with the LIST function, the result is an association list that contains dotted pairs. To create an association list containing the design parameters for the gear shown in Figure 4–3, the following expression would be used.

```
(SETQ       gear_information
    (LIST (CONS 1 8)
        (CONS 2 18)
        (CONS 3 20)
        (CONS 4 2.250)
        (CONS 5 2.101)
        (CONS 6 0.393)
        (CONS 7 0.196)
        (CONS 8 1.9375)
    )
)
```

Resulting in:

```
((1 . 8) (2 . 18) (3 . 20) (4 . 2.25) (5 . 2.101) (6 . 0.393) (7 . 0.196)
    (8 . 1.9375))
```

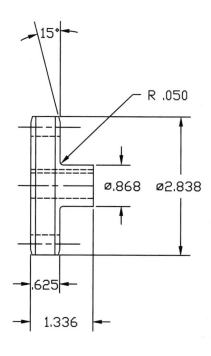

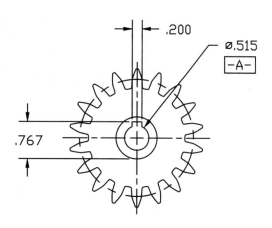

SPUR GEAR DATA	
DIAMETRAL PITCH	8
NUMBER OF TEETH	18
PRESSURE ANGLE	20°
PITCH DIAMETER	2.250
BASE CIRCLE DIAMETER	2.101
CIRCULAR PITCH	.393
CIRCULAR THICKNESS	.196
ROOT DIAMETER	1.9375

5. PROFILE TOLERANCE .003

4. PITCH TOLERANCE .003

3. ALL TOOTH ELEMENT SPECIFICATIONS ARE FROM DATUM A

2. INTERPRET GEAR DATA PER ANSI Y14.7.1-1979.

1. INTERPRET DIMENSIONS AND TOLERANCES PER ANSI Y14.5M-1982.

NOTES:

Figure 4–3 *Information about spur gear*

or:

```
(SETQ    gear_information
    (LIST (CONS "DIA_PIT" 8)
        (CONS "NUM_TEE" 18)
        (CONS "PRE_ANG" 20)
        (CONS "PIT_DIA" 2.250)
        (CONS "BAS_CIR" 2.101)
        (CONS "CIR_PIT" 0.393)
        (CONS "CIR_THI" 0.196)
        (CONS "ROO_DIA" 1.9375)
    )
)
```

Resulting in:

```
(("DIA_PIT" . 8) ("NUM_TEE" . 18) ("PRE_ANG" . 20) ("PIT_DIA" . 2.25)
    ("BAS_CIR" . 2.101) ("CIR_PIT" . 0.393) ("CIR_THI" . 0.196) ("ROO_DIA"
    . 1.9375))
```

Retrieving Elements from an Association List

Like a regular list, the information embedded within an association list can be accessed using the CXXXX functions. For example, given the association list ((1 . 8) (2 . 18) (3 . 20) (4 . 2.25) (5 . 2.101) (6 . 0.393) (7 . 0.196) (8 . 1.9375)), the second element of the dotted pair (1 . 8) can be obtained with the following expression:

```
(SETQ example_list '((1 . 8) (2 . 18) (3 . 20) (4 . 2.25) (5 . 2.101) (6
    . 0.393) (7 . 0.196) (8 . 1.9375)))
(CDR (CAR example_list))
```

Resulting in:

```
8
```

Recall from the introduction that the main advantage of an association list is that the exact position of the elements in relation to the list is of no importance. This is because the group code that is assigned to each sub-list can be used as an index or reference, allowing the programmer to scan and retrieve information from that list more efficiently. To scan an association list for a specific group code and retrieve that information, the AutoLISP ASSOC function (ASSOC element alist) must be used. This function, when supplied with a group code and a list, searches the specified list looking for the first occurrence of the provided group code. Once a match has been confirmed, the function returns the sub-list associated with the specified group code. If a match cannot be made, then the function returns nil.

The first argument supplied to this argument is the group code associated with a sub-list. The second argument supplied is the association list this function is to search. For example, given the list ((1 . 8) (2 . 18) (3 . 20) (4 . 2.25) (5 . 2.101) (6 . 0.393) (7 . 0.196) (8 . 1.9375)), the sub-list (7 . 0.196) can easily be retrieved using the following expression:

```
(SETQ gear_information
    '((1 . 8) (2 . 18) (3 . 20) (4 . 2.25) (5 . 2.101) (6 . 0.393)
    (7 . 0.196) (8 . 1.9375))
)
(ASSOC 7 gear_information)
```

Resulting in:

```
(7 . 0.196)
```

By adding the CDR function to the previous expression, the second element of the returned sub-list can be obtained. For example:

```
(CDR (ASSOC 7 gear_information))
```

Resulting in:

```
0.196
```

Substituting a Sub-List in an Association List

Often when working with association lists, it becomes necessary to replace a sub-list or an element within a sub-list with another. This can be accomplished using a combination of the CXXXX, CONS, and LIST functions. If this method is used, then all sub-lists within the association list will have to be extracted. A new list is then constructed containing the new sub-list along with the remaining sub-lists. At first the process might seem simple, but this way would require several expressions, an IF statement or two, and a loop, making this approach difficult and somewhat unreliable. There is an efficient alternative provided in AutoLISP for these situations: the SUBST function (SUBST newitem olditem list). This function, when supplied with the necessary information, searches a list for a particular item and returns that list with the new item substituted for every occurrence of the old item. The first argument supplied to this function is the new item, the second argument is the old item to be replaced, and the third is the list where the old item resides. To illustrate how this function works, in the following example the sub-list (7 . 0.196) is replaced by the sub-list (7 . 0.2) in the association list ((1 . 8) (2 . 18) (3 . 20) (4 . 2.25) (5 . 2.101) (6 . 0.393) (7 . 0.196) (8 . 1.9375)).

```
(SETQ gear_information
         '((1 . 8) (2 . 18) (3 . 20) (4 . 2.25) (5 . 2.101) (6 .
            0.393) (7 . 0.196) (8 . 1.9375))
   old       (ASSOC 7 gear_information)
   new       (CONS 7 0.2)
   gear_information (SUBST new old gear_information)
   )
```

Resulting in:

```
((1 . 8) (2 . 18) (3 . 20) (4 . 2.25) (5 . 2.101) (6 . 0.393) (7 . 0.2)
   (8 . 1.9375))
```

AUTOCAD AND ASSOCIATION LISTS

As stated earlier, whenever AutoLISP requests information about an AutoCAD entity, that information is returned in the form of an association list. Before an association list of an AutoCAD entity can be extracted, the entity name must first be obtained. In AutoLISP this can be accomplished through one of three different functions: ENTSEL, ENTNEXT, and ENTLAST. While the ultimate result of these functions is the same, the technique that these functions employ in order to complete their assigned task differs. It is essential that the programmer becomes efficient in the use of all three functions.

Acquiring an Entity Name Using ENTSEL

The first method used to obtain the entity name assigned to an AutoCAD object is to physically select that entity from the graphics editor using the ENTSEL (ENTity SELect) function (ENTSEL [msg]). This function allows the user to select a single entity by either picking the entity or entering the coordinates of a point contained within that entity. Upon examination of the syntax for this function, it is apparent that the only argument that may be supplied to this function is an optional message or prompt. Even though this argument is not required, it should be supplied. The message should be constructed in such a way that the end user will know exactly what the program is expecting of them. For example, suppose that a program is constructed that will allow the user to change the scale of a block by simply selecting the block. If a message argument is supplied to the ENTSEL function, any misunderstandings about what is excepted will be omitted.

```
(ENTSEL "\nSelect block in which the scale is to be changed : ")
```

Once the entity has been selected, the function returns an association list containing the entity name of the object selected and the coordinates (in relation to the current UCS) of the point used to select the entity. In the following examples, the entity name of an AutoCAD line is needed so that the data associated with the line can be edited. In the first example, the line is selected by the entering of a set of coordinates for a point contained on the entity, while in the second example, the entity is selected by the user.

> Command: **(ENTSEL "\nSelect or enter the coordinates of the line to change : ")** (ENTER)
> Select or enter the coordinates of the line to change : **6.2861,3.2801,0** (ENTER)
> (<Entity name: 18ef558> (6.2861 3.2801 0.0))
> Command: **(ENTSEL "\nSelect or enter the coordinates of the line to change : ")** (ENTER)
> Select or enter the coordinates of the line to change :
> (<Entity name: 18ef558> (6.74292 4.10512 0.0))

 Note: If an entity is not selected, the function returns nil.

Acquiring an Entity Name Using ENTNEXT

The second method for acquiring an entity name assigned to an AutoCAD object is to page through the entities names within the AutoCAD drawing database using the ENTNEXT (ENTity NEXT) function (ENTNEXT [ename]). Unlike the ENTSEL function, which requires the user to select the object before the entity name can be extracted, this function searches the AutoCAD database looking for the first non-deleted entity and returns that entity name. If the function is supplied with an entity name, it returns the next non-deleted entity from the AutoCAD database that follows the one specified. If the entity name supplied to the function is the last non-deleted entity within the AutoCAD drawing database, then the function returns nil. For example, the AutoCAD graphics screen shown in Figure 4–4 contains three circles; using the ENTNEXT function without supplying an argument returns the entity name of the first non-deleted entity created, which in this case is the circle created with a hidden line type. If the entity name returned is set to a variable and that variable is supplied once again to the ENTNEXT function, then the value returned would be the next non-deleted entity. This would be the circle created with the center line type. Again, if the entity name in the previous execution of the ENTNEXT expression is supplied to another call of the ENTNEXT function, then the result would be the entity name for the circle created with the dotted line type. Finally, if the entity name from the last ENTNEXT function is supplied to another ENTNEXT function, then the result returned by this function would be equal to nil. This is because there are no other non-deleted entities contained within the drawing's database. An example of this follows.

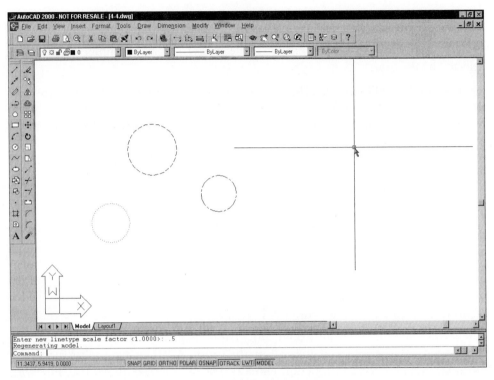

Figure 4–4 *AutoCAD graphics screen containing three circles*

Command: **(SETQ variable1 (ENTNEXT))** (ENTER)
<Entity name: 3410500>
Command: **(SETQ variable2 (ENTNEXT variable1))** (ENTER)
<Entity name: 3410508>
Command: **(SETQ variable3 (ENTNEXT variable2))** (ENTER)
<Entity name: 3410510>
Command: **(SETQ variable4 (ENTNEXT variable3))** (ENTER)
nil

Acquiring an Entity Name Using ENTLAST

The last of the three ENTXXXX functions, ENTLAST (ENTLAST), returns the entity name of the last non-deleted entity contained with in the AutoCAD database. This function can be extremely useful if a program requires the entity name of the last object created or if a program creates an object and then needs to ascertain information regarding that object.

Accessing Entities within a Complex Object

The functions illustrated in the previous sections allow the programmer to obtain the entity name of an object contained in an AutoCAD drawing. If the object that is selected is a complex entity (blocks, polylines, three-dimensional objects) the name that is returned is that assigned to the parent object. For example, a block containing the three circles shown in Figure 4–4, when selected with one of the previous entity selection functions, would return the entity name for the block and not the name of any of the individual components. The entity name for a component within a complex entity can be obtained through the ENTNEXT, NENTSEL (NENTSEL [msg]), or NENTSELP (NENTSELP [msg]) function. These functions allow the programmer to obtain the entity name of an object embedded within a complex entity. While the syntax for two of these functions is identical to that of the ENTSEL function, the amount of information that is returned is quite different. For example, using these functions to select the block containing the three circles shown in Figure 4–4 would produce the following results:

```
Command: (ENTSEL) (ENTER)
Select object: 4.13357,5.98631 (ENTER)
(<Entity name: 27805c0> (4.13357 5.98631 0.0))
```

In this example, the entity name as well as the coordinates used to select that entity are returned. However, this is the name that is assigned to the block and not to one of the circles contained within.

```
Command: (NENTSEL) (ENTER)
Select object: 4.13357,5.98631 (ENTER)
(<Entity name: 2780598> (4.13357 5.98631 0.0) ((1.0 0.0 0.0) (0.0 1.0 0.0) (0.0
0.0 1.0) (3.91991 4.42609 0.0)) (<Entity name: 27805c0>))
```

The NENTSEL function in the second example also returns the entity name of the parent object selected and the point used to select that object, but this time the name of the actual entity selected (that is embedded within the block) is now returned.

```
Command: (NENTSELP) (ENTER)
Select object: 4.13357,5.98631 (ENTER)
(<Entity name: 2780598> (4.13357 5.98631 0.0) ((1.0 0.0 0.0 3.91991) (0.0 1.0
0.0 4.42609) (0.0 0.0 1.0 0.0) (0.0 0.0 0.0 1.0)) (<Entity name: 27805c0>))
```

This is also the case for the third example, in which the NENTSELP function is used. Both functions return more information than just the name of the nested entity selected, the point used to select that object and the entity name of the parent object. The NENTSEL function also returns the model to world matrix. This matrix is a set of four sub-lists that can be used to translate the entities definition data points from a Model Coordinate System (an internal coordinate system) to a World Coordinate System. The NENTSELP function also returns a matrix, but it is a different matrix than that supplied by the NENTSEL function. The matrix supplied by the NENTSELP function is a 4x4-transformation matrix where the first three columns specify scaling and rotation, and the last column is the translation vector. This matrix is similar to those used by ObjectARX functions (this will be covered in more detail later).

If either the NENTSEL or NENTSELP function is used to select a non-complex object, then the results returned are identical to that returned by the ENTSEL function. These results would be the entity name of the object selected and the coordinates of the point used to select that object. This is illustrated in the example below, in which the ENTSEL, NENTSEL and NENTSELP functions are used to retrieve the entity name of the line shown in Figure 4–5.

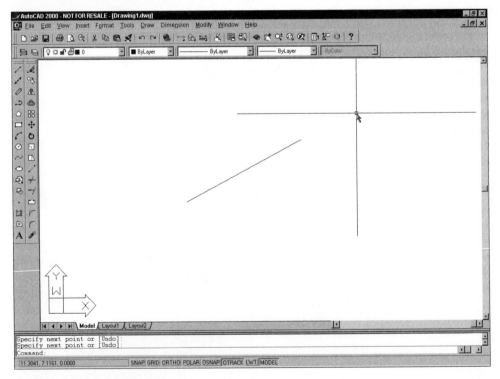

Figure 4–5 *Line in AutoCAD*

Command: **(ENTSEL)** (ENTER)
Select object: **8.86704,3.48996** (ENTER)
(<Entity name: 27805c8> (8.86704 3.48996 0.0))

```
Command:
Command: (NENTSEL) (ENTER)
Select object: 8.86704,3.48996 (ENTER)
(<Entity name: 27805c8> (8.86704 3.48996 0.0))
Command:
Command: (NENTSELP) (ENTER)
Select object: 8.86704,3.48996 (ENTER)
(<Entity name: 27805c8> (8.86704 3.48996 0.0))
Command:
```

Once a complex entity has been selected through either the NENTSEL or NENTSELP function and its entity definition data returned, then additional sub-entities can be accessed with the ENTNEXT function. Recall that when this function is supplied with an entity name, it returns the next non-deleted entity in the AutoCAD database. If this function is supplied with a sub-entity name, then the function returns the entity name of the next entity contained with the complex object.

Removing an AutoCAD Entity Using ENTDEL

In AutoCAD, an entity can be removed from the graphics screen through the ERASE command. When objects have been removed with this command, their definition remains in the drawing database until the current session is terminated, at which time the definitions are purged. This allows the user to recall objects that were removed during the current drawing session using either the UNDO or the OOPS command. AutoLISP also provides a function for deleting an entity from the graphic screen. That is the ENTDEL (ENTDEL ename) function. Just as with the ERASE command, an entity's definitions remain in the database until the current drawing session is terminated. Unlike the ERASE command, this function can also be used to restore an entity that has been deleted in the current drawing session. If the name of an entity that has been deleted with the ERASE command is known, then that entity can be restored with the ENTDEL function. However, entities that have been removed with the ENTDEL function cannot be restored with the AutoCAD OOPS command. They must be restored using either the UNDO command or by reissuing the ENTDEL function. When this function is evaluated, it returns the entity name of the object deleted. This function is limited to deleting one entity at a time. If a group of entities are to be deleted using this function, the entity name of each object must be supplied to the function. For example, to delete the first circle created in Figure 4–4, the following syntax would be used:

```
Command: (SETQ ent (ENTDEL (ENTNEXT))) (ENTER)
<Entity name: 2780500>
Command:
```

To restore the entity deleted in Figure 4–4 the following syntax is used:

```
Command: (ENTDEL ent) (ENTER)
<Entity name: 2780500>
```

or

```
Command: UNDO (ENTER)
Command:
```

 Note: If the UNDO command is used, the object will be restored, but changes made by an AutoLISP program to the drawing will be lost. This process is not recommended. This function cannot be used to remove nested entities (object inside a block) or sub-lists (vertex of a polyline).

Returning a List Associated with an Entity

Once the entity name of the object has been obtained, then the association list for that particular object can be accessed through the ENTGET (ENTGET ename [applist]) function. This function, when supplied with a valid entity name, returns the association list for that object. Values from this association list may be extracted and/or the list may be modified. For example, to obtain the association list for the polyline shown in Figure 4–2 the following code would be used:

```
(SETQ entname (CAR (ENTSEL)))        ;Allows the user to select an
                                     ;entity. Once the entity has been
                                     ;selected the name is extracted
                                     ;using the CAR function and set to
                                     ;the variable entname
(ENTGET entname)                     ;Returns the association list for
                                     ;the entity selected
```

Resulting in:

```
((-1 . <Entity name: 162c958>) (0 . "LWPOLYLINE") (330 . <Entity name:
   162c8f8>) (5 . "2B") (100 . "AcDbEntity") (67 . 0) (301 . "Model") (8
   . "0") (100 . "AcDbPolyline") (90 . 2) (70 . 0) (43 . 0.0) (38 . 0.0)
   (39 . 0.0) (10 4.20059 2.19533) (40 . 0.0) (41 . 0.0) (42 . 0.0) (10
   10.3626 5.71121) (40 . 0.0) (41 . 0.0) (42 . 0.0) (210 0.0 0.0 1.0))
```

or

```
(ENTGET (CAR (ENTSEL)))              ;Allows the user to select an entity.
                                     ;Once the entity has been selected the
                                     ;name is extracted using the CAR
                                     ;function and passed to the ENTGET
                                     ;function, where the association list
                                     ;is returned for the entity selected.
```

Resulting in:

```
((-1 . <Entity name: 162c958>) (0 . "LWPOLYLINE") (330 . <Entity name:
   162c8f8>) (5 . "2B") (100 . "AcDbEntity") (67 . 0) (301 . "Model") (8
   . "0") (100 . "AcDbPolyline") (90 . 2) (70 . 0) (43 . 0.0) (38 . 0.0)
   (39 . 0.0) (10 4.20059 2.19533) (40 . 0.0) (41 . 0.0) (42 . 0.0) (10
   10.3626 5.71121) (40 . 0.0) (41 . 0.0) (42 . 0.0) (210 0.0 0.0 1.0))
```

The second argument to this function is an optional argument that, when supplied, enables extended entity data that may be attached to the entity to be retrieved (extended entity data is covered in more detail in Chapter 5). If the entity has extended data attached, the information is returned along with the association list produced by the ENTGET function. If the programmer had attached extended data to the polyline in Figure 4–2, then that data could be accessed by the following expression:

```
(ENTGET (CAR (ENTSEL)) '("*"))        ;the "*" instructs AutoLISP to
                                      ;obtain all registered applications
                                      ;that are associated with that
                                      ;entity.
```

Resulting in: (extended data shown in bold italic)

```
((-1 . <Entity name: 162c958>) (0 . "LWPOLYLINE") (330 . <Entity name:
   162c8f8>) (5 . "2B") (100 . "AcDbEntity") (67 . 0) (301 . "Model") (8
   . "0") (100 . "AcDbPolyline") (90 . 2) (70 . 0) (43 . 0.0) (38 . 0.0)
   (39 . 0.0) (10 4.20059 2.19533) (40 . 0.0) (41 . 0.0) (42 . 0.0) (10
   10.3626 5.71121) (40 . 0.0) (41 . 0.0) (42 . 0.0) (210 0.0 0.0 1.0))
   (-3 ("NEWDATA" (1000 . "This is an example of extended data"))
   ("NEWDATA1" (1000 . "The possibilities are endless")))))
```

 Note: The '("*") argument will not cause AutoLISP to return an error if no extended data is present for that entity. The function still returns the association list for the entity selected. Once the extended data has been retrieved, information from that list can be accessed by the ASSOC and CXXXX functions.

Working with AutoCAD Association Lists

As indicated earlier, once the association list for an AutoCAD entity has been obtained, the programmer can not only retrieve information from that list but also modify, delete, and even create new entities in ways that were not possible with ordinary AutoCAD commands. The AutoLISP functions ENTMOD, ENTMAKE, ENTMAKEX, and ENTUPD are used to accomplish these changes. When these functions are combined with the other list and association list functions previously discussed, the strata and the complexity of programming possibilities are greatly increased.

Modifying an AutoCAD Entity using the ENTMOD Function

AutoLISP provides two different ways that an AutoCAD entity maybe modified. The first method is by using the COMMAND (COMMAND [argument]….) function. This function allows the programmer to interface with the AutoCAD graphic environment using ordinary AutoCAD commands. The commands supplied to this function must be contained in quotes and arranged in the order in which they would normally be used. This method of interfacing with the AutoCAD graphics editor is constrained by the AutoCAD commands it employs, thus limiting the type and complexity of the program designed to interact with the AutoCAD entity.

The second method, modifying an entity's association list, is often preferred because it is able to perform complex procedures and operations not possible using ordinary AutoCAD commands. This method is generally accomplished in four steps. They are:

1. Obtain entity name

2. Aquire entity association list

3. Extract and modify entity association list

4. Update entity definition in AutoCAD drawing file

The first step of this process involves ascertaining the entity name of the AutoCAD object to be modified through any of the entity selection functions previously discussed (ENTSEL, ENTNEXT, ENTLAST, NENTSEL, or NENTSELP). Once the entity name is secured, the second step is to acquire the association list for that entity using the ENTGET function. After this has been completed, the programmer is free to extract, modify and replace data within the association list using any of the CXXXX, SUBST, CONS, LIST, and QUOTE functions, or any other list processing functions previously discussed. The changes that are made will not take effect until the database has been updated using the ENTMOD (ENTMOD ename) function. This function, when supplied with a valid association list of a graphic or non-graphic entity, updates that object's definition in the drawing's database. This is illustrated in the following program, where the user is allowed to change the layer of an entity by simply selecting another entity.

```
;;;****************************************************************
;;; Program Purpose: This program allows the user to change the layer
;;;                    of an entity by simply selecting that entity.
;;; Written By: James Kevin Standiford
;;; Date: 03/12/99
;;;****************************************************************
(DEFUN c:match ()
```

```
    (SETQ entity1 (ENTSEL "\nSelect entity to change : "))
    (IF (/= entity1 NIL)                    ;Program proceeds only
                                            ;If an object is selected
      (PROGN
        (SETQ elist1 (ENTGET (CAR entity1))
                                            ;Association list for entity1
            entity2 (ENTSEL "\nSelect entity to match : ")
        )
        (IF (/= entity2 NIL)                ;Program proceeds only
                                            ;If an object is selected
          (PROGN
            (SETQ elist2   (ENTGET (CAR entity2))
                                            ;Association list for entity2
                old_layer (ASSOC 8 elist1)  ;layer to be changed
                new_layer (ASSOC 8 elist2)  ;layer to change to
                elist1    (SUBST new_layer old_layer elist1)
            )                               ;Old value replaced with new
            (ENTMOD elist1)                 ;Drawing database updated
          )                                 ;End of second progn
        )                                   ;End of second if
      )                                     ;End of first progn
    )                                       ;End of first if
    (PRINC)
  )                                         ;End of program
 (PRINC
   "\nEnter match at the command prompt to start program : "
 )
 (PRINC)
```

If the program shown above is modified, entities that are embedded within complex objects may also be modified. However, the changes that are made to the entity will not appear in the graphics editor unless the user either regenerates the drawing by using the REGEN command or updates the graphics screen using the ENTUPD (ENTUPD ename) function. This function, when supplied with an entity name, causes AutoCAD to regenerate that object, thus displaying any modifications that were made to the object. If the function is successful in updating the graphics screen, the name of the entity is returned. Initially this function was intended for complex objects such as polylines or blocks, but it may be used with any entity contained within an AutoCAD drawing. Adding a few more lines of code to the previous program gives the user the capability of changing the layer of a nested entity (without first exploding that entity) by selecting an entity containing the desired properties. Once the update has been successful, the entity is then updated on the graphics screen. In the following code, additions or changes to the previous code are shown in bold.

```
;;;****************************************************************
;;;; Program Purpose: This program allows the user to change the layer
;;;;                  of an entity by simply selecting that entity.
;;;; Written By: James Kevin Standiford
;;;; Date: 03/12/99
;;;****************************************************************
;;;
(DEFUN c:match ()
 (SETQ    entity1 (NENTSEL "\nSelect entity to change : ")
   ename  (nth 3 entity1)
 )
 (IF (/= ename NIL)                          ;Test for complex objects
   (SETQ ename (CAR ename))                  ;If object is complex get main
                                             ;header
   (SETQ ename (CAR entity1))                ;if object is not get main header
 )
 (IF (/= entity1 NIL)                        ;Program proceeds only
                                             ;If an object is selected
   (PROGN
    (SETQ elist1 (ENTGET (CAR entity1))
                                             ;Association list for entity1
       entity2 (NENTSEL "\nSelect entity to match : ")
   )
    (IF (/= entity2 NIL)                     ;Program proceeds only
                                             ;If an object is selected
     (PROGN
      (SETQ elist2   (ENTGET (CAR entity2))
                                             ;Association list for entity2
           old_layer (ASSOC 8 elist1)    ;layer to be changed
           new_layer (ASSOC 8 elist2)    ;layer to change to
           elist1    (SUBST new_layer old_layer elist1)
     )                                       ;Old value replaced with new
      (ENTMOD elist1)                        ;Drawing database updated
      (ENTUPD ename)                         ;Updates the graphic screen
     )                                       ;End of second progn
    )                                        ;End of second if
   )                                         ;End of first progn
 )                                           ;End of first if
 (PRINC)
)                                            ;End of program
(PRINC
 "\nEnter match at the command prompt to start program : "
)
(PRINC)
```

Note: Although the ENTUPD function was written for polylines and blocks, the function does not always update the graphics screen. When this update fails to occur, the REGEN command may be used to update the graphics screen. Inserting the expression (COMMAND "REGEN") into a program ensures that the graphics screen will always be updated. Nevertheless, inserting the REGEN command into a program is not recommended because large, complex objects require extensive regeneration time.

Creating an Entity

In addition to modifying an AutoCAD entity, AutoLISP can be used to create both graphic and non-graphic AutoCAD entities. In AutoCAD, a graphic or graphical entity is defined as the visible objects used to create a drawing (lines polylines, arc, circles, text, etc). A non-graphic or non-graphical entity is defined as any object that is not visible to the AutoCAD user (layers, linetypes, dimension styles, text styles, etc). To create these entities in AutoCAD using AutoLISP, the ENTMAKE or ENTMAKEX functions must be used. These functions may be used to create both simple and complex entities. While the syntax for both functions is the same ((ENTMAKE elist) and (ENTMAKEX elist), their results are quite different.

Using ENTMAKE Function to Create an AutoCAD Entity

For an object to be created using ENTMAKE, a valid association list containing all the necessary information must be supplied. If an invalid list or one missing vital information is supplied, then the function returns nil and the entity is not created. Otherwise the function creates the entity and returns that object's association list of definition data. If optional data is omitted, such as a layer, line type, font, or color, the function uses the default (current) information for those omissions and the object is created. The association list supplied to the ENTMAKE function must contain the entity type as either the first or second sub-list in the association list. If the entity type is supplied as the second sub-list of the association list, then it must be preceded by the entity name, in which case the name is ignored and the entity is created. If the association list contains handles for the new object, then those handles are also ignored and the entity is created. For example, to create a line starting at 0,0,0 and ending at 2,5,0 the following expression would be used.

Command: **(SETQ f (LIST (CONS 0 "line")(CONS 10 '(0 0 0))(CONS 11 '(2 5 0))))** (ENTER)
((0 . "line") (10 0 0 0) (11 2 5 0))
Command: **(ENTMAKE f)** (ENTER)
((0 . "line") (10 0 0 0) (11 2 5 0))

Since the layer and linetype were not specified, the ENTMAKE function automatically uses the current values for these setting. This can be confirmed by the ENTLAST and ENTGET functions.

Command: **(ENTGET (ENTLAST))** (ENTER)
Select object: ((-1 . <Entity name: 2780618>) (0 . "LINE") (5 . "43") (100 .
"AcDbEntity") (67 . 0) **(8 . "0")** (100 . "AcDbLine") (10 0.0 0.0 0.0) (11 2.0
5.0 0.0) (210 0.0 0.0 1.0))

 Note: If a layer that does not currently exist is specified, it is created when the ENTMAKE
expression is evaluated.

When complex objects such as polylines and blocks are created through the ENTMAKE
function, the ENTMAKE function must be used multiple times. Each time the function
is called, the next sub-entity contained within the complex object must be supplied.
When complex objects are created by the ENTMAKE function, a temporary file is
created containing the supplied information. Information is fed into this file until
the process is either interrupted or the seqend or endblk sub-entities are issued. Then
the information is checked to determined whether or not it is valid and consistent. If
the information checks out, then the temporary file is deleted and the object is added
to the drawing. If the information is not valid or the creation process is interrupted,
then the temporary file is deleted and the object is not created. To create a red
polyline on a new layer called poly that starts at 0,0,0 proceeds to 3,5,0 and ends at
7,9,0 the following syntax would be used:

```
(ENTMAKE (LIST (CONS 0 "POLYLINE")        ;Entity type to create
              (CONS 8 "POLY")             ;Layer to create entity on
              (CONS 66 1)                 ;Entity color
              (CONS 62 1)                 ;Vertices are to follow
       )                                  ;End list
)                                         ;Completes first pass of ENTMAKE
 (ENTMAKE (LIST (CONS 0 "VERTEX")         ;Entity type
              (CONS 10 '(0 0 0))          ;Starting point
       )                                  ;End list
)                                         ;Completes second pass of ENTMAKE
 (ENTMAKE (LIST (CONS 0 "VERTEX")         ;Entity Type
              (CONS 10 '(3 5 0))          ;Starting point
       )                                  ;End list
)                                         ;Completes third pass of ENTMAKE
 (ENTMAKE (LIST (CONS 0 "VERTEX")         ;Entity
              (CONS 10 '(7 9 0))          ;Starting point
       )                                  ;End list
)                                         ;Completes fourth pass of ENTMAKE
 (ENTMAKE (LIST (CONS 0 "SEQEND")         ;Sequence end
       )                                  ;end list
 )                                        ;Completes fifth pass of ENTMAKE
```

Note: The seqend sub-entity is used to mark the end of a polyline. The endblk sub-entity is used to mark the end of a block definition.

If a block name is specified and that name is currently assigned to an existing block, ENTMAKE will redefine the original block's definition.

Using ENTMAKEX Function to Create an AutoCAD Entity

AutoCAD graphic and non-graphic entities may also be created using the ENTMAKEX function. While this function is similar to the ENTMAKE function in both syntax and argument requirements, it differs in the fact that when an entity is created using this function, an owner is not assigned. The function does assign a handle and an entity name to the object created. In AutoCAD, all entities created must have an owner assigned to them or the object will not be saved to the drawing file, nor will it be written to a DXF file. Therefore, when this function is used, the owner must be established before the drawing session is completed or a DXF file is created. Once the expression containing the function has been evaluated, the entity name of the object created is returned and the object is created. If the entity cannot be created, the function then returns nil.

LOOPS

Often, when an application is developed it becomes necessary for the program to repeat a series of operations. This should have become apparent in the previous chapter in the writing of the series circuit program. In that application the programmer had to know exactly how many resistors were contained in each circuit, and the program could not exceed that number of resistors. If that number was exceeded, then modifications had to be made to include those resistors in the calculations, thereby making the program limited and impractical. To remedy this situation, a process must be introduced that allows the developer to continually execute certain portions of the program until a test condition returns a value of nil (false), at which time the cycle is broken. This process is known as a *Loop* or *Looping*. A *Loop* or *Looping* is nothing more than the continuous execution of a portion or portions or a program. Inserting a loop into the series program gives the application the ability to calculate the equivalent resistance for an infinite number of resistors.

AutoLISP provides two means of looping expressions in a program. They are the WHILE function and the REPEAT function. Either function can be used to create a loop, and both functions return the value of the last expression evaluated, but the WHILE function has one advantage over the REPEAT function: the WHILE function's loop is constructed around a test condition, whereas the REPEAT function is structured around a specified number of cycles. Typically the WHILE function is the primary function used for the construction of a loop; however, both functions should be mastered to give the programmer a variety of methods that can be used.

CREATING A LOOP USING THE WHILE FUNCTION

The WHILE function is most often used when constructing a loop. In application and syntax this function is similar to another AutoLISP function, the IF function. Like the IF function, the WHILE function contains a test expression that when evaluated, if a value other than nil is returned, causes the remaining expressions grouped under the WHILE function to be executed. The grouped expressions will continue to be evaluated until the test expression returns nil, at which time the program continues with the first expression following the loop. Unlike the IF function, the WHILE function does not contain an ELSE expression. If the test expression is first evaluated as nil, then the program skips the loop and resumes execution starting with the first expression following the WHILE loop. The syntax for this function is (WHILE testexpr [expr...]). The *testexpr* argument is the test expression(s) used to determine if the loop is to be executed. This argument can either be single or multiple test expressions. If multiple test expressions are used, then the functions AND and/or OR must also be employed. The *expr* arguments are the expressions that are to be looped. The quantity and type of expressions used in a WHILE loop are not limited, but one of the expressions must reset the value of the variable used in the test expression; otherwise the loop will not break. For example, the following illustrates how the series program from Chapter 3 can be modified to enable the user to enter an undetermined number of resistors. In this program, the user is prompted to enter a resistance. If the user enters a numeric value, then that value is added to the list of resistors and the program begins the loop. This cycle is continued until the user fails to enter a value for the resistance, at which time the program breaks out of the loop and the first expression following the loop is evaluated.

```
;;;********************************************************************
;;;; Program Purpose: This program allows the user to calculate the
;;;;                  voltage drop across a group of series
;;;;                  resistors.
;;;; Written By: James Kevin Standiford
;;; Date: 03/12/99
;;;********************************************************************
(defun c:series    ()
  (setvar "cmdecho" 0)              ;Command echo off
  (setq    voltage (getreal "\nEnter voltage : ")
                                    ;Prompts the user to enter a voltage
                                    ;and setq the value to the variable
                                    ;voltage
        cnt    1                    ;Sets the Variable cnt 1
        resis    (getreal "\nEnter resistance for resistor #1 :"
                                    ;Prompts the user to
        )                           ;enter the resistance for resistor
                                    ;number one
```

```
                                           ;End GETREAL function
      )                                    ;End SETQ function
      (if (/= resis nil)                   ;Start If statement, if user enters a
                                           ;value
                                           ;for resistance then
        (progn                             ;Start PROGN function
          (setq resistor_list (list resis))
                                           ;Creates a list and sets to the variable
                                           ;resistor_list
          (while (/= resis nil)
                                           ;Start WHILE loop, continue
                                           ;looping until user fails
                                           ;to enter a resistance
            (setq cnt  (1+ cnt)            ;Sets cnt to 1 + previous value
                  resis (getreal (strcat   ;Prompts user to enter the
                                           ;resistance for the current
                                           ;resistor
                                           ;The current resistor is the same as
                                           ;the loop number.
                            "\nEnter resistance for resistor #"
                            (itoa cnt)
                            " : "
                          )                ;End STRCAT function
                 )                         ;End GETREAL function
            )                              ;End SETQ function
            (if (/= resis nil)
              (setq resistor_list (append resistor_list (list resis)))
            )                              ;Start If statement, if user enters a
                                           ;value
                                           ;for resistance then append value to
                                           ;resistor list.
          )                                ;End WHILE loop
        )                                  ;End PROGN function
      )                                    ;End IF function
    )                                      ;End program
```

Nesting WHILE Functions

In Chapter 3, the concept of nesting was introduced with the IF function. IF expressions are not the only ones that can be nested; loops can also subjected to this concept. While the theory of nesting functions is simple and can provide the programmer with a valuable means of gathering and processing data, extra care should be taken to provide clear and concise comments concerning critical steps of each loop so that debugging does not become a laborious task. Even when sufficient comments are provided, nested loops can be confusing and frustrating for beginning programmers to construct and debug. For this reason, it is recommended that the nesting of WHILE loops should be avoided when possible. If it is absolutely necessary that one or more loops be nested, then a flowchart can prove to be a valuable asset for constructing and debugging nested loops.

When a WHILE loop is nested within another, the original loop is evaluated until the nested loop is encountered, at which time the nested loop is evaluated and its contents are executed until the loop is broken. Control of the program is once again reestablished by the first loop and the process is repeated until the test expression for the external loop returns nil.

CREATING A LOOP USING THE REPEAT FUNCTION

A loop can also be constructed using the REPEAT function (REPEAT int [expr...]). This function, when supplied with the necessary arguments, repeats a sequence of expressions a predetermined number of times. The first argument that must be supplied is an integer. Its purpose is to instruct the REPEAT function on how many times the second argument (one or more expressions) is to be evaluated. If any other data type is supplied as the first argument, then an error is returned. This makes it somewhat limited compared to the WHILE function, which can loop indefinitely. To illustrate how this function could be used, the program below prompts the user to enter the number of times the expressions supplied to the REPEAT function are to be executed:

```
(DEFUN c:repeatfunction   ()
  (SETQ    num (GETINT "\nEnter number of time to loop: ")
                                ;Prompts user to enter number of times
                                ;to loop expressions
     cnt 0                      ;Sets the variable cnt to 0
  )                             ;End of SETQ function
  (REPEAT num                   ;Start REPEAT function
    (SETQ cnt (1+ cnt))         ;Setq the variable cnt to current value
                                ;plus one
    (PRINC (STRCAT "\nCycle # " (RTOS cnt)))
                                ;Converts the variable cnt into a string
                                ;and merges it with another test string
  )                             ;End of REPEAT function
  (PRINC)                       ;Prints blank space at the end of the
                                ;program
  )                             ;End of program
```

When the program is executed from the command prompt of AutoCAD, the following results are displayed:

```
Command: REPEATFUNCTION (ENTER)
Command: 9 (ENTER)
Cycle # 1.0000
Cycle # 2.0000
Cycle # 3.0000
Cycle # 4.0000
Cycle # 5.0000
Cycle # 6.0000
Cycle # 7.0000
Cycle # 8.0000
Cycle # 9.0000
Command:
```

Again, if a non-integer value is supplied to the function as the first argument, then an error is returned. Removing the variable *num* in the previous example and substituting the expression (< cnt num) causes the following error to be displayed.

```
Command: REPEATFUNCTION (ENTER)
Enter number of time to loop: 4 (ENTER)
; *** ERROR: bad argument type: fixnump: T
```

If an expression is supplied as the first argument, then that expression must return an integer value. For example if the variable *num* is an integer, then the following examples would return a valid result.

```
(REPEAT (+ 25 num) expressions)
(REPEAT (/ num 5) expressions)
(REPEAT (* num 2) expressions)
```

If the variable *num* is a real number, then it must first be converted or the function returns an error. For example, if the variable *num* is equal to the real number 20.5, then the following example would return a valid result.

```
(REPEAT (FIX num) expressions)
(REPEAT (FIX (/ 5 num)) expressions)
(REPEAT (FIX (EXPT num 5)) expressions)
```

SUMMARY

When AutoLISP requests information about an AutoCAD entity, that information is returned in the form of a list known as an association list. Association lists are the primary method used by AutoCAD to maintain an entity's definition data. An association list contains elements that are somewhat linked by an index (group code). When a list is created that contains only two elements and those elements are separated by a period, that list is known as a dotted pair. Dotted pairs can be incorporated into association lists.

The easiest way to create either a dotted pair or an association list is with the QUOTE function. This function returns any arguments supplied to it without evaluating the arguments. Dotted pairs can also be created using the CONS function.

Information contained within an association list can be retrieved through the Cxxxx functions. Although AutoLISP does supply a function that allows the developer to utilize the unique qualities of the association list and therefore retrieve information by using the association list's group code, to retrieve information using an association lists group code, the ASSOC function must be employed.

Once information has been retrieved from an association list, the developer can substitute current values associated with the list with new values using the AutoLISP SUBST function. This function, when supplied with the necessary information, searches a list for a particular item and returns that list with the new item substituted for each occurrence.

All information regarding AutoCAD entities is returned to AutoLISP in the form of an association list. Before an entity's association list can be extracted from AutoCAD, the entity's name must first be known. In AutoLISP an entity's name can be determined using one of the three ENTxxx functions. While the basic information returned by these functions is the same, the circumstances in which these functions are used differ. Once an entity's name has been determined, the association list that contains that entity's information can be extracted using the ENTGET function.

When AutoLISP requests information regarding complex AutoCAD entities (blocks, polylines, three-dimensional objects) the association list that is returned by AutoCAD is the association list for the parent object. To obtain the entity name of the complex object's components, the AutoLISP NENTSEL function must be used.

Once changes have been made to an entity's association list, those changes have to be updated in the drawing's database (dwg file). In AutoLISP this is accomplished by the ENTMOD function. To update the object in the AutoCAD graphics editor, the AutoLISP ENTUPD function is used.

In addition to modifying an AutoCAD entity, AutoLISP can also be used to create both graphic and non-graphic AutoCAD entities. In AutoCAD, a graphic entity is any visible object used to create a drawing, where a non-graphic entity is any object not visible to the user. To create an AutoCAD entity using AutoLISP, the ENTMAKE and ENTMAKEX functions must be used.

To prevent the programmer from having to repeat code in a program, all programming languages employ the concept of loops. A loop is nothing more than the continuous execution of a portion or portions of a program. In AutoLISP, a loop can be constructed using either the WHILE or REPEAT function.

REVIEW QUESTIONS AND EXERCISES

1. Define the following terms:
   Association List
   Dotted Pair
   Extended Entity Data
   Graphic Entities
   Loop
   Non-graphic Entities

2. What is the difference between a list and an association list?

3. What is the difference between an association list and a dotted pair?

4. What is the difference between ENTMAKE and ENTMAKEX?

5. How are loops created in AutoLISP?

6. What is the difference between ENTSEL and NENTSEL?

7. What AutoLISP function is used to create a dotted pair? Give an example of how it is used.

8. What AutoLISP function is used to substitute one element in an association list with another? Give an example how it is used.

9. What AutoLISP function is used to delete and restore entities from the AutoCAD graphics screen?

10. True or False: The only way to create an association list in AutoLISP is by using the QUOTE function. (If the answer is false, explain why.)

11. True or False: An AutoCAD entity association list can not be modified using AutoLISP. (If the answer is false, explain why.)

12. What is the difference between the AutoLISP WHILE function and REPEAT function?

13. True or False: In AutoLISP a loop cannot be nested inside another loop. (If the answer is false, explain why.)

14. True or False: In AutoLISP a loop cannot be nested inside an IF or COND function. (If the answer is false, explain why.)

15. True or False: In AutoLISP a loop is limited to evaluating only one expression. (If the answer is false, explain why.)

What is the value returned by the following expressions?

16. (cons 3 "The value is 4")

17. (list (cons 4 0.65)(cons 5 .08))

18. (list (cons 4 0.65)(cons 5 0.08))

19. (entget (car (entsel)))

20. (entlast)

21. (entget (entlast))

22. (entnext)

23. (entget (entnext))

24. (entdel (cdr (assoc -1 (entget (entlast)))))

25. (setq exampleList (list "this is an example" "list"))

26. (subst (list "list") (list "Neat List") exampleList)

Using the following association list, what value is returned by the following expressions?

List

(setq ent (entget (car (entsel))))

((-1 . <Entity name: 15a4160>) (0 . "LINE") (330 . <Entity name: 15a40f8>) (5 . "2C")
    (100 . "AcDbEntity") (67 . 0) (410 . "Model") (8 . "0") (100 . "AcDbLine") (10
    9.84791 5.59698 0.0) (11 5.12027 2.85854 0.0) (210 0.0 0.0 1.0))

Expressions

27. (cdr (assoc -1 ent))

28. (car (assoc 5 ent))

29. (subst (cons 5 "2B")(assoc 5 ent) ent)

30. (assoc 10 ent)

31. (entmod (subst (cons 10 9.9 8.7 0.0)(assoc 10 ent) ent))

32. (entmod (subst (cons 10 (list 9.9 8.8 0.0))(assoc 10 ent) ent))

33. (entupd ent)

34. (entupd (cdr (assoc -1 ent)))

Debug the following loops:

35. (Setq num 10)

    (While (> num cnt)

    (setq numberOne (getreal "\nEnter first number : ")

        numberTwo (getreal Enter second number :)

    (if (= numberOne numberTwo)

        (Princ "\nThe two number entered are the same : ")

        (setq answer (* numberOne numerTwo))

    (princ "The products of the two numbers are" (rtos answer))

)

36. (while (<= loop_cnt list_cnt)

        (setq resistor (nth loop_cnt resistor_list))

        (if (/= resistor nil)

        (setq   equilivent_resistance

                (+ equilivent_resistance resistor)

)

)

        (setq loop_cnt (1- loop_cnt))

)

37. Write an application that will calculate the btu/hr heat for up to ten different material types using the formula btu/hr = Area * U * (difference in temperature), U = 1/R-value.

38. Write an application that will allow the user to change the location of a line by selecting new starting and ending points.

39. Append the program in question 37 so that the results can either be saved to a file or the AutoCAD graphics screen.

CHAPTER 5

Extended Entity Data, Xrecords, Symbol Tables, Dictionaries, and Selection Sets

OBJECTIVES

Upon completion of this chapter the reader will be able to:

- Define the terms **Extended Entity Data** and **Xrecords** and describe how these two concepts differ from one another
- Describe what application names are and how they are used with extended entity data
- Describe the process used to register an application name using the REGAPP function
- Define the purpose of the Application Identification Table
- Describe how to discover if an application name has already been registered using the REGAPP and TBLSEARCH functions
- Use the ENTMOD function to append extended entity data to an AutoCAD object and modify it
- Use the ENTGET function to retrieve extended entity data from an AutoCAD object
- Construct a selection set using the SSGET function
- Search a symbol table for a specific entity using the TBLSEARCH function
- Page through a symbol table using the TBLNEXT function
- Use the TBLOBJNAME function to search a particular table and return the entity name of specific object

KEY WORDS AND AUTOLISP FUNCTIONS

Application Identification Table	Table
Application Name	TBLNEXT
Dictionary	TBLOBJNAME
Extended Entity Data	TBLSEARCH
Filters	Xdata
REGAPP	XDROOM
Register Application	XDSIZE
SSGET	Xrecord

INTRODUCTION TO EXTENDED ENTITY DATA

Chapters 3 and 4 introduced list manipulation and processing, which is the true power of AutoLISP programming. Even though these capabilities have greatly extended the potential of AutoLISP programming, one drawback still exists concerning non-AutoCAD entity information: Unless the information is either assigned to an AutoCAD attribute or saved in an external file, the program is unable to recall any of the information once the current drawing session has been terminated. This limits the applications where these capabilities can be used. For example, if the resistance program illustrated in Chapter 4 used attributes contained within blocks to store information concerning the resistance and voltage drop for each resistor, then extreme care would have to be taken to ensure that these blocks are not exploded. If one or more of the blocks were exploded, then this would result in the program's ignoring the values that were assigned to those entities, thereby increasing the voltage drop for the remaining resistors. If this were to happen and the situation were to go unnoticed, then the results obtained from running the program would produce a situation in which the actual circuit could fail. Therefore the programmer is forced to rely on constructing a program that would either prevent the user from exploding the attributed blocks or one that warns the user of the potential danger involved if this process is continued. On the other hand, if the program had saved the information to an external file, then a similar situation could result if the file were deleted. AutoLISP provides two possible solutions to this problem. They are *Extended Entity Data* (or *Xdata* as it is sometimes referred to) and *Xrecord*. While both methods are similar in concept and application, each method has its advantages and disadvantages. The programmer should be well versed in the use of both of these methods. The intent of these methods is to provide the programmer with a means of storing and managing information in an object's association list using DXF codes. Attached information can be relevant or non-relevant to the particular entity.

USING EXTENDED ENTITY DATA

Extended entity data is really nothing more than an extension of the object's regular entity data. Through the procedures and techniques previously discussed for handling and manipulating the association list of an AutoCAD entity, along with a few specifically designed Xdata functions, extended entity data can be assigned, retrieved, modified, and even saved to an object with very little effort. Once the information has been saved to an object's definition, then it may be recalled at a later time, long after the initial drawing session has been terminated. This information remains attached to the entity until either the information is removed by a process similar to the way it was attached or the entity is erased from the drawing. Extended entity data does have one major limitation: the amount of information that can be attached to an entity is limited to 16 Kbytes.

Attaching Extended Entity Data to an AutoCAD Object

Extended entity data is attached to an object's definition data using the DXF codes 1000 to 1071. Because of the limited range of these DXF codes and the fact that Xdata can be set by any application, Autodesk has devised a method for keeping the information unique to a particular program. This reduces the possibility of information stored by one program affecting the data or even the operations of other applications. AutoCAD requires that all extended entity data be assigned an application name. That application name may only be used once in a drawing. This helps ensure that information assigned by more than one program to the same AutoCAD entity is unique as long as the application names used by these programs are different.

Registering an Application Name

Before AutoCAD will allow a developer to set extended entity data to an AutoCAD object, the application name must first be registered in the AutoCAD application identification (APPID) table. This table is used by AutoCAD to store the names of all the applications that have been registered in an AutoCAD drawing. To add an application name to the APPID table, the REGAPP function (REGAPP application) must be used. This function, when supplied with the name of the application, first checks the APPID table to determine if the application name is already in use. If the application name is not used, then the function adds the name to the table and returns the name that was added. If the application name already exists, then the function returns nil. The nil in this particular case means only that the application name is already registered. This is illustrated in the following examples, in which the application name "Visual_LISP" is registered:

```
Command: (REGAPP "Visual_LISP")
"VISUAL_LISP"
Command: (REGAPP "Visual_VLISP")
nil
Command:
```

In the first example, the application name VISUAL_LISP is added to the APPID table. In the second example, the REGAPP function returns nil because the application name already exists in the APPID table. Also notice in the first example that the application name returned by the REGAPP function is displayed as all uppercase letters. When a name is supplied to the REGAPP function, all letters contained in that name are converted to uppercase and are stored in the APPID table as such. Application names are limited to 31 characters in length and can be made up of letters, numbers or a combination of the two. They can even contain special characters such as dollar signs ($), hyphens (-), and underscores (_), but they cannot contain less than and greater than (<>), forward slashes and backslashes (∧), quotation marks ("), question marks (?), colons (:), asterisks (*), vertical bars (|), commas (,), equal signs (=), backquotes (`) and semicolons (;). Table 5–1 provides examples of valid and invalid application names.

Table 5–1 Examples of Valid and Invalid Application Names

Valid Application Names	Invalid Applications Names
VISUAL_LISP	VISUAL*LISP
VISUAL_1_LISP	VISUAL<>1_LISP
VISUAL$LISP	VISUAL?LISP
$VISUAL_LISP	VISUAL:LISP
1VISUAL_LISP	VISUAL'LISP
$1-VISUAL_LISP	VISUAL=LISP

Although the REGAPP function takes some precautions to ensure that an overlap of application names does not occur, there is a possibility that an overlap can happen. To guarantee that this does not occur, the AutoLISP TBLSEARCH function (TBLSEARCH table-name symbol [setnext]) can be used. This function, when supplied with the name of the table to search along with the symbol name to check for, returns either nil, if the symbol is not present, or an association list containing the symbol's name and table, if the symbol exists. To check the application identification table to determine if the application VISUAL_1$_LISP-EXAMPLE has been registered, and if it is not, register it, the following syntax would be used:

```
(SETQ ans (TBLSEARCH "appid" "VISUAL_1$_LISP-EXAMPLE"))
                            ;Searches the APPID table for the
                            ;symbol VISUAL_1$_LISP-EXAMPLE and
                            ;returns the results to the
                            ;variable ans.
    (IF (/= ans nil)        ;Tests the value of the variable
                            ;ans
      (REGAPP "VISUAL_1$_LISP-EXAMPLE") ;If ans is not equal to nil then
```

```
                                        ;the symbol VISUAL_1$_LISP-EXAMPLE
                                        ;is added to the APPID table using
                                        ;the REGAPP function
  )                                     ;End of IF statement
  (IF (= ans NIL)                       ;Tests the value of the variable
                                        ;ans
    (PRINC                              ;If ans is equal to nil then a
                                        ;message indicating that the
                                        ;symbol is already registered is
                                        ;displayed.
      "\nApplication VISUAL_1$_LISP-EXAMPLE is already registered : "
    )
  )                                     ;End of IF statement
```

Assigning Extended Entity Data to an AutoCAD Object

Once the application name for the extended entity has been registered in the application identification table, extended data may now be assigned to an AutoCAD object. This is usually accomplished by appending the extended data to the object definition data using one of the many list functions provided in AutoLISP. Then the entity data is updated in the AutoCAD drawing database using ENTMOD function and assigned the extended information. Before extended data can be appended to an existing entity, it must be contained within an association list that starts with the code −3 and is followed by the name of the application with which the data is associated. Extended data can be assigned only to existing AutoCAD objects. The entity must either already exist or be created before the extended data can be assigned. An object can be assigned multiple application names and extended entity data. The following program illustrates how extended data can be assigned to an AutoCAD object.

```
(DEFUN c:extended_entity_data ()
  (SETQ select_entity (ENTGET (CAR (ENTSEL))))
                                        ;Gets the association list of
                                        ;definition data for the entity
                                        ;selected.
  (REGAPP "EXAMPLE_1_EXTENDED_DATA")    ;Registers the application name
                                        ;EXAMPLE_1_EXTENDED_DATA
  (SETQ     extend_data                 ;Sets the variable extend_data
                                        ;equal
          '((-3
              ("EXAMPLE_1_EXTENDED_DATA"
                                        ; to the new extended data, which
                                        ; is a text string and two
                                        ; scaleable values 1041 and 1042.
                (1000 . "Imagine what can be saved to extended
  entity data")
```

```
                         (1041 . 4.5)
                         (1042 . 10.0)
                         )
                       )                        ; End of association list
         )                                      ; End of QUOTE function
     new_entity_def
                   (APPEND select_entity extend_data)
    )
                                                ; Appends the extended entity data
                                                ; to the object definition data.
     (ENTMOD new_entity_def)                    ; Updates the entity defination in
                                                ; the AutoCAD database with the new
                                                ; definition data containing the
                                                ; extended entity data.

  )
```

Retrieving Extended Entity Data from an AutoCAD Object

When an object has been assigned extended entity data, it cannot be accessed through traditional AutoCAD object-listing or entity-modifying commands (LIST, DDMODIFY, PROPERTIES, etc.). This alone assures the developer that the information assigned to the object cannot be tampered with by the user. The only way the information can be extracted once it has been assigned is with the use of the ENTGET function. Recall from Chapter 4 that when this function is supplied with any entity's name, it returns the definition data for the specified object. However, after a closer examination of the function's syntax, one important aspect emerges: This function, when supplied with an application name in addition to the entity name, returns the entity definition along with the extended entity data that is associated with the specified application. For example, to obtain the extended entity data that was assigned to an object by the previous program, the following syntax would be used:

Command: **(SETQ entity_data (ENTGET (CAR (ENTSEL))**
'("EXAMPLE_I_EXTENDED_DATA"))) (ENTER)
Select object:
((-I . <Entity name: 2770500>) (0 . "LINE") (5 . "20") (100 .
"AcDbEntity") (67 . 0) (8 . "0") (100 . "AcDbLine") (10 2.19416 5.16515 0.0) (11
5.15258 2.37318 0.0) (210 0.0 0.0 1.0) *(-3*
("EXAMPLE_I_EXTENDED_DATA" (1000 . "Imagine what can be saved to
extended entity data") (1041 . 4.5) (1042 . 10.0))))
Command:

In order for the Xdata assigned to each application associated with an object to be retrieved, all application's names must be supplied to the ENTGET function. To retrieve the extended entity data appended to an object's definition data by the application names

EXAMPLE_1_EXTENDED_DATA and EXAMPLE_2_EXTENDED_DATA, the following syntax would be used:

> Command: **(SETQ entity_data (ENTGET (CAR (ENTSEL))**
> **'("example_1_extended_data" "example_2_extended_data"))** (ENTER)
> Select object:
> ((-1 . <Entity name: 2770500>) (0 . "LINE") (5 . "20") (100 . "AcDbEntity") (67 . 0)
> (8 . "0") (100 . AcDbLine") (10 2.32565 2.06113 0.0) (11 6.31952 4.44252 0.0)
> (210 0.0 0.0 1.0) *(-3 "EXAMPLE_1_EXTENDED_DATA" (1000 . "The*
> *possibilities are endless") (1041 . 14.5) (1042 . 12.0))*
> *("EXAMPLE_2_EXTENDED_DATA" (1000 . "Imagine what can be saved to*
> *extended entity data") (1041 . 14.5) (1042 . 12.0))))*
> Command:

The wild card character asterisk (*) is an easier method that can be employed to retrieve multiple applications assigned to an AutoCAD object. When this character is supplied in place of an application name, ENTGET returns all applications associated with the selected object. In other words, the same results that were returned in the previous example can also be achieved using the following syntax.

> Command: **(SETQ entity_data (ENTGET (CAR (ENTSEL)) '("*"))** (ENTER)
> ((-1 . <Entity name: 2770500>) (0 . "LINE") (5 . "20") (100 . "AcDbEntity") (67 . 0)
> (8 . "0") (100 . AcDbLine") (10 2.32565 2.06113 0.0) (11 6.31952 4.44252 0.0)
> (210 0.0 0.0 1.0) *(-3 "EXAMPLE_1_EXTENDED_DATA" (1000 . "The*
> *possibilities are endless") (1041 . 14.5) (1042 . 12.0))*
> *("EXAMPLE_2_EXTENDED_DATA" (1000 . "Imagine what can be saved to*
> *extended entity data") (1041 . 14.5) (1042 . 12.0))))*
> Command:

Modifying Extended Entity Data

Once the extended entity data has been retrieved with the ENTGET function, the programmer is free to modify any portion or all of the information using the list and association list functions previously discussed. At this point the extended entity data is treated just as the association list data was in Chapter 4. The dotted pair containing the old data is first obtained using the ASSOC function. Next, the new dotted pair is created using the CONS function and the two are switched using the SUBST function. Finally, the changes are recorded in the drawing database using the ENTMOD function. Using the extended entity data from the previous example, the following expressions would be used to replace the dotted pair (1041 . 14.5) with the dotted pair (1041 . 0.0125):

```
(SETQ      old_value    (ASSOC 1041
                         (CDR (CAR (CDR (ASSOC -3
                              new_entity_def)))))
```

```
                               ;This expression extracts the value to
                               ;be replaced (1041 . 14.5)
new_value      (cons 1041 0.568)
                               ;This expression creates the new value
                               ;(1041 . 0.0125)
ent            (CDR (CAR (CDR (ASSOC -3 new_entity_def))))
                               ;This expression retrieves the
                               ;association list assigned to the
                               ;application EXAMPLE_1_EXTENDED_VALUE
                               ;((1000 . "The possibilities are
                               ;endless") (1041 . 14.5) (1042 . 12.0))
ent1           (assoc -3 new_entity_def)
                               ;This expression retreives the extended
                               ;entity data attached to the select
                               ;object (-3 ("EXAMPLE_1_EXTENDED_DATA"
                               ;(1000 . "The possibilities are
                               ;endless") (1041 . 14.5) (1042 . 12.0)))
ent2           (car (cadr ent1))
                               ;This expression obtains the name of the
                               ;application where the extended data
                               ;is registered
                               ;"EXAMPLE_1_EXTENDED_DATA"
ent3           (list -3
                    (append (list ent2) (subst new_value
                    old_value ent)))
                               ;This expression substitutes the new
                               ;value created earlier with the existing
                               ;value in the association list that is
                               ;registered to the application
                               ;"EXAMPLE_1_EXTENDED_DATA, thus
                               ;returning the new association list (-3
                               ;("EXAMPLE_1_EXTENDED_DATA" (1000 .
                               ;"The possibilities are endless") (1041
                               ;. 0.0125) (1042 . 12.0)))
ent4           (subst ent3 (assoc -3 new_entity_def)
                    new_entity_def))
                               ;This expression substitutes the new
                               ;extended association list with the old
                               ;extended association list in the entity
                               ;definition list of the object selected.
                               ;This returns ((-1 . <Entity name:
                               ;2760500>) (0 . "LINE") (5 . "20") (100
                               ;. "AcDbEntity") (67 . 0) (8 . "0") (100
                               ;. "AcDbLine") (10 1.71753 4.06478 0.0)
```

```
;(11  5.38268  6.15055  0.0)  (210  0.0  0.0
;1.0)  (-3  ("EXAMPLE_1_EXTENDED_DATA"
;(1000  .  "The  possibilities  are
;endless")  (1041  .  0.0125)  (1042  .  1
;2.0))))
```

(entmod ent4)

```
;This  expression  updates  the  entity
;definition  in  the  drawing  database
;thus  returning  ((-1  .  <Entity  name:
;2760500>)  (0  .  "LINE")  (5  .  "20")  (100
;.  "AcDbEntity")  (67  .  0)  (8  .  "0")  (100
;.  "AcDbLine")  (10  1.71753  4.06478  0.0)
;(11  5.38268  6.15055  0.0)  (210  0.0  0.0
;1.0)  (-3  ("EXAMPLE_1_EXTENDED_DATA"
;(1000  .  "The  possibilities  are
;endless")  (1041  .  0.0125)  (1042  .
;12.0))))
```

The same process of modifying the extended entity data could have been accomplished by simply redefining the extended entity data using the same program that assigned the data to the object in the first place, but this time inserting the new values in the appropriate place. For example, by substituting the dotted pair (1041 . 0.568) into the original program and then rerunning that application, the same results could have been achieved. If an object has extended entity data already assigned to it, that data can be redefined if the program uses the same application name as the one already registered to that object.

Managing Extended Entity Data

Extended entity data has one main limitation that could cause an application that uses extended entity data to fail. Extended entity data is currently limited in size (the amount of information that can be stored) to 16 KB. This should be a major concern with developers because Xdata can be used by multiple applications for the same AutoCAD object. AutoLISP does offer two functions that provide a means for checking the amount of available space remaining for a particular AutoCAD entity, as well as the amount of space required to store a specified extended entity association list. These functions are XDROOM (XDROOM ename) and XDSIZE (XDSIZE 1st). The XDROOM function, when supplied with an entity name, returns the amount of available extended entity data space remaining for the specified object. The result is returned in the form of an integer that represents the available space in bytes. The other function, XDSIZE, when supplied with an extended entity association list that is to be appended to an AutoCAD object, returns the amount of space required to contain that list. This result is also returned in the form of an integer representing the number of bytes required to store the information. By using these two functions in a program, the developer has a better chance of controlling the Xdata associated with an AutoCAD entity.

The previous section discussed how to modify extended entity data using the ENTMOD function. If the modifications made to the Xdata had exceeded the size limitation, then ENTMOD would have been unable to make the necessary changes to the drawing database. The ability to check the remaining amount of extended entity data space and compare it to the amount required to store a particular amount of data gives the developer the opportunity to decide what action to take if the amount required is greater than the amount available. If the supplied information exceeds the available space, then the developer has the option of not making the changes, removing any non-essential information from the Xdata list, or attaching it to another entity in the drawing file. To check the amount of available space that an entity can receive and compare it to the amount of space required to append a particular amount of data, the following expressions would be added to the previous program segment:

```
(SETQ remaining (XDROOM (CDR (ASSOC -1 new_entity_def)))
                              ;This expression extracts the entity
                              ;name then passes it to the XDROOM
                              ;function where the amount of available
                              ;remaining space is returned. Afterward
                              ;the information is set to the variable
                              ;remaining.
    use      (XDSIZE ent3)    ;This expression returns the amount of
                              ;space required to store the modified
                              ;xdata and sets the results to the
                              ;variable use.
)
  (IF (> remaining use)       ;This expression tests the variable
                              ;remaining and if it is greater than the
                              ;variable use then the following
                              ;expressions are carried out.
    (PROGN
```

XRECORDS

Another type of extended entity data that is available to the AutoLISP developer is Xrecords. Xrecords are similar to Xdata in that they are used to store and manage miscellaneous non-AutoCAD information in an association list constructed from DXF group codes (1 through 369). Unlike Xdata, Xrecords are not limited in size or order, thus allowing the developer to store more information. Xrecords are designed to not offend earlier versions of AutoCAD such as R13c0 through R13c3. If a drawing containing Xrecords is loaded into a release prior to R13c4 level of AutoCAD, then those Xrecords disappear. The following example illustrates how Xrecords are created:

```
(DEFUN C:create_xrecord ( / xrec xname )
  (setq xrecord_list '((0 . "XRECORD")(100 . "AcDbXrecord")
                       (1 . "The possibilities are endless.")
              (10 3.0 4.0 0.0)
                       (40 . 3.15)
                       (50 . 3.15)
                       (62 . 1)
                       (70 . 180)
                       )
  )
  (SETQ xrecord_name (ENTMAKEX xrecord_list))
  (DICTADD (NAMEDOBJDICT) "XRECLIST" xname)
  (PRINC)
)
```

SYMBOL TABLES AND DICTIONARIES

It has been demonstrated that the AutoCAD entity database contains very detailed information regarding the graphic objects contained within a drawing. This database retains information concerning non-graphic entities (layers, linetypes, etc.) as well. Non-graphic entities are stored as either dictionaries or symbol table objects. While these objects are similar in concept, the functions used to handle these objects are different. All objects, graphic and non-graphic, support the ENTGET, ENTMOD, ENTDEL, and ENTMAKE functions; however, the type associated with the individual object dictates the extent to which these functions can be used.

SYMBOL TABLES

The symbol table can be thought of as a filing cabinet. It is used to store information regarding non-graphic AutoCAD objects: Layers, Linetypes, Styles, Viewports, Views, Dimension Styles, User Coordinates Systems and Applications. The symbol table itself contains sub-tables where data associated with each category is contained. Before a non-graphic entity can be appended to a symbol table using AutoLISP, certain restrictions and rules must be adhered to. They are:

- The guidelines and restrictions that apply to graphic objects also apply to non-graphic objects.

- The ENTMAKE function can be used to create symbol table entries if a valid description is passed to the function. The function cannot be used to create symbol tables.

- Once an entry has been made in the symbol table, it cannot be updated using the ENTUPD function.

- No entry names can conflict with one another among all sub-tables except the VPORT table.

- Once an entry has been made to the symbol table, it cannot be deleted with the ENTDEL function.

- The entity name of a symbol table entry may be obtained using the TBLOBJNAME function. Once the name has been acquired, its data may be returned using the ENTGET function.

- Neither the ENTMAKE nor the ENTMOD function will allow the developer to specify a handle using DXF codes 5 or 105 in the creation or modification of an entity. These two functions also ignore the DXF code 70 when they are supplied with valid record lists.

- All symbol table entries except ones that are contained within the APPID table can be modified using the ENTMOD function only if the entity's name is furnished. The entity name of an entry in a table can be obtained using the ENTGET function and not the TBLSEARCH or TLBNEXT function.

- Symbol table entries can be renamed. However, the new name cannot duplicate an existing name except for entries contained in the VPORT symbol table.

- The table entries *ACTIVE from the VPORT table and CONTINUOUS from the LINETYPE table cannot be modified or renamed. The entries STANDARD from the STYLE and DIMSTYLE tables and *MODEL_SPACE and PAPER_SPACE from the BLOCKS table may be modified but cannot be renamed.

- Entries contained within the APPID table cannot be renamed.

When a request is made by AutoLISP regarding the data associated with a non-graphic object, that information is returned in the form of an association dotted pair list. Once this information has been obtained, it can be modified only if the rules previously mentioned have been satisfied and then by using the list and association list functions previously discussed. Before the information regarding a non-graphic object can be manipulated, it must first be retrieved from the symbol table. In AutoLISP, this can be achieved by using one of the three symbol-handling functions provided. These functions are TBLSEARCH, TBLNEXT, and TBLOBJNAME.

The TBLSEARCH Function

The TBLSEARCH function (TBLSEARCH table-name symbol [setnext]) searches a symbol table for a specific entry name (symbol). If the symbol is contained within the specified sub-table, then the TBLSEARCH function returns an association dotted pair for that symbol. The sub-table that the TBLSEARCH function is directed to search is specified by the *table-name* argument and must be a valid sub-table name. If an invalid *table-name* argument is supplied, then an error is returned. The *symbol* argument is the entry where the TBLSEARCH function is to search. If the symbol does not exist, then the function returns nil. Finally, the optional argument *setnext* is used to reset the entry counter used by the TBLNEXT function. If the argument supplied is not equal to nil, then the counter is reset so that the TBLNEXT function, when used, returns the next entry following the one supplied to the TBLSEARCH function. If this argument is equal to nil, then the counter is not affected. To search the sub-table layer for the layer 2 the following syntax would be used:

Command: **(TBLSEARCH "layer" "2" T)**
((0 . "LAYER") (2 . "2") (70 . 0) (62 . 7) (6 . "Continuous")) *(Value returned by the TBLSEARCH function.)*
Command: **(TBLSEARCH "layer" "2" e)**
((0 . "LAYER") (2 . "2") (70 . 0) (62 . 7) (6 . "Continuous")) *(Value returned by the TBLSEARCH function.)*
Command: **(TBLSEARCH "layer" "2")**
((0 . "LAYER") (2 . "2") (70 . 0) (62 . 7) (6 . "Continuous")) *(Value returned by the TBLSEARCH function.)*

In the first example, the optional argument *setnext* is furnished with the symbol **T**. Because this symbol retains a value not equal to nil, the entry counter used by the TBLNEXT function is reset. In the second example, the argument is supplied with the symbol **e**. If this symbol retains a value of nil, then the counter is not affected. Finally, in the third example the *setnext* argument is not furnished; thus the entry counter is not affected. In all three examples, the results returned by the TBLSEARCH function are identical.

The TBLNEXT Function

The TBLNEXT function (TBLNEXT table-name [rewind]), when supplied with a valid sub-table name, returns the next entry residing in that table. The function is similar to the TBLSEARCH function in one respect: a valid sub-table name must be specified or the function will return an error. Unlike the TBLSEARCH function arguments, the optional argument associated with this function is used to rewind the function to the first entry residing in the specified sub-table. If the function has not previously been executed or the entry counter has not been reset by the TBLSEARCH function, then the first entry contained in the sub-list is returned. Once the function has been issued, repeated use of this function will return the next corresponding entry. Once the last entry in a sub-table has been returned, repeated use of the TBLNEXT function returns the value nil, unless the rewind argument is supplied (as a non-nil value), at which time the first entry is returned. For example:

Command: **(tblnext "layer")**
((0 . "LAYER") (2 . "0") (70 . 0) (62 . 7) (6 . "Continuous"))

Command: **(tblnext "layer")**
((0 . "LAYER") (2 . "1") (70 . 0) (62 . 5) (6 . "Continuous"))
Command: **(tblnext "layer")**
((0 . "LAYER") (2 . "2") (70 . 0) (62 . 2) (6 . "Continuous"))
Command: **(tblnext "layer")**
((0 . "LAYER") (2 . "3") (70 . 0) (62 . 4) (6 . "Continuous"))
Command: **(tblnext "layer")**
((0 . "LAYER") (2 . "4") (70 . 0) (62 . 1) (6 . "Continuous"))
Command: **(tblnext "layer")**
nil
Command: **(tblnext "layer" t)**
((0 . "LAYER") (2 . "0") (70 . 0) (62 . 7) (6 . "Continuous"))

In the first five examples, repeated use of the TBLNEXT function returns the data associated with each corresponding entry as they appear in the layer sub-table. Once all entries have been accessed and the TBLNEXT function is reissued, a value of nil is returned as in the sixth example. However, in the seventh example a non-nil rewind argument is supplied and the first entry is returned.

Coordinating Table Searches Using the TBLSEARCH and TBLNEXT Functions

As stated earlier, when a non-nil argument is supplied to the TBLSEARCH function, then the next time the TBLNEXT function is used the entry following the one specified in the TBLSEARCH function is accessed. This feature is extremely useful in working with the VPORT symbol table. In the VPORT symbol table, all viewports for a particular configuration have the same name.

 Note: Any changes made to the VPORT symbol table while TILEMODE is turned off will not be visible until TILEMODE is reactivated.

DICTIONARY OBJECTS

Dictionary objects are similar in concept to symbol tables; their purpose is to store non-graphic information that cannot be contained in a symbol table. Unlike symbol tables, dictionary objects can be deleted using the ENTDEL function. Dictionary objects consist of a key (which is a text name) plus an ownership handle (which references to an entity). Before a non-graphic entity can be added to a dictionary using AutoLISP, certain restrictions and rules must be adhered to. They are:

- Access to entries contained in a dictionary is made through either the DICTSEARCH or the DICTNEXT function.

- Dictionary objects may be examined using the ENTGET function and their Xdata modified with ENTMOD. Entries cannot be altered with the ENTMOD function.

- All Dictionary entries that begin with ACAD* cannot be renamed.

- Dictionary entries may be removed by directly passing entry object names to the ENTDEL function.

- The text name key uses the same syntax and valid characters as symbol table names.

AutoLISP does provide five functions that are designed just for working with dictionary objects. These functions are DICTADD, DICTRENAME, DICTREMOVE, DICTSEARCH, and DICTNEXT. The last two functions are similar to their symbol table counterparts TLBSEARCH and TBLNEXT. The DICTSEARCH (DICTSEARCH ename symbol [setnext]) function searches a dictionary for a specific dictionary object, and the DICTNEXT (DICTNEXT ename [rewind]) function returns the next dictionary object in a dictionary. The other three functions are unique to dictionaries and do not have a symbol table counterpart. These functions allow the developer to add (dictadd ename symbol newobj), rename (dictrename ename oldsym newsym) and remove (dictremove ename symbol) items from a specified dictionary.

SELECTION SETS

Up to this point, when an application needed to edit an AutoCAD entity, one of the entity selection (ENTXXX) functions was used. However, these functions do have a limitation: they can only select one entity at a time. If more than one entity is to be edited, then the process of selecting the entities must be repeated until all entities have been changed, and the program can quickly become a hassle instead of an asset. For example, in AutoCAD Release 14.01, if the user wanted to change the contents of several strings of text using either the DDMODIFY or DDEDIT command, then each separate entity had to be selected and the new string retyped each time. Having to retype the same string each time makes these commands tedious. AutoLISP does offer a function that allows the developer to work around situations where multiple entities can be edited by using selection sets. A selection set is nothing more than a group of entities that may be edited as a whole or each individual entity may be extracted and edited separately. Before a selection set can be used in an AutoLISP program, there are several functions that must be examined.

BUILDING A SELECTION SET IN AUTOLISP

Just as in AutoCAD, before a selection set can be used it must first be created. In AutoLISP this is accomplished by the SSGET function (ssget [sel-method] [pt1

[pt2]] [pt-list] [filter-list]). This function has several optional arguments that may be supplied; however, if no arguments are furnished, then the user is free to select a type of entity contained within the drawing using any of the standard selection techniques. For example, to prompt the user to create a selection set of objects contained in an AutoCAD drawing the following expressions would be used:

```
(PRINC "\nSelect objects to change : ")    ;Displays the prompt
                                           ;"select objects to
                                           ;change : ".
(SETQ selection_set (SSGET))               ;Returns a selection
                                           ;Set and sets it
                                           ;to the variable
                                           ;selection_set
```

The first argument associated with this function is the *sel-method* or selection mode argument. This is a string value that can be used to specify the method by which objects are selected. Objects can be selected using regular or crossing windows, previous objects selected, last object created, fence, crossing polygon, or a windowed polygon. When one of these modes is used, then the programmer must also supply the argument(s) *pt1* and/or *pt2*. These arguments are the points associated with the particular selection mode specified. In other words, if the developers use a crossing window selection mode, then two points (the diagonal corners of the window) must be specified. To set the SSGET function to a crossing window selection mode, the following expressions would be used:

```
(SETQ pt (GETPOINT :\nSelect first corner of crossing window : "))
                                        ; Prompts the user to
                                        ;select the first
                                        ;corner defining the
                                        ;crossing window to be
                                        ;used with the SSGET
                                        ;crossing window
                                        ;select mode option.
(SETQ pt1 (GETPOINT :\nSelect diagonal corner of crossing window: "))
                                        ;Prompts the user to
                                        ;select the diagonal
                                        ;corner defining the
                                        ;crossing window.
(SETQ selection_set (SSGET "C" pt pt1))  ;Creates a selection
                                        ;set using the
                                        ;crossing window mode
                                        ;starting with the
                                        ;first point specified
```

```
;through the second
;point. The selection
;set is then set to
;the variable
;selection_set.
```

Table 5–2 provides valid selection mode options.

Table 5–2 Selection Mode Options

Option	Description
C	Crossing window
CP	Crossing polygon
F	Fence
I	Implied
P	Previous created selection set
W	Window
WP	Windowed polygon
X	Entire drawing database
:E	Everything within the cursor's object selection
:N	Call SSNAMEX for additional information on container blocks and transformation matrices for any entities selected during the SSGET operation.

 Note: Because selection sets can contain paper and model space entities, only the entities that are contained in the current space mode are used when an operation is invoked. When a selection set is created that contains complex objects, only the main entities of the objects selected are stored in the selection set.

When the selection mode options CP, F, and WP are specified, then argument *pt-list* or point list is supplied in place of the arguments *pt1* and *pt2*. This is a list of points that will be used by these options to carry out their intended operation. To create a selection set using the crossing polygon option, the following syntax would be used:

```
(SETQ pt1 (GETPOINT "\nSelect first point of the polygon : ")
      pt2 (GETPOINT "\nSelect second point of the polygon : ")
      pt3 (GETPOINT "\nSelect third point of the polygon : ")
      pt4 (GETPOINT "\nSelect fourth point of the polygon : ")
      pt_list (list pt1 pt2 pt3 pt4)
      selection_set (SSGET "_CP" pt_list)
)
```

The final argument that can be supplied to this function is the *filter-list* option. This is an association list used as a filter where only objects matching specified criteria are added to a selection set. This option allows the developer to perform either broad or narrow searches of the drawing database. To create a selection set of all string data contained in a drawing, the following expression would be used:

```
(SETQ selection_set (SSGET "X" (LIST (CONS 0 "text"))))
```

The selection can be narrowed even further with the addition of more criteria to the filter list option. To create a selection set of only yellow text contained on the layer test, the following expression would be used:

```
(SETQ selection_set (SSGET "x" (LIST (CONS 8 "TEST")(CONS 0 "text")(CONS
    62 2)))
```

 Note: If a selection set is to be created by specifying an association list of properties, then the "X" must also be used.

Adding Entities to a Selection Set

Once a selection set has been created, additional entities may be appended to the selection set through the SSADD function (SSADD [ename [ss]]). This function, when supplied with an entity name, appends the object to a specified selection set. To append a line to the selection set created in the previous example, the following expressions would be used:

```
(SETQ entity_name (CAR (ENTSEL)))        ;Extracts the entity
                                         ;name of the object to
                                         ;be appended to the
                                         ;selection set.
(SETQ selection_set (SSADD entity_name selection_set))
                                         ;Adds the entity
                                         ;to the selection set.
```

Removing Entities from a Selection Set

If an entity can be appended to a selection set once the selection set has been defined, then an entity can also be removed from a selection set. In AutoLISP this is accomplished by the SSDEL function (SSDEL [ename [ss]]). Just like the SSADD function, this function, when supplied with an entity name and a previously created selection set, removes that object from the specified set. To remove a line from the selection set created in the previous example, the following expressions would be used:

```
    (SETQ entity_name (CAR (ENTSEL)))          ;Extracts the entity
                                               ;name of the object to
                                               ;be removed from the
                                               ;selection set.
    (SETQ selection_set (SSDEL entity_name selection_set))
                                               ;Removes the entity
                                               ;from the selection
                                               ;set.
```

Determining if an Entity Belongs to a Selection Set

From time to time it is necessary to determine if an entity is already member of a selection set. The SSMEMB (SSMEMB ename SS) function can be used for this purpose. This function, when supplied with an entity name and a selection set, returns either the entity name, if the entity is a member of the selection set, or nil if it is not. To determine if the entity selected in the previous example is already a member of the selection set before it is added, the following expressions would be used:

```
    (SETQ entity_name (CAR (ENTSEL)))          ;Extracts the entity
                                               ;name of the object to
                                               ;tests as a member of
                                               ;a selection set.
    (IF (= (SSMEMB entity_name selection_set) nil)
                                               ;Test the entity
                                               ;selected in the
                                               ;previous expression
                                               ;against the selection
                                               ;set selection_set.
    (PROGN
        (SETQ selecton_set (SSADD entity_name selection_set))
        (PRINC "\nEntity added to selection set ")
                                               ;IF the selected object is not
                                               ;a member of the selection
                                               ; set then the entity is
                                               ;appended and the message
                                               ;"Entity added to selection
                                               ;set" is displayed.
    )
        (PRINC "\nEntity already belongs to selection set)
                                               ;If the object is already a
                                               ;member of the selection set
                                               ;then the message "Entity
                                               ;already belongs to select
                                               ;set" is displayed.
    )
```

WORKING WITH SELECTION SETS

Once a selection set has been created, the developer can then either modify the contents of the selection set one at a time or modify the selection set as a whole. If the selection set is to be dealt with as a whole (such as by changing the layer of the selected entities), then the modification can be achieved through standard AutoCAD commands (in conjunction with the AutoLISP COMMAND function). If the entities are to be modified in a way that is not possible with any of the AutoCAD commands, then the entity name of each object embedded within the selection set must be obtained.

Determining the Number of Entities Contained Within a Selection Set

Before the entity name of an object contained within a selection set can be extracted, the actual location of that object in relation to the selection set must be known. Often this is accomplished by first determining the number of entities held by a selection set and then using a loop that is terminated once the process has evaluated the last entity in that set. To determine the number of entities within a selection set, the AutoLISP SSLENGTH function (SSLENGTH selection-set) is used. This function, when supplied with a valid selection set, returns an integer value representing the number of objects contained within that set. To create a selection set containing all the text shown in Figure 5–1 and then determine the number of entities contained within the selection set, the following expressions would be used:

```
(SETQ sel (SSGET "x" (LIST (CONS 0 "text"))))   ;Creates a selection
                                                ;set containing all
                                                ;the text within the
                                                ;drawing.
(SETQ len (SSLENGTH sel))                       ;Determines the number
                                                ;of entities contained
                                                ;within the selection
                                                ;set. When these
                                                ;expressions are
                                                ;applied to Figure 5-1
                                                ;the SSLENGTH function
                                                ;returns the integer
                                                ;value of 17.
```

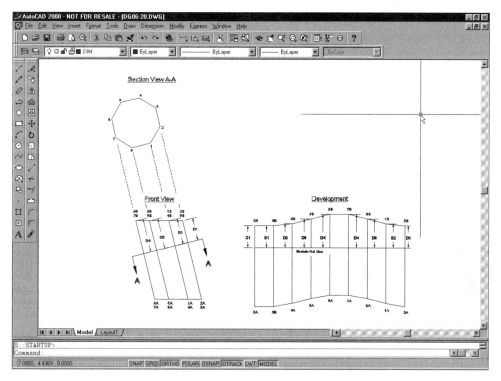

Figure 5–1   *Determining the number of objects within a selection set*

Extracting Entities from a Selection Set

In AutoLISP the extraction of an entity name from a selection set is accomplished by the SSNAME function (SSNAME selection-set index), but before an entity name can be extracted, an index number must be supplied. Index numbers are integer values that range from 0 to 32767 and represent the position of the object within the selection set. To extract the entity name of the first object contained within the selection set from the last example, the following expression would be used:

```
(SSNAME sel 0)
```

If the SSNAME function is successful, then the entity name is returned and the association list of the object's properties may be obtained through the ENTGET function. If the function is not successful (for example, if the specified index number is greater than the position of the last entity contained within the selection set), then nil is returned.

Extracting Entity Names from Selection Sets that Contain more than One Entity

By now it should be clear that one advantage of using the SSGET as opposed to any of the ENTXXX selection functions is that the developer is able to construct programs where groups of entities may be manipulated at one time. To accomplish this once the selection set has been created, the programmer must employ a loop. This is illustrated in the following example, where a loop is constructed to extract the remaining entity names from the selection set constructed in the previous section.

```
(SETQ sel (SSGET "x" (LIST (CONS 0 "text"))))     ;Creates a selection
                                                  ;set consisting of
                                                  ;all the text
                                                  ;contained within the
                                                  ;drawing and sets it
                                                  ;to the variable sel.
      len (SSLENGTH sel)                          ;Determines the length
                                                  ;of the selection set
                                                  ;previously created.
      cnt 0)                                      ;Sets the variable cnt
                                                  ;to zero.
(WHILE (< cnt (- len 1))                          ;Begins looping
                                                  ;following the
                                                  ;expression
                                                  ;while cnt is less
                                                  ;than length of
                                                  ;the selection set
                                                  ;minus one.
      (SETQ entity_name (SSNAME sel cnt))         ;Extracts the entity
                                                  ;name of the object
                                                  ;specified by the
                                                  ;index number supplied
                                                  ;by the cnt variable
                                                  ;and sets it to the
                                                  ;variable entity_name.
    (IF (/= entity_name nil)(SETQ ent_data (ENTGET entity_name)))
                                                  ;If the variable
                                                  ;entity_name is not
                                                  ;equal to nil then the
                                                  ;association list for
                                                  ;the specified entity
                                                  ;is returned.
      (SETQ cnt (1+ cnt))                         ;Increases the value
                                                  ;of cnt by one.
)
```

APPLICATION – CONSTRUCTING AND ATTACHING EXTENDED ENTITY DATA TO A SIMPLE GEAR TRAIN

In Chapter 3 the concepts of list manipulation and processing were addressed, in Chapter 4 association lists were introduced, and in Chapter 5 the subject of extended entity data has been covered. In reality these concepts overlap and in many cases are all employed in the same program. To illustrate this point, the following program uses both association lists to construct a simple gear train and extended entity data to append design criteria to the appropriate entities once they have been created.

```
;;;********************************************************************
;;;
;;; Program Name: VL05.lsp
;;;
;;; Program Purpose:This program allows the user to construct a
;;;                 simple gear train, from the Pinion Gear RPM,
;;;                 the gear ratio, the diametral pitch, shaft
;;;                 diameter, and number of teeth on the pinion
;;;                 gear. The two gears are constructed as blocks,
;;;                 and the rpms, number of teeth and gear ratio
;;;                 of both gears are saved to the newly constructed
;;;                 entities in the form of extended entity data.
;;;
;;; Program Date: 2/31/99
;;;
;;; Written By: James Kevin Standiford
;;;
;;;********************************************************************
;;;********************************************************************
;;;
;;;              Main Program
;;;
;;;********************************************************************
(DEFUN c:gear (/    rpm        number_teeth
        gear_ratio dia_pitch  shaft_dia   PT
        pitch_dia  gear_teeth gear_out     Pinion_out
        gear_rpm   dist       pt_x                    pt_y
        pinion     gear       shaft_pinion
        Shaft_gear lastent    exdata1       newent1
        exdata     newent
      )
  (SETQ   rpm          (GETREAL "\nEnter RPM of Pinion Gear : ")
    number_teeth (GETREAL "\nEnter Number of Teeth : ")
    gear_ratio   (/ 1 (GETREAL "\nEnter Gear Ratio : "))
    dia_pitch    (GETREAL "\nEnter Diametral Pitch : ")
```

```
    shaft_dia    (GETREAL "\nEnter Shaft Diameter : ")
    PT           (GETPOINT "\nSelect Point : ")
;;;~~~~~~~~~~~~~~~~~~~~~~~~~~~~~~~~~~~~~~~~~~~~~~~~~~~~~~~~~~~~~
;;; Calculates the pitch diameter, number of teeth on the gear,
;;; the outside diameter of the gear the outside diameter of the
;;; pinion, the distance between the pinion and the gear and the
;;; position of the gear
;;;~~~~~~~~~~~~~~~~~~~~~~~~~~~~~~~~~~~~~~~~~~~~~~~~~~~~~~~~~~~~~
    pitch_dia    (/ number_teeth dia_pitch)
    gear_teeth   (* gear_ratio number_teeth)
    gear_out     (+ pitch_dia (* 2 (/ 1 (/ gear_teeth pitch_dia))))
    pinion_out   (+ pitch_dia (* 2 (/ 1 (/ number_teeth pitch_dia))))
    gear_rpm     (* gear_ratio rpm)
    dist         (+ (* 0.5 pinion_out) (* 0.5 gear_out))
    pt_x         (+ dist (CAR pt))
    pt_y         (CADR pt)
    pinion       (LIST (CONS 0 "CIRCLE")
                       (CONS 10 pt)
                       (CONS 40 (/ pinion_out 2))
                       (CONS 8 "pinion")
                 )
    gear         (LIST (CONS 0 "CIRCLE")
                       (CONS 10 (LIST pt_x pt_y))
                       (CONS 40 (/ gear_out 2.0))
                       (CONS 8 "gear")
                 )
    shaft_pinion (LIST (CONS 0 "CIRCLE")
                       (CONS 10 pt)
                       (CONS 40 (/ shaft_dia 2.0))
                       (CONS 8 "pinion")
                 )
    shaft_gear   (LIST (CONS 0 "CIRCLE")
                       (CONS 10 (LIST pt_x pt_y))
                       (CONS 40 (/ shaft_dia 2.0))
                       (CONS 8 "pinion")
                 )
  )
;;; ~~~~~~~~~~~~~~~~~~~~~~~~~~~~~~~~~~~~~~~~~~~~~~~~~
;;; Begins the construction of the pinion block
;;; ~~~~~~~~~~~~~~~~~~~~~~~~~~~~~~~~~~~~~~~~~~~~~~~~
  (ENTMAKE (list (cons 0 "block")
           (CONS 2 "pinion")
           (cons 10 (LIST pt_x pt_y))
           (cons 70 64)
       )
```

```
  )
  (ENTMAKE  pinion)
  (ENTMAKE  shaft_pinion)
  (ENTMAKE  (list (cons 0 "endblk")))
  (ENTMAKE  (list (CONS 0 "INSERT")
            (cons 2 "pinion")
            (cons 10 (LIST pt_x pt_y))
      )
  )
;;;  ~~~~~~~~~~~~~~~~~~~~~~~~~~~~~~~~~~~~~~~~~~~~~
;;; Modifies the entity created by attaching XDATA
;;;  ~~~~~~~~~~~~~~~~~~~~~~~~~~~~~~~~~~~~~~~~~~~~~
  (setq lastent (entget (entlast)))
  (regapp "pinion")
  (setq     exdata1
     (LIST
       (LIST "pinion"
             (CONS 1000 (STRCAT "Pinion's RPM " (RTOS rpm)))
             (CONS 1041 gear_ratio)
             (CONS 1042 number_teeth)
      )
     )
  )
  (setq     newent1
     (append lastent (list (append '(-3) exdata1)))
  )
  (entmod newent1)
;;;  ~~~~~~~~~~~~~~~~~~~~~~~~~~~~~~~~~~~~~~~~~~~~~
;;; Begins the construction of the gear block
;;;  ~~~~~~~~~~~~~~~~~~~~~~~~~~~~~~~~~~~~~~~~~~~~~
  (ENTMAKE (list (cons 0 "block")
            (CONS 2 "gear")
            (cons 10 (LIST pt_x pt_y))
            (cons 70 64)
      )
  )
  (ENTMAKE  gear)
  (ENTMAKE  shaft_gear)
  (ENTMAKE  (list (cons 0 "endblk")))
  (ENTMAKE  (list (CONS 0 "INSERT")
            (cons 2 "gear")
            (cons 10 (LIST pt_x pt_y))
      )
  )
;;;  ~~~~~~~~~~~~~~~~~~~~~~~~~~~~~~~~~~~~~~~~~~~~~
```

```
;;; Modifies the entity created by attaching XDATA
;;; ~~~~~~~~~~~~~~~~~~~~~~~~~~~~~~~~~~~~~~~~~~~~~~~
(setq lastent (entget (entlast)))
(regapp "gear")
(setq    exdata
   (LIST
     (LIST "gear"
             (CONS 1000 (STRCAT "Mating Gear's RPM " (RTOS rpm)))
             (CONS 1041 gear_ratio)
             (CONS 1042 gear_teeth)
     )
    )
)
(setq    newent
    (append lastent (list (append '(-3) exdata)))
 )
 (entmod newent)
 (princ)
)
```

SUMMARY

Non-AutoCAD program-specific information can be saved to a drawing file as either extended entity data or as an Xrecord. The information is saved to a drawing by attaching it to an entity's association list using DXF codes. For extended entity data, the DXF codes range from 1000 to 1071, while the DXF codes for Xrecords range from 1 to 369. Both extended entity data and Xrecords are intended to allow the developer to store information within a drawing. Xrecords do have one advantage over extended entity data: extended entity data is limited to 16 KB while Xrecords have no size limit.

Before the developer can set extended entity data to an AutoCAD object, the application name that will be associated with the extended entity must first be registered. In AutoLISP to register an application name, the REGAPP function must be used. Once the application name has been registered, the extended entity data may be attached to an AutoCAD object through the ENTMOD function. To retrieve extended entity data that has been attached to an AutoCAD object, the ENTGET function must be used and the application name must be supplied to the function as an argument. To retrieve all extended entity data attached to an AutoCAD object, the wild card character asterisk may be used. Once extended entity data has been retrieved from an object, the data may be manipulated using any of the AutoLISP list handling functions.

Because extended entity data is limited to 16 KB, it is often necessary to check the amount of space available before attempting to attach or even modify existing extended entity data. AutoLISP supplies two functions for managing extended entity data: XDROOM and XDSIZE. The XDROOM function returns the amount of space an object has available for extended entity data, while the XDSIZE function returns the amount of space required to attach extended entity data to an object.

In AutoCAD information regarding non-graphic entities is stored as either symbol table objects or dictionary objects. Before a non-graphic entity can be appended to an AutoCAD symbol table, certain rules and restrictions must be adhered to. They are:

- The guidelines and restrictions that apply to graphic objects also apply to non-graphic objects.

- The ENTMAKE function can be used to create symbol table entries if a valid description is passed to the function. The function cannot be used to create symbol_tables.

- Once an entry has been made in the symbol table, it cannot be updated using the ENTUPD function.

- No entry names can conflict with one another among all sub-tables except the VPORT table.

- Once an entry has been made to the symbol table, it cannot be deleted with the ENTDEL function.

- The entity name of a symbol table entry may be obtained using the TBLOBJNAME function. Once the name has been acquired its data may be returned using the ENTGET function.

- Neither the ENTMAKE nor the ENTMOD function will allow the developer to specify a handle using DXF codes 5 or 105 in the creation or modification of an entity.. These two functions also ignore the DXF code 70 when they are supplied with valid record lists.

- All symbol table entries except ones that are contained within the APPID table can be modified using the ENTMOD function only if the entity's name is furnished. The entity name of an entry in a table can be obtained using the ENTGET function and not the TBLSEARCH or TLBNEXT function.

- Symbol table entries can be renamed. However, the new name cannot duplicate an existing name except for entries contained in the VPORT symbol table.

- The table entries *ACTIVE from the VPORT table and CONTINUOUS from the LINETYPE table cannot be modified or renamed. The entries STANDARD from the STYLE and DIMSTYLE tables and *MODEL_SPACE and PAPER_SPACE from the BLOCKS table may be modified but cannot be renamed.

- Entries contained within the APPID table cannot be renamed.

- Dictionary objects are similar to symbol table objects, but their purpose is to store non-graphic information that cannot be stored in a symbol table.

- If multiple entities are to be manipulated using AutoLISP, then those entities can be selected and their entity names stored as a selection set for later processing. AutoLISP selection sets are similar to AutoCAD selection sets in that entity names can be added and subtracted from the selection set.

REVIEW QUESTIONS AND EXERCISES

1. Define the following terms:
   Symbol Table
   Application Identification Table
   Application Name
   Dictionary
   Extended Entity Data
   Filters

2. What is an application name and what role does it play in defining extended entity data?

3. Describe the process used to register an application name using the REGAPP function.

4. What is the purpose of the Application Identification Table?

5. Describe how to find out if an application name has already been registered using the REGAPP and TBLSEARCH functions.

6. True or False: The ENTMOD function cannot be used to append and/or modify extended entity data that is attached to an AutoCAD object. (If the answer is false, explain why.)

7. True or False: The ENTGET function can be used to retrieve extended entity data from an AutoCAD object. (If the answer is false, explain why.)

8. What is the purpose of extended entity data and how does it differ from an association list?

9. What AutoLISP function is used to create a selection set? Give an example of how the function would be used.

10. What is the difference between the TBLSEARCH and TBLNEXT functions?

What is the value returned by the following expressions?

11. (ssget "x" (list (cons 8 "line")))

12. (ssget "x" (list (cons 0 "line")))

13. (ssget "x" (list (cons 8 "0")(cons 410 "model")))

14. (ssget 'x (list (cons 410 "model")))

15. (ssget "c" (getpoint "\nSelect first corner : ")(getpoint "\nSelect second corner : "))

16. (ssget "c" (setq pt (getpoint "\nSelect first corner : "))(getpoint pt "\nSelect second corner : "))

17. (ssget "x")

18. (ssget)

19. (ssget "p")

20. (setq entity_data (entget (car (entsel))'("*"))

21. (sslength (ssget "X"))

22. (setq len (sslength (setq sel (ssget "X"))))

23. (ssname sel 0)

24. (entget (ssname sel -1))

25. (entget (ssname sel 0) '("*"))

26. (DEFUN C:create_xrecord (/ xrec xname)
```
                (setq xrecord_list '((0 . "XRECORD")(100 . "AcDbXrecord")
                    (1 . "The possibilities are endless.")
        (10 3.0 4.0 0.0)
        (40 . 3.15)
        (50 . 3.15)
        (62 . 1)
        (70 . 180)
                )
  )
  (SETQ xrecord_name (ENTMAKEX xrecord_list))
  (DICTADD (NAMEDOBJDICT) "XRECLIST" xname)
  (PRINC)
  )
```

27. (tblsearch "layer")

28. (tblsearch "layer" "0")

29. (ssadd entity_name selection_set)

30. (xdsize (cdr (assoc -1 (entget (car (entsel))))))

31. Using the heating load application developed in the Chapter 3 exercises, modify the program to allow the user to save the BTU/hr information as extended entity data.

32. Write an application that allows the user to change a group of entities from one layer to another simply by selecting an object that resides on the target layer.

CHAPTER 6

Diesel and Dialog Boxes

OBJECTIVES

Upon completion of this chapter the reader will be able to:

- Describe the difference between DIESEL and AutoLISP
- Use the MODEMACRO system variable to configure the AutoCAD status bar
- Use DIESEL expressions to make the status bar reflect the internal state of AutoCAD
- Incorporate DIESEL expressions in an AutoLISP program
- Use DIESEL expressions in an AutoCAD menu file
- Describe the difference between Dialog Control Language and AutoLISP
- Use the proper format when constructing a dialog box
- Describe the components of a dialog box
- Define the following terms: **Dialog Box**, **Tile**, **Attribute**, **Parent**, and **Child**
- Place the following components in a dialog box: Buttons, Sliders, Images, Image Buttons, Toggles, Radio Buttons, Edit Boxes, and Text
- Describe why aesthetics and ergonomics are important issues to consider when developing a dialog box

KEY WORDS AND AUTOLISP FUNCTIONS

Attribute	image	radio_button
AutoCAD PDB	image_button	radio_column
boxed_column	Julian Date	radio_row
boxed_radio_column	Julian Time	row
boxed_radio_row	list_box	slider
boxed_row	MENUCMD	spacer
button	MODEMACRO	spacer_0
Children	ok_cancel	spacer_I
column	ok_cancel_help	Status Line
concatenation	ok_cancel_help_errtile	Subassemblies
DCL	ok_cancel_help_info	text
dialog	ok_only	text_part
Dialog Box	paragraph	Tile
DIESEL	Parent	toggle
edit_box	popup_list	Tree Structure
errtile	Prototypes	

INTRODUCTION TO DIESEL

An important feature of Windows 95, 98, and NT and Windows 2000 is the status line. This feature allows an application to display vital information that has been generated without disturbing the user productivity. For example, in AutoCAD, the user can visually obtain the coordinates of the current position of the cursor by simply glancing at the AutoCAD status line (also known as the AutoCAD status bar) without having to activate menu options or dialog boxes. To modify the AutoCAD status bar, the *D*irect *I*nterpretively *E*valuated *S*tring *E*xpression *L*anguage (also known as *DIESEL*) is used in conjunction with the MODEMACRO system variable. DIESEL expressions, unlike AutoLISP expressions, are limited to accepting and generating only string data and results. When an AutoLISP application must transmit information to a DIESEL expression, the AutoCAD system variables USERS1, USERS2, USERS3, USERS4, and USERS5 can prove to be a valuable asset in accomplishing this task.

USING THE MODEMACRO SYSTEM VARIABLE TO CONFIGURE THE AUTOCAD STATUS LINE

All changes made to the AutoCAD status bar are done through the MODEMACRO system variable. This variable, when supplied with a string, displays that string in the left-aligned panel of the AutoCAD status bar (see Figures 6–1 and 6–2). The system variable can be set from the AutoCAD command prompt or through the AutoLISP SETVAR function. The length of the string that can be passed to the MODEMACRO system variable is limited by restrictions that are imposed by AutoLISP and the

AutoLISP to AutoCAD communication buffer (255 characters). However, the number of characters that can be displayed in the status bar is limited only by the size of both the AutoCAD window and the monitor. Any information that is passed to the MODEMACRO system variable is lost once AutoCAD has been restarted. The value of the MODEMACRO system variable is not saved in either the AutoCAD configuration or drawing files. If the status bar is to be modified to reflect a particular setting produced by an application each time AutoCAD is activated, then the s::startup function must be defined in the acad2000.lsp file (this file is normally located in the c:\programs files\acad2000\support directory). Without the use of "macro expressions" (DIESEL programming language) any changes to the internal state of AutoCAD do not affect the status bar. The value assigned to the MODEMACRO system variable will not change until the value of the system variable has been reassigned.

Note: When multiple drawings are open in AutoCAD 2000, a change made to the status bar in one drawing is reflected in the others (see Figures 6–3 and 6–4). These changes will remain in effect until either the value of the MODEMACRO system variable is changed or AutoCAD has been terminated. If an application is run where the status bar is changed in one drawing, and then run again in another drawing, the status bar will reflect the changes made in the last drawing where the status bar was changed.

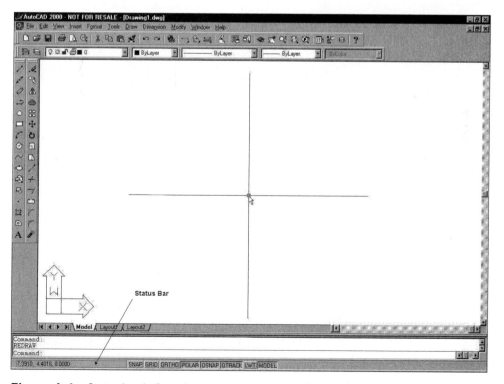

Figure 6–1 *Status bar before changes are made using the* MODEMACRO *system variable*

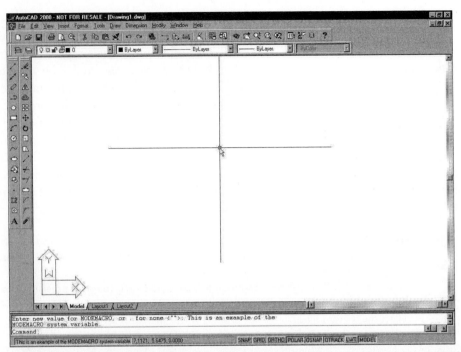

Figure 6–2 *Status bar after changes have been made using the* MODEMACRO *system variable*

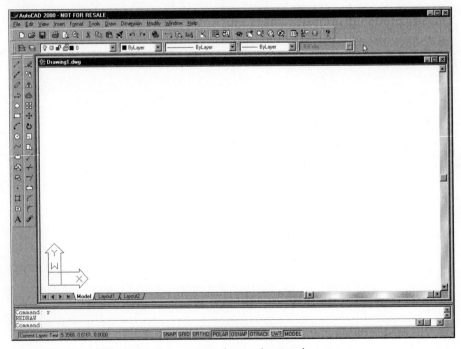

Figure 6–3 *AutoCAD dual screen – status bar in drawing 1*

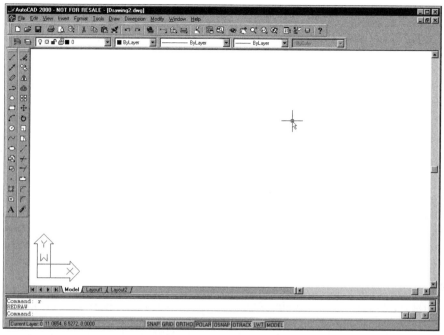

Figure 6–4 *AutoCAD dual screen – status bar in drawing 2*

MAKING THE STATUS BAR REFLECT AUTOCAD INTERNAL STATE USING DIESEL

As mentioned earlier, the status bar is controlled by the MODEMACRO system variable. This variable can be set by either entering MODEMACRO at the command prompt or by using the AutoLISP SETVAR function. In either case changes made to AutoCAD internal state are not automatically displayed in the status bar unless a "macro" expression (DIESEL) is inserted into the panel. To display the current text style in the AutoCAD status bar, the AutoLISP expression (SETVAR "modemacro" (GETVAR "cmlstyle")) would be used. If the current text style is changed again, then the status bar will no longer reflect the current text style until the previous expression has been reevaluated. If the desired result is to automatically display the current text style without having to reevaluate an AutoLISP expression every time, then the DIESEL expression $(getvar, cmlstyle) could be inserted into the previous AutoLISP expression.

DIESEL expressions all follow the same basic syntax, which is $(somefun, arg1, arg2, ...). Compare this to the basic syntax for an AutoLISP expression (function argument1 argument2 ...) and their differences as well as similarities become apparent. Both languages list the name of the function first, followed by the argument(s) that are to be used, and both languages use parentheses to form expressions. DIESEL expressions always start with a dollar sign while AutoLISP expressions do not. AutoLISP expressions are limited to nine data types, while DIESEL is limited to only one, strings. Both DIESEL and AutoLISP expressions can be nested within other expressions, thus allowing the developer to create complex applications and expressions. DIESEL expressions use commas to separate function name and argument while AutoLISP uses a single space. In other words, DIESEL expressions are comma-delimited while AutoLISP expressions are space-delimited. Finally, DIESEL functions in many cases resemble their AutoLISP counterpart function in appearance, performance and even values returned.

When an expression is evaluated in DIESEL, the characters used to make up the expression are copied directly into output until either a dollar sign or quotation mark is encountered. Quotation marks are typically used to suppress the evaluation of characters. When a series of characters is enclosed within quotation marks, those characters are not evaluated as a DIESEL expression but are instead passed along as output, without the quotation marks. When quotation marks are needed as part of the final output to the status bar, the use of two adjacent quotation marks will produce the required result. For example, from the AutoCAD command prompt:

Command: **MODEMACRO** (ENTER)
Enter new value for MODEMACRO, or . for none <"">: **"The Current Text style is """"$(getvar, cmlstyle)""""** (ENTER)

Figure 6–5 shows the result.

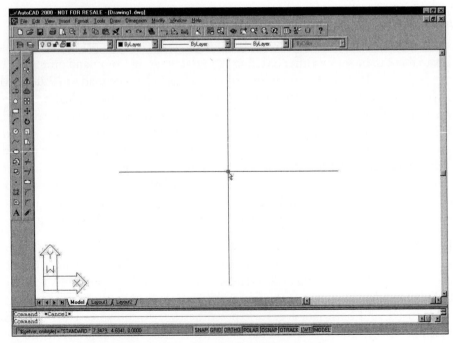

Figure 6–5 *Current text style displayed in the AutoCAD status bar*

Using the AutoLISP SETVAR function produces the same result:

```
(SETQ current_t_style " \"\"\"$(getvar, cmlstyle) = \"\"\"$(getvar,
   cmlstyle) \"\"\" ")
(SETVAR "modemacro" current_t_style)
```

Using DIESEL Expressions in AutoLISP Programs

By incorporating DIESEL expressions into AutoLISP applications, the developer is able to control the appearance and functionality of the AutoCAD status bar, thus allowing them to display important information to the user. One way this can be done is by using the STRCAT function to combine DIESEL functions, punctuation and string data into a DIESEL expression. Once the expression has been constructed, it can then be supplied to the MODEMACRO system variable using either the AutoLISP SETVAR or COMMAND function. The following AutoLISP program is used to redesign the AutoCAD status bar so that the current text style, dimension style, and save time settings are displayed using DIESEL. A DIESEL expression is also used to evaluate the current setting of the CMDECHO and CMDNAMES system variables. If the CMDECHO system variable is set to zero and the CMDNAMES variable is not equal to " ", then the active command is displayed in the status bar. Otherwise the last point selected is displayed. Although this application is not that practical in the type of information displayed, it does serve as a good example of how AutoLISP and DIESEL can be used to control the status bar.

```
;;;****************************************************************
;;; Program Name : example_chapter_6
;;;
;;; Program Purpose: Illustrates how DIESEL can be incorporated
;;;                  into an AutoLISP program.
;;;
;;; Written By: James Kevin Standiford
;;;
;;; Date: 05/07/99
;;;****************************************************************
;;;****************************************************************
;;; Main Program
;;;****************************************************************
(defun c:example_chapter_6 ()
  (setq cmd (getvar "cmdecho"))     ;Sets the current cmdecho system
                                    ;variable to the variable cmd.
  (setvar "cmdecho" 0)              ;Turns the command echo off
  (setvar                          ;Sets the system variable
```

```
                              ;MODEMACRO to the following
                              ;values.
     "modemacro"
     (strcat
      "Text = $(getvar, textstyle), "        ;Current Text Style
      "Dim = $(getvar, dimstyle), "          ;Current Dim Style
      "Save Time = $(getvar, savetime) "     ;Current Layer
;;;************************************************************
;;;
;;; The following IF statement compares the current setting of the
;;; cmdecho system variable and the current command, and if the
;;; the cmdecho variable is set to 1 and the current command is
;;; equal to "" then the last point selected is displayed, otherwise
;;; the current command is displayed.
;;;
;;;************************************************************
      "$(if, $(and, $(=, $(getvar, cmdecho), 1),"
           "$(eq, $(getvar, cmdnames), \"\")), \""
           "Last Point Picked : \" $(index, 0, $(getvar, lastpoint))"
           "\" \" $(index, 1, $(getvar, lastpoint))"
           "\" \" $(index, 2, $(getvar, lastpoint)),"
           "[\"Current Command = \" $(getvar, cmdnames)]"
      ")"
     )
   )
  (setvar "cmdecho" cmd)                      ;resets the command echo
  (princ)                                     ;to its previous setting
 )
```

Figures 6–6 and 6–7 show the two resulting displays.

| Text = Standard, Dim = Standard, Save Time = 120 Last Point Picked : 4.70984076 1.50343678 0 1.9457 -3.0441 0.0000   SNAP GRID ORTHO POLAR OSNAP OTRACK LWT |

Figure 6–6 *Status bar – current text style, current dimension style, save time, and last point selected*

| Text = Standard, Dim = Standard, Save Time = 120 Current Command = Line 9.6085, 1.4499, 0.0000   SNAP GRID ORTHO POLAR OSNAP OTRACK LWT MODEL |

Figure 6–7 *Status bar – current text style, current dimension style, save time, and current command*

The application can be modified so that the last time the drawing was saved is now displayed in the status bar instead of the current text style or current dimension style.

This is achieved by the replacement of:

```
"Text = $(getvar, textstyle), "        ;Current Text Style
"Dim = $(getvar, dimstyle), "          ;Current Dim Style
```

with the single expression:

```
"LSave $(edtime, $(getvar,tdupdate), H:MMam/pm) "
```

Figure 6–8 shows the resulting display.

LSave = 9:48PM ETime 0:07 ASave = 120 Last Point = 10.27843689 6.20491803 0 10.8420, 4.5930, 0.0000   SNAP GRID ORTHO POLAR OSNAP OTRACK LWT MODEL

Figure 6–8 *Status bar – last time saved*

Although AutoLISP functions cannot be used in a DIESEL expression, DIESEL expressions can be used in AutoLISP applications through the MENUCMD AutoLISP function (MENUCMD string), where the string argument is used to specify a menu area and a value assigned to the menu area. Allowable values for menu areas are B1–B4 for buttons 1–4, A1–A4 for AUX menus 1–4, P0–P16 for drop-down menus 1–16, I for image tile menus, S for screen menus, T1–T4 for tablet menus, and M for DIESEL expressions. To use the DIESEL function EDTIME to return the current date and time in a format other than Julian time, the following AutoLISP/DIESEL expression would be used.

```
(SETQ time_string (MENUCMD "M=$(edtime, $(getvar, date), DDDD MONTH DD
   YYYY HH:MM:SS AM/PM)"))
```

> **Note:** The arguments DDDD and MONTH return the current day and month spelled out. The arguments DD and YYYY return the current day in a two digit format and the current year in a four digit format. The HH:MM:SS arguments return the current hour, minute, and second. For example, Monday May 10 1999 08:28:10 PM.

To illustrate how this can be used in a program, the following AutoLISP application displays an AutoCAD alert dialog box containing the current time, last time saved, time between saves and auto save time.

```
;;;***************************************************************
;;; Program Name : Example_chapter_6_C
```

```
;;;
;;; Program Purpose: To illustrate how DIESEL expressions can be used
;;;                  in an AutoLISP program
;;;
;;; Written By: James Kevin Standiford
;;;
;;; Date : 05/10/99
;;;****************************************************************
;;;
;;; Main Program
;;;
;;;****************************************************************
;;; Notes: The control codes are used to display each string on a
;;;        separate line. If the control codes were not used, the
;;;        STRCAT function would cause each line to follow the
;;;        next resulting in one long string
;;;****************************************************************
(DEFUN c:example_chapter_6_C ()
  (SETQ    time_string   (MENUCMD
"M=\nCurrent Time : $(edtime, $(getvar, date), DDDD MONTH DD YYYY
   HH:MM:SS AM/PM)"
              )
    time_string_1  (MENUCMD
"M=\nLast Time Saved : $(edtime, $(getvar,tdupdate), DDDD MONTH DD YYYY
   HH:MM:SS AM/PM)"
              )
    time_string_2  (MENUCMD
"M=\nTime Between Saves : $(edtime, $(-, $(getvar, date), $(getvar,
   tdupdate)), HH:MM:SS)"
              )
    time_string_3  (MENUCMD
                 "M=\nAuto Save Time : $(getvar, savetime) min"
              )
  )
;;;****************************************************************
;;; Notes: The AutoLISP ALERT function when supplied with a string
;;;        value displays an AutoCAD message dialog containing the
;;;        string and an OK button.
;;;****************************************************************
  (ALERT
    (STRCAT time_string
```

```
            time_string_1
            time_string_2
            time_string_3
      )
    )
    (PRINC)
  )
  (PRINC)
)
```

Figure 6–9 shows the resulting AutoCAD message box.

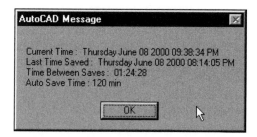

Figure 6–9   *AutoCAD message box displaying current time, last time saved, time between saves and auto save time*

Note: The AutoCAD system variables CDATE and DATE share a common purpose: they both use real numbers to store the current date and time. But this is where the similarities of these two variables end. The CDATE variable uses a format in which the year, month, and day are displayed as integers and the hours, minutes, seconds, and milliseconds are displayed as a decimal value. The first four digits represent the year, the fifth and sixth digits represent the month, the seventh and eighth digits represent the day, the ninth and tenth digits represent the hour, the eleventh and twelfth digits represent the minutes, the thirteenth and fourteenth digits represent the seconds, and the fifteenth and sixteenth digits represent the milliseconds. For example, May 10, 1999 12:13:59 PM would be displayed as 19990510.12135900. The DATE system variable uses a real number to display the current data and time in Julian format. For example, May 10, 1999 at 12:13:59 PM would be displayed as 2451309.50979248 in Julian Format. The integer portion of this example represents the number of days that have passed since the beginning of the first Julian cycle, January 1, 4713 BC. (There are 7,980 years in one cycle. The next cycle will begin January 22, 3268.) The fraction portion of this format represents the portion of a 24-hour day (Hours, Minutes, Seconds, etc. starting at midnight) that has elapsed. Julian time can be calculated using the formula (H + (M + S/60)/60)/24. For example to convert 07:42:34 p.m. into Julian time, the equation (19 + (42 + 34/60)/60)/24 would be used, thus resulting in 0.821226852. In this equation the hour portion is expressed in terms of a 24-hour clock, therefore 7 p.m. is equal to 19, or 1900 hours.

USING DIESEL EXPRESSIONS IN AUTOCAD MENU FILES

Although it is not within the scope of this book to cover AutoCAD menu customization, DIESEL can be used in menu files, allowing the developer to construct menus that reflect the internal state of AutoCAD. By inserting the following DIESEL expression into one of the AutoCAD drop-down menus contained in the acad.mns file, the developer has created a menu option that shows the current setting for the system variable savetime. Just like the status bar, this menu option is updated each time the variable is changed.

The DIESEL expression:

```
[$(eval,"AutoSAVE: " $(getvar,savetime))]^c^csavetime
```
The inquiry sub-menu of the drop-down menu POP6:

```
***POP6
**TOOLS
ID_MnTools      [&Tools]
ID_Spell        [Sp&elling]'_spell
ID_Qselect      [&Quick Select...]^C^C_qselect
ID_MnOrder      [->Display &Order]
ID_DrawordeF      [Bring to &Front]^C^C^P(ai_draworder "_f") ^P
ID_DrawordeB      [Send to &Back]^C^C^P(ai_draworder "_b") ^P
          [--]
ID_DrawordeA      [Bring &Above Object]^C^C^P(ai_draworder "_a") ^P
ID_DrawordeU      [<-Send &Under Object]^C^C^P(ai_draworder "_u") ^P
ID_Inquiry      [->Inquir&y]
ID_Dist         [&Distance]'_dist
ID_Area         [&Area]^C^C_area
ID_Massprop       [&Mass Properties]^C^C_massprop
          [--]
ID_List         [&List]^C^C_list
ID_Id         [&ID Point]'_id
          [--]
ID_Time         [&Time]'_time
        [$(eval,"AutoSAVE: " $(getvar,savetime))]^c^csavetime
```

 Note: The $ is a special character used in menus to either load a menu section or designate a conditional DIESEL macro expression.

Figures 6–10 and 6–11 show the two different settings for AutoSAVE time.

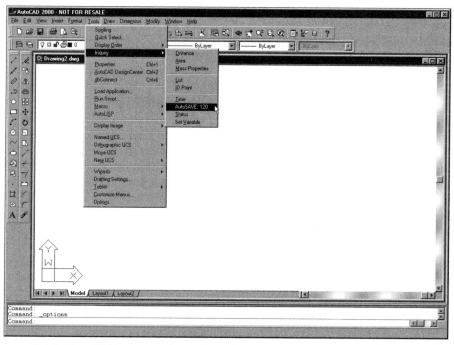

Figure 6–10 *AutoSAVE time set for 120 minutes*

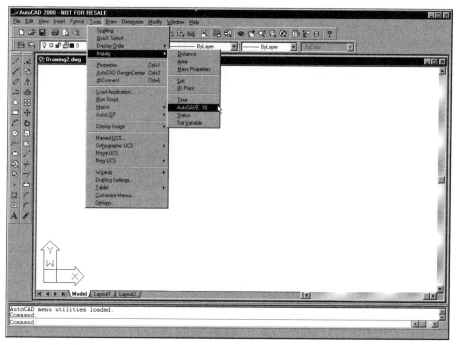

Figure 6–11 *AutoSAVE time set for 10 minutes*

Tip: Before making changes to any of the AutoCAD support files, first make a backup copy. Before testing the changes made to a menu file, be sure to delete the filename.mnc file for that menu. This will cause AutoCAD to recompile the menu source file and display the changes.

DIESEL FUNCTIONS

Currently, the DIESEL programming language is comprised of 28 functions. It is from these functions that DIESEL expressions are created. All DIESEL functions are limited to ten parameters that can be supplied, which includes the function name. Table 6–1 lists the functions associated with the DIESEL language, along with a description of each and their syntax.

Table 6–1   DIESEL Functions

Function Name	Description	Syntax
+	Returns the sum of two or more numbers.	$(+, val1 [, val2, …, val9])
-	Returns the difference between two or more numbers.	$(-, val1 [, val2, …, val9])
*	Returns the product of two or more numbers.	$(*, val1 [, val2, …, val9])
/	Returns the quotient of two or more numbers.	$/, val1 [, val2, …, val9])
=	Compares the numeric value of *val1* and *val2*. If these arguments are the same, then the function returns 1. If the values are not the same, then the function returns 0.	$(=, val1 val2)
<	Compares the numeric value of *val1* and *val2*. If *val1* is less than *val2*, then the function returns 1.	$(<, val1, val2)
>	Compares the numeric value of *val1* and *val2*. If *val1* is greater than *val2*, then the function returns 1.	$(>, val1, val2)
!=	Compares the numeric value of *val1* and *val2*. If these arguments are the same, then the function returns 0. If the values are not the same, then the function returns 1.	$(!=, val1, val2)

Function Name	Description	Syntax
<=	Compares the numeric value of *val1* and *val2*. If *val1* is less than *val2* or equal to *val2*, then the function returns 1.	$(<=, val1, val2)
>=	Compares the numeric value of *val1* and *val2*. If *val1* is greater than or equal to *val2*, then the function returns 1.	$(>= val1, val2)
AND	Returns the bitwise logical AND of the integers *val1* through *val9*.	$(AND, val1[, val2, …, val9])
ANGTOS	Returns the angle value of the specified number in a predetermined format. These formats are degrees - 0, Degrees /Minutes/Seconds - 1, Grads - 2, Radians - 3 and Surveyor's units - 4.	$(ANGTOS, value [, mode, precision])
EDTIME	Returns a date and time based on a given format phrase(s) (picture). These phrases are D – day, DD – day, DDD - abbreviation for week day, DDDD – week day spelled out, M – month, MO – two-digit month, MON – abbreviation for month, YY – two-digit year, YYYY – four-digit year, H – hour, HH – two-digit hour, MM – two-digit minute, SS – two-digit second, MSEC – three-digit millisecond, AM/PM – uppercase am or pm, am/pm – lowercase am or pm, A/P – uppercase A for am, a/p – lowercase a for am.	$(EDTIME, time, picture)
EQ	Compares the value of string *val1* and string *val2*. If the values are the same, then the function returns 1.	$(EQ, val1, val2)
EVAL	Passes the string *str* to the DIESEL evaluator and returns the result.	$(EVAL, str)
FIX	Converts a real number to an integer by truncating the fractional portion.	$(FIX, value)

Table 6–1   DIESEL Functions *(continued)*

Function Name	Description	Syntax
GETENV	Returns the value of the specified AutoCAD environment variable.	$(GETENV, varname)
GETVAR	Returns the value of the specified AutoCAD system variable.	$(GETVAR, varname)
IF	Conditionally evaluates expressions.	$(IF, expr, dotrue [, dofalse])
INDEX	Extracts the specified member from a comma-delimited string.	$(INDEX, which, string)
LINELEN	Returns an integer value representing the longest length in character that the status line can display to the user.	$(LINELEN)
NTH	Evaluates and returns the arguments selected by which. If which is 0, nth returns arg0, and so on.	$(NTH, which, argo [, arg1, .., arg7])
OR	Returns the bitwise logical OR of the integers *val1* through *val9*.	$(OR, val1 [, val2, .., val9])
RTOS	Returns the real value in the format and precision specified.	$(RTOS, value [, mode, precision])
STRLEN	Returns the length of string in characters.	$(STRLEN, string)
SUBSTR	Returns the substring of a string, starting at the character start and extending through the length characters.	$(SUBSTR, string, start [, length])
UPPER	Returns the string converted to uppercase according to the rules of the current locale.	$(UPPER, string)
XOR	Returns the bitwise logical XOR of the integer val1 through val9.	$(xor, val1 [, val2 .., val9])

Courtesy of the AutoLISP Developers Guide

INTRODUCTION TO DIALOG BOXES

Ever since the introduction of the Graphical User Interface in the 1970s by the Xerox's PARC laboratory, end users have demanded that computer programs be equipped with this technology. As stated in Chapter 1, the use of graphics in a computer program enables the person working with the application to increase productivity by spending more time using the application for its intended purpose instead of learning commands. This push for computer programs to become more graphically oriented has not escaped the realm of AutoCAD; in fact it has become a tremendous asset for the AutoCAD developers. Programmers can develop applications that are user friendly, run faster and input data in ways that might not be possible without the use of graphics. When an AutoCAD application is being developed, the use of graphics or graphical user interface should be an integral part of the program design when possible.

DIALOG BOX FILE FORMAT

The source code for dialog boxes do share one thing in common with AutoLISP: they must first be written and saved in the ASCII format. This is where the similarities between these two customization tools end. Dialog boxes do not use the AutoLISP or Visual LISP programming languages for their creation, but instead they use these languages as a means of managing the data received by and sent to the actual dialog box. Dialog boxes are created using a language referenced to as *DCL* or *Dialog Control Language*. DCL is a relatively easy language to understand and follow. Its code is written in a tree-structure format. The functions, or *Tiles* as they are called, are written in standard English, making them easy to interpret and understand. The arguments, or *Attributes*, are also written in the same fashion.

COMPONENTS OF A DIALOG BOX

A dialog box consists of two main components, the box and the tiles. Tiles, which are predefined in the *Programmable Dialog box Facility*, can be arranged in a variety of ways to create complex tiles referred to as *Subassemblies*. A subassembly is a grouping of tiles into rows and columns. Subassemblies can be either enclosed or not enclosed by a border, depending upon the developer of the application and the task. When a subassembly is defined in a dialog box, the row or column used in the subassembly is referred to as a *Cluster*. Therefore, the definition of a subassembly can be rewritten to a grouping of tiles and/or clusters that can be used in many different dialog boxes. When a subassembly is created, it is treated as a single tile, with each individual tile referred to as a *Child*, and together they are referred to as *Children*. Although most of the tiles that are used to create a dialog box are predefined in the PDF, the developer can define custom tiles to be used in a dialog box. These are referred to as *Prototypes*. A prototype is typically defined when the developer wants to keep a certain consistency among several different dialog boxes or when a particular arrangement of tiles is to be used over and over again in different dialog boxes.

All dialog boxes are comprised of the same basic classification of tiles. They are *Predefined Active Tiles, Tile Clusters, Decorative and Information Tiles, Text Cluster Tiles, Dialog Box Exit Buttons and Error Tiles*, and *Restricted Tiles*.

Predefined active tiles are used to create the vast majority of the functioning portion of a dialog box. This category includes `button`, `edit_box`, `image_button`, `popup_list`, `list_box`, `radio_button`, `slider`, and `toggle`.

As indicated earlier, clusters are rows or columns used in a subassembly; cluster tiles are the tiles used to create these clusters. These tiles include `boxed_column`, `boxed_radio_column`, `boxed_radio_row`, `boxed_row`, `column`, `dialog`, `radio_column`, `radio_row`, and `row`.

The overall appearance of the dialog box can be controlled with decorative and information tiles. These tiles are used to add images, text, and spacing to dialog boxes. This category of tiles includes `image`, `text`, `spacer`, `spacer_0`, and `spacer_1`.

When more flexibility is needed for the format of text in a dialog box, text cluster tiles are often employed. This category of tiles includes `concatenation`, `paragraph`, and `text_part`.

All dialog boxes, regardless of design and functionality, must contain a means for the end user to exit the dialog once it has been loaded and executed. This is why all dialog boxes must contain at least one tile from the exit buttons and error tiles category. This category includes `ok_only`, `ok_cancel`, `ok_cancel_help`, `ok_cancel_help_errtile`, and `ok_cancel_help_info`.

The last of the classifications of tiles is the restricted tiles (`cluster` and `tile`.) These tiles should not be used in a DCL file.

Figure 6–12 shows the common components of a dialog box.

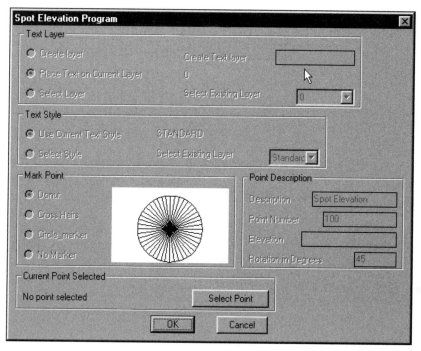

Figure 6–12 *Common components of a dialog box*

DIALOG BOX STYLE AND SYNTAX

Similar to AutoLISP application, the preparation of a dialog box requires planning at the front end. Before any code is ever typed into the computer, the appearance and functionality of the dialog box should be defined. Because dialog boxes are constructed in a tree-structure format (similar to the tree structure used by the Windows operating system), the preparation of a dialog box is best done using flowcharts instead of Pseudo-Code. Compare the tree structure of a Windows directory listing shown in Figure 6–13 to the flowchart for the same dialog box in Figure 6–14.

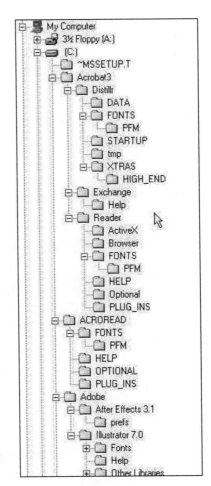

Figure 6–13 *Tree structure of the dialog box*

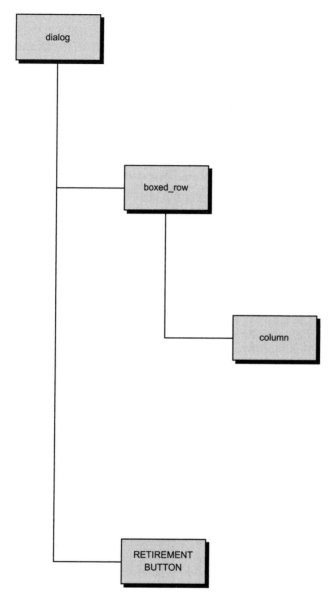

Figure 6–14 Flowchart of the dialog box

In AutoLISP, functions are grouped together with other functions and/or expressions using parentheses. The expressions are then grouped together to form an application or program. In DCL, curly brackets are used instead of parentheses to define a dialog box or tile. This is illustrated in the following example, where an AutoLISP program is used to display two messages to the user. This is followed by an example of a dialog box definition that displays the same information to the user.

The AutoLISP program:

```
;;;*********************************************
;;;
;;;
;;;Program name: first.lsp
;;;
;;;Purpose: To illustrate how AutoLISP differs from
;;;        from DCL
;;;
;;;
;;;*********************************************
(DEFUN c:print_message ()
  (TEXTPAGE)
  (PRINC "\nThis AutoLISP program is designed to illustrate")
  (PRINC "\nthe differences between DCL and AutoLISP")
  (GETSTRING "\nOk")
  )
```

The DCL equivalent:

```
//*********************************************
//
//
// Dialog Box name: first.dcl
//
// Purpose: To illustrate how dcl differs from
//        from AutoLISP
//
//
//*********************************************
first : dialog {
    label = "Dialog Box Example #1";
      : text {
        label = "This is a simple dialog box designed to illustrate";
          }
      : text {
        label = "the differences between DCL and AutoLISP";
          }
```

```
        ok_only;
          }
```

Figures 6–15 and 6–16 show the resulting messages, in the text screen and in the dialog box.

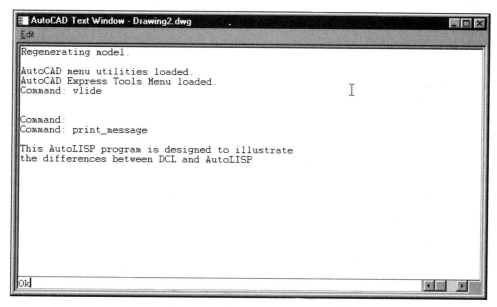

Figure 6–15 *AutoCAD text screen displaying two messages to the user*

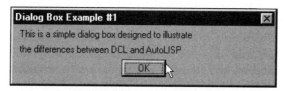

Figure 6–16 *AutoCAD dialog box displaying two messages to the user*

To declare a comment in AutoLISP, the semicolon is used. A semicolon must proceed the comment in order for AutoLISP to ignore the statement when the program is being executed. If the comment is placed in an expression, it must also be followed by a semicolon in order for the remaining portion of the expression to be evaluated. To place a comment in a DCL file, the developer uses two forward slashes //. When a comment is embedded within a line of DCL code, it is separated from that code with a /* at the beginning of the comment and a */ at the end.

Example of a comment embedded in a line of AutoLISP code:

```
(SETQ ang (GETANGLE "\nSelect angle ")) ;Prompts for an angle
(SETQ num ;Calculates the sin of the angle;(SIN (* (/ 180 3,14) ang)))
```

Example of a comment embedded in a line of DCL code:

```
: text {
        label = /*Displays text*/ "the differences between DCL and
    AutoLISP";
          }
```

All attributes are assigned in DCL using the equal sign and terminated using the semicolon. This can be seen in the previous example, where the attribute label is set equal to the string "the difference between DCL and AutoLISP". Tiles, on the other hand, always start with a colon. Again, examining the previous example, the reader will note that the tile text is preceded by a colon. Unlike AutoLISP, DCL is case sensitive with all attributes and tiles being written in lowercase. The tiles TEXT, Text, and text are all treated differently in DCL. Extra care should be observed when writing a DCL program as to the case of all tiles and attributes; one mistake and the dialog box will fail to open.

 Note: The style used for creating comments in DCL /* */ is the same style as used in the C programming language.

Alignment and Spacing

Just because DCL files are constructed in a tree-structure format does not mean that the programmer is required to indent statements or provide extra spaces between characters. However, just as in AutoLISP, this practice does make the file much easier to interpret by the programmer or anyone else who might be affiliated with the project. Once the file has been loaded and called upon by an AutoLISP application, the indentation is ignored. The practice of lining up statements and providing comments will save the developer of the application many hours of debugging.

Tiles

The actual components (radio buttons, edit boxes, text, toggles, etc.) of a dialog box are defined using tiles. Although each tile is unique and serves a specific purpose, all tiles follow the same basic format.

```
name : item1 [ : item2 : item3 ... ] {
        attribute = value;
```

```
        } 
```
          . . .

The tile *name* can be either a name used to call a dialog box or a name for a predefined tile. When this option is used to call a dialog box, then the same guidelines used for defining an AutoLISP function name also apply. The name can be all uppercase or lowercase, and it can contain all letters or a combination of letters and numbers, but it must always start with a letter. Finally, a name can contain underscores, but again it must always start with a letter. The *item* attribute in this arrangement is a predefined tile. The attributes needed for each tile (*item*) are grouped to that tile using opening and closing curly brackets and placed in sequence with the tile. This arrangement allows for some tiles to be nested. Nested tiles are contained within the curly brackets of the tile where they are nested. For example:

```
name : tile {                       /*Defines a tile called name*/
    Attribute = "value";            /*Attribute associated with name*/
    Attribute = "value";            /*Attribute associated with name*/
    : tile1 {                       /*Tile nested within the name*/
                                    /*tile*/

      attribute = "value";          /*Attribute associated with tile1*/
      attribute = "value";          /*Attribute associated with tile1*/
      : tile2 {                     /*Tile nested within tile1*/
        attribute = "value";        /*Attribute associated with tile2*/
      }                             /*Closing curly bracket for tile2*/
    }                               /*Closing curly bracket for tile1*/
}                                   /*Closing curly bracket for tile*/
                                    /*name*/
```

Attributes

All tiles that are used to define a dialog box have attributes. Attributes are used in DCL to define a tile function and layout. They can even be compared to an AutoLISP variable because they consist of a name and a value. Attributes have four data types associated with them: integers, real numbers, text (set in quotes), and reserved words. The first three data types are bound by the same rules associated with their AutoLISP counterparts. An integer cannot contain a decimal point, real numbers between 1 and −1 must have a leading zero, and text must be placed inside quotes and can contain special control characters (\" embedded quote, \\ backslash, \n new line, and \t horizontal tab). Reserved words are unique: they are not contained within quotes, and they consist of alphanumeric characters that define special options that can be used with many of the attributes. Examples of reserved words include left, centered, right, horizontal, and vertical, just to name a few.

Attributes may be user defined, but their name cannot conflict with any of the predefined attributes already provided by DCL. The name assigned to the attribute must start with a letter, although the remaining portion can contain numbers and underscores.

 Note: The predefined attributes **type**, **horizontal_margin**, **vertical_margin** are restricted and should not be used. The list of predefined words can be obtained from the Visual LISP Developers Guide, pages 406-408.

CHILD /PARENT RELATIONSHIP

As mentioned earlier, DCL is based on a tree-structure format similar to the way directories are organized on a computer (see Figure 6–17). In a directory structure the C: drive is the main trunk with each directory being a branch off the main truck (C:). In DCL this arrangement is slightly different. Instead of calling a tile a main trunk and/or branch, they are referred to as a parent and/or child. For example, the tiles RETIREMENT BUTTON and boxed_row are children of the tile dialog (see Figure 6–18). The tile column is a child of the tile boxed_row. The concept of parent/child relationship is an important concept, because it helps determine the behavior of the components of a dialog box.

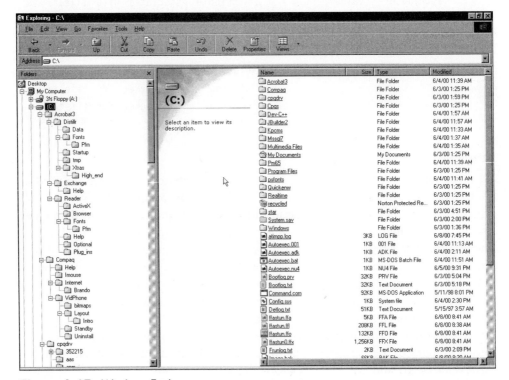

Figure 6–17 *Windows Explorer*

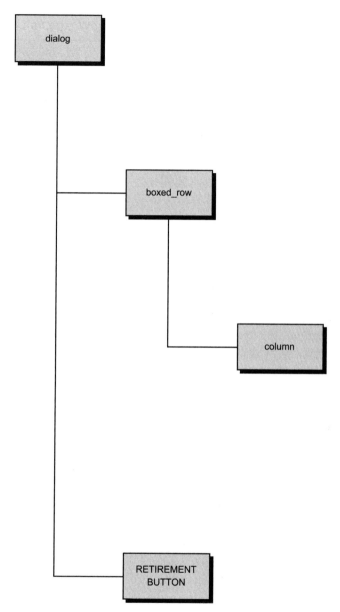

Figure 6–18 *Child/parent relationship*

DEFINING A DIALOG BOX

When a dialog box is defined, all tiles and corresponding attributes that make up that dialog box are grouped together or defined using the dialog tile. This tile is slightly different from its AutoLISP counterpart function DEFUN in the respect that the dialog name used to call the dialog box precedes the tile separated by a colon and ending with an open curly bracket. The tile is concluded with a closing curly bracket only after the attributes associated with it and other tiles and their attributes, which define the components of the dialog box, have been listed. For example:

```
spot_elevation : dialog {          /*Defines the dialog spot_elevation*/
    tile {
            attribute;
            attribute;
    }
    tile {
            attribute;
            attribute;

    }
    }                              /*Closing curly bracket for dialog*/
```

The attributes that are specific to this tile are label, initial_focus, key, and value. Any combination of these attributes may be used with this tile, depending upon the particular application for which the dialog box is designed. The label attribute, when supplied with a valid text value, displays a message in the dialog title bar. The initial_focus attribute, when supplied with a text string specifying the key or tile that is initially highlighted, will be activated once the user presses ENTER. The value attribute can also be used to specify a dialog box's title; however this attribute is a runtime attribute and its value isn't inspected at layout time. The developer should take this into consideration and make sure that the number of characters assigned to this attribute is not greater than the overall width of the dialog box. Truncation of the message will result if the number of characters to be displayed exceeds the overall width of the dialog box. The following example shows how the initial_focus and label attributes may be used when a dialog box is defined:

```
cp_rotate : dialog {
    initial_focus = "select_object"; /*Allows the user to activate the*/
                                      /*select_object tile by pressing*/
                                      /*return*/
    label = "Copy_rotate";            /*Displays the label copy_rotate*/
                                      /*on the dialog box title bar*/
        : boxed_column {
        fixed_width = true;
        label = "Operations";
        : edit_box {
                label = "Rotation Angle";
```

```
        key = "rotation";
        width = 5;
    }
    : button {
        label = "Base Point";
        key = "base_point";
        width = 15;
    }
    : button {
        label = "To Point";
        key = "to_point";
        width = 15;
    }
    : button {                          /*Activates when the user*/
                                        /*presses Enter */

        label = "Select Objects";
        key = "select_object";
        width = 15;
    }
```

Figure 6–19 shows the resulting dialog box.

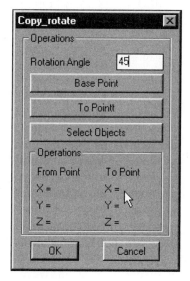

Figure 6–19 *Copy_rotate dialog box*

Using Buttons to Provide the User with Choices

One of the advantages of using a graphical interface with a program is that it allows the user to make a choice by selecting graphic objects. One of the primary objects used for this type of operation is the button. Currently, DCL has four types of buttons that can be incorporated into the design of a dialog box: They are toggles, radio buttons, buttons, and image buttons. Each type of button has it own unique characteristics and one particular button may not be desirable for all applications.

Toggles `:toggle { attributes; }`

Toggles are among the most basic type of buttons provided in DCL (see Figure 6–20). They can be compared to a single pole light switch: they only have two states, either on or off. When the toggle has been activated, an X appears in the box; when the toggle is deactivated, the box appears clear. The values that can be supplied to and extracted from a toggle are the text strings 1 or 0 (1 means on and 0 means off). A single toggle can be used in a dialog box or several can be grouped together. When toggles are grouped together, each toggle in the grouping will return a separate value depending upon the state of the toggle. Grouping several of these tiles together is an effective means of providing the user with several choices.

 Note: Each time a toggle is selected, the state of the toggle is changed.

```
: toggle {
    key = "create";        // Key associated with this tile
    label = "Create";      // Text displayed in dialog box
    mnemonic = "C";        // Allows the user to change the focus of
                           // the dialog box by pressing "h"

}
```

Figure 6–20 *Toggle*

Radio Buttons `:radio_button { attributes; }`

Radio buttons are similar to toggles in the respect that the value supplied to and extracted from these tiles is either 1 or 0. However, when radio buttons are grouped together, the user is permitted to make only one selection. If a second radio button is selected from the same group (cluster) of radio buttons, then the previous button is deactivated and a value of 0 is returned for that button. These buttons are used primarily when a single selection must be made from a list of options (see Figure 6–21).

```
: radio_button {
    key = "current_layer";          // Key associated with
                                     // this tile
    label = "Place Text on Current Layer";  // Text displayed in
                                     // dialog box
}
```

Figure 6–21 *Radio button*

Buttons `: button { attributes; }`

Similar to toggles, buttons work independently of one another. If a button is selected, it will not reset the value of the other buttons contained within the dialog box (see Figure 6–22). Buttons also return a value of 1 or 0 and are often used when the user must either interface with the graphic screen or with another dialog box. All dialog boxes must contain at least one button that allows the user to escape from the dialog box once it has been evaluated. Normally, when a button is selected the same action is performed, unlike a toggle that may be associated with a decision expression (IF statement).

```
: button {
    label = "Select Point";    // Text displayed in dialog box
    key = "s_point";           // Key associated with this tile
}
```

Figure 6–22 *Button*

Image Buttons `: image_button { attributes; }`

A button can contain a graphic image instead of a label (see Figure 6–23). These buttons are referred to as image buttons. When an image button is selected, the program obtains the coordinates of the point selected and returns those coordinates in the form of a text string. The coordinates that are returned can then be used in a variety of different ways.

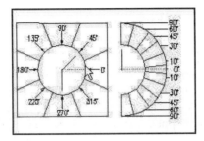

Figure 6–23 *Image button*

Using Lists to Provide the User with Choices

At times the user may be required to choose an option from a list that varies in content and length from drawing to drawing or even session to session. A dialog box that allows the user to select a layer or text style from a list cannot be accomplished easily or effectively with the use of buttons. A different strategy must be developed; in DCL this is accomplished by the list handling and processing tiles `list_box` and `popup-list`.

List Box : `list_box { attributes; }`

A list box is a component that is used to display text strings arranged in a vertical row (see Figure 6–24). The list, as well as the list box, can be variable in length or a fixed length. This is often the case where space is a consideration. When a fixed list box is created that contains more rows than can be displayed in that box, a scroll bar is automatically generated on the right hand side, thus allowing users to scroll down the list to complete their selection. A list box will typically allow the user to make only one selection; however, applications can be created that allow the user to make more than one selection.

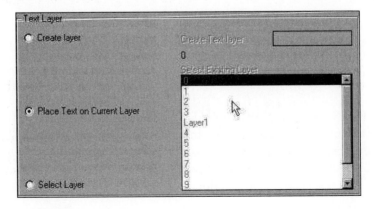

Figure 6–24 *List box*

Popup List : popup_list { attributes; }

The popup list, or popups as they are often referred to, are similar to list boxes in functionality. These components, when first displayed, will look like an edit box or button that has a downward-pointing arrow displayed to its right. By selecting the down arrow the remaining or preset amount of the list is then displayed (see Figure 6–25). Once the selection has been made, the list returns to the collapsed state, with the selection the user made appearing in the display field. Popup lists do not allow for multiple selections.

```
: popup_list {
    label = "Select Layer";      // Text displayed in dialog box
    key = "lyr_pop";             // Key associated with this tile
    is_enabled = false;          // Disables the tile on startup
    width = 27;                  // Sets the width of the pop up list
}
```

Figure 6–25 *Popup list*

Using an Edit Box to Request User Input

When text, numeric, or a combination of the two is required from the user, DCL provides a means of acquiring such information through an edit box.

Edit Box : edit_box { attributes; }

An edit box provides a field in the dialog box where the user is allowed to enter information (see Figure 6–26). All data returned from an edit box is in the form of text. If numeric data is required, then once the information has been obtained from the dialog box, it must be converted to either an integer or a real number through either the ATOI or ATOF AutoLISP function. When a label is assigned to an edit box, it appears on the left hand side. If the data entered by the user contains more characters than the length of the edit box, then the text is scrolled horizontally to the right.

```
: edit_box {
    label = "Text Height";        // Text displayed in dialog box
    mnemonic = "h";               // Allows the user to change the focus of
                                  // the dialog box by pressing "h"
    key = "enthe";                // Key associated with this tile
    edit_width = 15;              // Sets the width of the edit box
}
```

Figure 6–26 *Edit box*

Providing Visual Effects Using Images

The phrase "a picture is worth a thousand words" can also be applied to dialog boxes. In many cases it is easier to provide the user with an illustration of a particular operation or process rather than describe it in words.

Images : image { attributes;}

In DCL, images are displayed in a dialog box with the image tile (see Figure 6–27). This tile displays a vector graphic image in a rectangle, where the height and width can be controlled by the specification of the height and aspect ratio, the width and aspect ratio, or both the height and the width. For example:

```
: image {
    key = "test";
    color = 0;
    width = 25;
    height = 5;
}
```

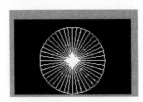

Figure 6–27 *Dialog box image*

Adding Text to a Dialog Box

Even though graphics can be used in a dialog box as a means of explaining a complex operation, sometimes it is necessary to also provide the user with written instructions or comments (aside from labels that are assigned to a tile with the label attribute.). When this is necessary, DCL has four tiles that are designed to handle these situations. They are text, text_part, paragraph, and concatenation.

Text : text { attributes; }

The text tile provides the developer with a means of displaying text in a dialog box. Depending upon how the dialog box is set up, the text can be either static, meaning that it does not change from session to session, or dynamic. For static text to be displayed in a dialog box, the label attribute must be set to the text that will be displayed. If the text is to be dynamic (see Figure 6–28), then the developer must supply the tile with a key attribute. The key is used as a link where information is provided to the dialog box through AutoLISP. This is covered in more detail in Chapter 7, "Interfacing and Managing Dialog Boxes."

Static text:

```
:text {
    label = "This is an example of static text";
    width = 75;
}
```

Dynamic text:

```
:text {
    key = "field_position";
    width = 35;
}
```

Figure 6–28 *Dynamic text – once the user has selected a point, the X, Y and Z coordinates of that point are displayed in the dialog box*

Text Part : `text_part { attributes; }`

The `text_part` tile is used to combine two or more text strings into a single piece of text (paragraph). The tile suppresses the margins of the supplied text so that it may be combined with other text parts using either the `paragraph` or the `concatenation` tile.

Paragraph : `paragraph { attributes; }`

To combine text parts into a paragraph, the `paragraph` tile must be employed. This tile allows the developer to combine text parts into a paragraph where the `text_part` tiles are arranged vertically. Depending upon the way the dialog box is set up, paragraphs can be either static or dynamic. The `paragraph` tile applies a single margin to the whole paragraph.

Concatenation –: `concatenation { attributes; }`

When the developer must supply a standard message to a dialog box where a portion of that message must change during runtime, then the `concatenation` tile should be used. A concatenation is a line of text made up of two or more concatenated `text_part` tiles. When a concatenation is used, a single border is applied to the whole.

Tile Clusters

Through the combining and arranging of buttons, images, text, and edit boxes, the possibilities are endless for the type and style of dialog boxes that can be created. Tiles can be grouped with or without borders, with or without labels, horizontally, vertically, or in any combination. The grouping of tiles in a dialog box is accomplished by the use of one or more tiles from a category of tiles known as *Tile Clusters*. A *Cluster* is a group of tiles arranged in rows or columns where the cluster (for layout purposes only) is treated as a single entity (tile). Tiles in this category include `column`, `boxed_column`, `row`, `boxed_row`, `radio_column`, `boxed_radio_column`, `radio_row`, and `boxed_radio_row`.

Columns : `column { attributes; tiles }`

When tiles are to be grouped together vertically without a border and with or without a label, the `column` tile is used (see Figure 6–29). When this tile is used to group other tiles together, the order in which the tiles are grouped in the dialog box is determined by the order in which they appear in the DCL file. For example:

```
: column {
    fixed_width = true;
    label = "Operations";
    : edit_box {
```

```
              label = "Rotation Angle";
              key = "rotation";
              width = 5;
          }
        : button {
              label = "Base Point";
              key = "base_point";
              width = 15;
          }
      : button {
              label = "To Point";
              key = "to_point";
              width = 15;
          }
      : button {
              label = "Slect Objects";
              key = "select_object";
              width = 15;
          }
      }
    }
```

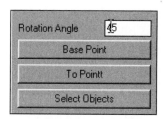

Figure 6–29 *Column*

Boxed Columns : boxed_column { attributes; tiles }

A boxed column is identical to a column in all aspects except one: a border is created around the group of tiles. For example, if the code is changed in the previous example from a column to a boxed column, the following visual effect is achieved (see Figure 6–30).

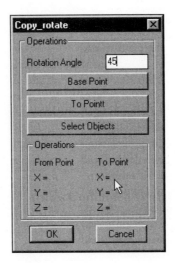

Figure 6–30 *Boxed column*

Rows

Tiles can also be grouped into horizontal arrangements called rows. The characteristics for row-grouping tiles are exactly the same as column-grouping tiles, with the obvious exception that the tiles are now grouped horizontally instead of vertically.

Radio Grouping : tile { attributes, tiles}

When radio buttons are employed in a dialog box, they are normally used to force the user into making a single choice from a list of options. However, it is a little more complex than just inserting a series of radio buttons in a DCL file. The series must be grouped together using one of the radio grouping tiles (radio_column, boxed-radio_column, radio_row, boxed_radio_row). These tiles behave just like the other grouping tiles with one exception: the developer can assign an action to these tiles.

```
: radio_column {
   key = "c1";
   label = "Text Layer";
   : radio_button {
         key = "create_layer";
         label = "Create layer";
   }
   : radio_button {
         key = "current_layer";
         label = "Place Text on Current Layer";
   }
```

```
    :radio_button {
        key = "select_layer";
        label = "Select Layer";
    }
}
```

Figure 6–31 shows a radio column.

Figure 6–31   *Radio column*

Providing a Means of Terminating a Dialog Box

As stated earlier, all dialog boxes must contain at least one button as a means of exiting the dialog once the dialog box is activated or the purpose of the dialog box has been fulfilled. DCL provides the designer with five different tiles for accomplishing this task. They are ok_only, ok_cancel, ok_cancel_help, ok_cancel_help_info, and ok_cancel_help_errtile. While the syntax for all five tiles is basically the same, the components they display vary, as illustrated in Figures 6–32 to 6–36.

Figure 6–32   *Ok_only*

Figure 6–33   *Ok_cancel*

Figure 6–34   *Ok_cancel_help*

Figure 6–35 *Ok_cancel_help_info*

Figure 6–36 *Ok_cancel_help_errtile*

Controlling the Spacing of Components in a Dialog Box

The alignment of components in a dialog box can be controlled by attributes. For example, by simply setting the fixed_width attribute to true, the width of a specified tile is not permitted to fill its available space. When spacing between tiles must be allowed, then the DCL tiles spacer, spacer_0 and spacer_1 must be used. These tiles have no effect on a dialog box whatsoever, except to provide a blank space for layout purposes. The only attributes permitted with a spacer tile are the standard layout attributes, fixed_height, fixed_width, alignment, height, and width.

DIALOG BOX DESIGN CONSIDERATIONS

When a dialog box for a program is designed, one objective must be fulfilled: the dialog box must serve the needs of the program for which it is intended. However, this is not the developer's only consideration; aesthetics and ergonomics, should also be taken into account. Aesthetics, the appearance of the dialog box, and ergonomics, the relationship between the dialog box and the end user, can play an important role in determining the value and functionality of the application in regard to the end user. If the dialog box is cluttered or poorly phrased, then the end user may have difficulties in interpreting exactly what data the program is requesting or in reading the results the program returns. Applications that call ineffective dialog boxes have a tendency to render the application useless regardless of how well the program is written. Issues that should be avoided when a dialog box is to be developed are unforgiving errors, not providing help, abbreviations, nested dialog boxes (use only when necessary), not handling keyboard input, and not providing user control. Whenever possible, related items should be grouped together with ample space provided between components. All components should be labeled using phrases and terminology that can easily be recognized by the user.

SUMMARY

An important feature of the Windows operating system is the status bar. The status bar allows developers to display important information regarding an application without disturbing user productivity. In AutoCAD, changes to the status bar are accomplished through the DIESEL programming language in conjunction with the MODEMACRO system variable. This system variable can be set from either the AutoCAD command prompt or through the AutoLISP SETVAR function. The number of characters that can be passed to the MODEMACRO system variable is limited to 255. However, the number of characters that can be displayed in the status bar is determined by the monitor and the size of the AutoCAD window.

To display changes to the internal state of AutoCAD, macro expressions (DIESEL) must be used. The basic syntax for a macro expression is similar to an AutoLISP expression, function name followed by the arguments. Unlike AutoLISP, DIESEL is limited to only one data type, strings. DIESEL expressions are also comma-delimited, meaning that a comma is used to separate the function from the arguments. DIESEL expressions can be nested within other expressions. This allows the developer to construct complex expressions and applications.

Another way a program can interact graphically with the user is by the incorporation of the use of dialog boxes. Dialog boxes are created through a programming language known as Dialog Control Language, or DCL. In DCL, the functions used to define a dialog box are referred to as tiles, while the arguments supplied to the functions are called attributes.

All dialog boxes consist of two main components: the box and the tiles. Tiles can be arranged in a variety of ways to create subassemblies. A subassembly is a grouping of tiles into rows and columns. When a subassembly is defined in a dialog box, the row or column used in the subassembly is known as a cluster. When a subassembly is created, it is treated as a single tile. The individual tiles that make up the assembly are treated as children.

All dialog boxes are comprised of the same basic classification of tiles. They are Predefined Active Tiles, Tile Clusters, Decorative and Information Tiles, Text Cluster Tiles, Dialog Box Exit Buttons and Error Tiles, and Restricted Tiles.

REVIEW QUESTIONS AND EXERCISES

1. Define the following terms:
   Attribute
   AutoCAD PDB
   boxed_column
   Children
   Concatenation
   DCL
   DIESEL
   Julian Date
   Julian Time
   Prototypes
   Status Line
   Subassemblies
   Tree Structure
   Parent

2. What is the difference between Dialog Control Language and AutoLISP?

3. What is the purpose of DIESEL? How is it different from Dialog Control Language? AutoLISP?

4. Why are aesthetics and ergonomics important issues to consider when developing a dialog box?

5. True or False: The length of the string that can be passed to the MODEMACRO system variable is limited by restrictions set by the operating system. (If the answer is false, explain why.)

6. What AutoCAD system variable controls the AutoCAD status bar?

7. True or False: The basic format for a DIESEL expression is $(someFun, arg1, arg2 ...). (If the answer is false, explain why.)

8. True or False: DIESEL expressions cannot be nested. (If the answer is false, explain why.)

9. True or False: In AutoCAD 2000 changes made to the status bar in one drawing are not reflected in another. (If the answer is false, explain why.)

10. True or False: DIESEL is limited to one data type, strings. (If the answer is false, explain why.)

11. What is the purpose of a prototype in DCL?

12. How are comments placed in a DCL source code file?

13. What is the purpose of the initial_focus attribute in DCL?

14. How do toggles differ from radio buttons in DCL?

15. What is the difference between a list box and a popup list in DCL? Give an example of when these would be used.

16. How is user input requested in DCL?

17. True or False: A dialog box defined in DCL must provide the user with a means of exiting; otherwise the user could become trapped. (If the answer is false, explain why.)

What is the value returned by the following expressions?

18. (setvar "modemacro" "This is an example or DIESEL")

19. (setvar "modemacro" (strcat "Drawing Name = $(getvar, dwgname"))

20. (setvar "modemacro" (strcat "Drawing Name = ($getvar dwgname), ""Dim = $(getvar, dimstyle)"))

21. (alert (menucmd "M=\nCurrent Time : $(edtime, $(getvar, date), DDDD MONTH DD YYYY HH:MM:SS AM/PM)"))

22. (getvar "time")

23. (getvar "cdate")

24. (alert (strcat "The current date is " (getvar "cdate")))

25. (alert (strcat "The current date is " (rtos (getvar "cdate"))))

26. (setvar "modemacro" "M=\nCurrent Time : $(edtime, $(getvar, date), DDDD
    MONTH DD YYYY HH:MM:SS AM/PM)")

27. (setvar "modemacro" "$(if, $(=, ans 1), The value of variable ans = 1")

CHAPTER 7

Interfacing with and Managing Dialog Boxes

OBJECTIVES

Upon completion of this chapter the reader will be able to:

- Use the AutoLISP function LOAD_DIALOG to load a dialog box definition into memory
- Initialize a dialog box once it has been loaded into memory using the NEW_DIALOG function
- Force an active dialog box to display using the START_DIALOG function
- Terminate a dialog box using the DONE_DIALOG function
- Specify the location for a dialog box to appear on the graphics screen
- Define a default action for a dialog box to execute
- Abort a series of nested dialog boxes using the TERM_DIALOG function
- Unload a dialog box definition from memory using the UNLOAD_DIALOG function

KEY WORDS AND AUTOLISP FUNCTIONS

Action Expression	START_DIALOG
Callback Function	Static
Dynamic	TERM_DIALOG
LOAD_DIALOG	UNLOAD_DIALOG
NEW_DIALOG	

INTRODUCTION TO DIALOG BOX MANAGEMENT

Chapter 6 introduced the concepts and techniques necessary for the development of dialog boxes using the DCL programming language. Defining the dialog box using DCL is only the beginning; special AutoLISP functions must be embedded within the application to allow the program to employ the dialog box. These functions load the dialog box definition into memory, assign it an identification number, activate the dialog box, initialize its tiles, load images, specify the dialog box's action, verify the data entered, end the dialog session, extract the data entered, and finally unload the dialog box's definition. While dialog boxes are unique in functionality, the general process used to manage a dialog box is the same.

LOADING A DIALOG BOX INTO MEMORY

The first step in using a dialog box in conjunction with an AutoLISP program is to load the dialog box's definition into memory. In AutoLISP this is accomplished by the LOAD_DIALOG (load_dialog dclfile) function. This function not only reads the content of the specified DCL file and stores it into memory, but it also assigns the definition an identification number. It is the identification number that will be used whenever subsequent function calls are made to the dialog box. This function requires the use of only one argument: the name of the dialog box's definition file, supplied as a text string and enclosed in quotation marks. If the path for the definition file is not supplied, then the function performs a search according to the AutoCAD library search path. If the file is located and successfully loaded into memory, then the function returns a positive integer. If the file is not located or the definition file contains an error, then the function returns a negative integer.

```
(SETQ dia_id (LOAD_DIALOG "c:/program files/
                          acad2000/autocad_lisp_programs
                          /example")    ; Loads the dialog box
                                        ; definition file
                                        ; example into memory.
```

The placement of the expression containing the LOAD_DIALOG function in the AutoLISP program is not critical as long as the expression is supplied before any other AutoLISP dialog functions are called. The function can also be used multiple times in a program to load multiple dialog box definitions. For example:

```
(SETQ dia_id (LOAD_DIALOG "c:/program files/
                          acad2000/autocad_lisp_programs
                          /example")    ; Loads the dialog box
                                        ; definition file
                                        ; example into memory.
```

```
(SETQ dia_id_1 (LOAD_DIALOG "c:/program files/
                            acad2000/autocad_lisp_programs
    /example_1")                        ; Loads the dialog box
                                        ; definition file
                                        ; example_1.dcl into
                                        ; memory_1.
(SETQ dia_id_2 (LOAD_DIALOG "c:/program files/
                            acad2000/autocad_lisp_programs
                    /example_2")   ; Loads the dialog box
                                   ; definition file
                                   ; example_2.dcl into
                                   ; memory_2.
```

 Note: A dialog definition file can contain an extension other than .dcl. However, it is recommended that all dialog box definition files contain the .dcl extension for purposes of consistency. If a dialog box definition file contains an extension other than .dcl, then the extension must be specified when the AutoLISP LOAD_DIALOG function is used; otherwise an error will result. If the dialog box definition file contains the .dcl extension, then the extension does not have to be specified when the definition is loaded into memory.

INITIALIZING THE DIALOG BOX

Once a dialog box has been loaded into memory and an identification number has been assigned, then the dialog box must be initialized. In AutoLISP this is accomplished through the NEW_DIALOG (new_dialog dlgname dcl_id [action [screen-pt]]) function. This function, when supplied with a dialog name and identification number, initializes the dialog box. If the initialization process is successful, then the function returns T. If the function is unable to initialize the dialog box, then the function returns nil. The function has two arguments that are required and two arguments that are optional. The two required arguments are *dlgname* and *dcl_id*. The *dlgname* argument is a string representing the name assigned to the actual dialog box by the DCL dialog tile. The *dcl_id* argument is the identification number assigned to the dialog's definition by the LOAD_DIALOG function. Examples of how this function can be used are provided below.

```
(if (null (new_dialog "examle_dia" dcl_id)) ; Initializes dialog
    (exit)                              ; box. If
                                        ; new_dialog returns nil
                                        ; the AutoLISP program is
                                        ; exited.
)                                       ; Close if expression.
```

```
(IF (NEW_DIALOG "example_dia" dcl_id)      ; Initializes dialog box.
   (PROGN                                   ; If NEW_DIALOG does not
                                            ; return nil then the
                                            ; remaining dialog
                                            ; functions are executed.
     (expression)                           ; Dialog Function Call
     (expression)                           ; Dialog Fuction Call
     (UNLOAD_DIALOG dcl_id)                 ; Unloads Dialog
                                            ; definition from memory.
   )                                        ; Closes this portion of
                                            ; if expression.
   (ALERT "Error: Unable to load EXAMPLE.DCL : ") ; If NEW_DIALOG
                                            ; returns nil then
                                            ; the message
                                            ; "Unable to load
                                            ; EXAMPLE.DCL" is
                                            ; displayed.
   )                                        ; Close if
                                            ; expression.
```

If the dialog box is to be assigned a default action, then the optional argument *action* must be supplied. This argument is a string consisting of an AutoLISP expression that is to be used for the dialog's default action. If a default action is not needed or the developer does not specify one, then an empty string ("") can be used or the option omitted altogether. If a default action is specified, then the action is evaluated when the user selects an active tile that does not have an action or callback assigned to it by the ACTION_TILE function or defined in the dialog box's definition.

The last argument associated with this function is the *screen-pt* argument. This argument can be used to specify the location on screen where the dialog box will appear. The argument specifies the two-dimensional location of the upper right corner of the dialog box. If the argument is not supplied, then by default, AutoLISP displays the dialog box in the center of the AutoCAD graphics screen.

Notes: If the *screen-pt* argument is used with the NEW_DIALOG function, then an *action* argument must also be specified.

In the Windows environment, 0,0 is located in the upper right corner of the active window.

DISPLAYING THE ACTUAL DIALOG BOX

Once a dialog box has been initialized through the NEW_DIALOG function, that dialog box becomes and remains active until either an action expression or callback function executes a DONE_DIALOG function. Therefore, an active dialog box can be forced to display with the AutoLISP START_DIALOG (start_dialog) function. This function does not have any arguments associated with it; however, it does return the optional status that has been passed to the dialog with the DONE_DIALOG function.

Any functions that require user input from the AutoCAD command line or that affect the display outside the active dialog box (including the text writing functions) cannot be used once a dialog box is displayed. Use of the SSGET function is permitted as long as the function does not require user input. Using these types of functions between a DONE_DIALOG and START_DIALOG function call can cause AutoCAD to terminate all dialog boxes and display the error message "AutoCAD rejected function".

Note: The default value returned by the START_DIALOG function is 1 if the user selects OK and 0 if the user selects Cancel.

Caution: In the Windows environment, the dialog box facility takes control of input when a call is made to the NEW_DIALOG function. This does not affect the techniques that the developer uses to construct a program. Invoking the NEW_DIALOG function from either the AutoCAD command prompt or the Visual LISP Console window might cause the screen to freeze. If the NEW_DIALOG function is to be invoked from the AutoCAD command prompt or the Visual LISP Console window, then it must be incorporated into a PROGN expression and be followed by the START_DIALOG function.

TERMINATING A DIALOG BOX

Once a dialog box has been displayed, control is then shifted from the graphics window or text screen to the dialog box. The dialog box remains in control until it is dismissed through a DONE_DIALOG (done_dialog [status]) function. When a DONE_DIALOG expression is used, it must be placed within an action expression or a callback function. An *action expression* associates an AutoLISP expression with a specific tile defined in the dialog box. The AutoLISP expression defines the action that is to be taken once the tile has been selected. Information relating to how the user has selected or modified a tile's content is known as a *callback function*.

If the optional *status* argument is not supplied to an action expression that a DONE_DIALOG function is embedded in, then the START_DIALOG function returns the string value 1 if the user selects OK and 0 if the user selects cancel. The *status* argument for the DONE_DIALOG function is an integer value that can be 0, 1 or any number greater than 1. If an integer greater than 1 is supplied, then the meaning of that integer will be determined by the application that contains the DONE_DIALOG function.

The only value returned by a DONE_DIALOG expression is the coordinates of a dialog box on the graphics screen. This information can be rather helpful if a dialog box must be reopened.

 Note: Dialog boxes that employ the predefined exit tiles automatically issue a DONE_DIALOG function.

 Caution: Extreme care should be exercised to ensure that a DONE_DIALOG expression is called when redefining the actions of any tile whose keys are either Accept or Cancel. If these tiles are redefined and the DONE_DIALOG expression is omitted, then the user can be trapped within the dialog box. When a user becomes trapped within a dialog box, often rebooting the system or terminating AutoCAD by pressing CTRL+ALT+DELETE is the only way to break out.

TERMINATING ALL CURRENT DIALOG BOXES

Dialog boxes can be nested within one another, as in the case of the LAYER command. When the command is executed, the user is presented with a dialog box. From this dialog box the user is able to control every aspect of the drawing's layers. One aspect is the layer's color. If the user chooses to change this attribute, then the user is presented with a second dialog box. This is known as a dialog box nested within a dialog box (see Figure 7–1). After the selection has been made, then the nested dialog box is terminated and the original dialog box once again reestablishes control. When working with nested dialog boxes, control should be reestablished by the parent dialog once the action is completed in a child. If an error should occur, it might become necessary to escape all active dialog boxes and return to the command prompt. The TERM_DIALOG (term_dialog) function should be used to accomplish this. The TERM_DIALOG function terminates all current dialog boxes, active or not.

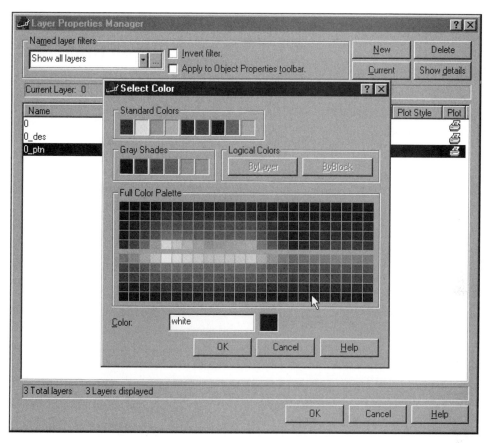

Figure 7–1 *AutoCAD* LAYER *command*

UNLOADING A DIALOG BOX

Because custom tiles can be defined and standard tiles can be redefined, it is often a good idea to unload a dialog box from memory once the program that uses the dialog box has finished executing. In AutoLISP this is accomplished by the UNLOAD_DIALOG (unload_dialog dcl_id) function. This function, when supplied with a dialog identification number, removes that dialog box's definition from memory. The UNLOAD_DIALOG function returns nil regardless of the function's success.

SETTING THE VALUE OF A TILE

The initial value of a tile contained within a dialog box can be set using the DCL attribute value. When this method is employed, the value assigned to a dialog box's tile is static and does not change from initialization to initialization. Assigning a static value to a dialog box's tile is not always desirable. When a dynamic value is to be assigned to a tile, then a different approach must be taken. In AutoLISP this is accomplished by the SET_TILE (set_tile key value) function. This function, when supplied with a key and a value, sets the specified value to the tile associated with the given key. The function returns the value assigned to the specified key once the function has been evaluated.

```
(set_tile "description" "Spot Elevation")
(set_tile "rotation" "45")
```

Note: Any calls to a SET_TILE function must be placed between the NEW_DIALOG and START_DIALOG function calls. The key argument is case sensitive and must exactly match the case used in the dialog box definition.

RETRIEVING THE VALUE OF A DIALOG BOX'S TILE

After a dialog box has been terminated, the information that was provided by the user must be extracted if it is to be used by the program. In AutoLISP this is accomplished by the GET_TILE (get_tile key) function. This function, when supplied with a tile's key, extracts the value retained by that tile. As with the SET_TILE function, the key supplied to the GET_TILE function is case sensitive and must exactly match the case of the key defined in the dialog box. But unlike the SET_TILE function, the GET_TILE function can follow a call made to the START_DIALOG function.

```
(setq description   (get_tile "description")
      elevation     (get_tile "elevation")
      rotation      (get_tile "rotation")
)
```

Another way information can be extracted from a dialog box is with the use of the AutoLISP GET_ATTR (get_attr key attribute) function. This function, when supplied with a key (specifying a tile) and an attribute (associated with the tile previously specified), returns the initial value of that attribute as it appears in the DCL definition.

ACTION EXPRESSIONS AND CALLBACK FUNCTIONS

As mentioned earlier, an action expression associates an AutoLISP expression with a specific tile defined in a dialog box. All action expressions used in an AutoLISP program must follow calls to the NEW_DIALOG function and precede calls to the START_DIALOG function. There are three ways that an action expression can be defined: defining the tile's action in the dialog box by using the DCL action attribute, using the NEW_DIALOG function, or using the ACTION_TILE function. If the action expression is defined using the NEW_DIALOG function, then the action is defined for the entire dialog box. If an action expression is to be defined for a particular tile, then the ACTION_TILE (action_tile key action-expression) function must be used. This function, when supplied with a key and an expression (a string defining an action to take once the specified tile has been selected), suspends the default action assigned to either the dialog box or the tile and replaces it with the action specified. In the AutoLISP expression below, the action expression is used to terminate the dialog box once the user has selected the tile s_point.

```
(action_tile "s_point" "(done_dialog 4)")
```

Notes: Because the DONE_DIALOG function is used in the action expression above and its status argument is set to 4, once the user exits the dialog box, the START_DIALOG function returns 4.

The AutoLISP COMMAND function cannot be called from an ACTION_TILE function. A tile can have only one action assigned to it. If a tile has an action assigned to it with the DCL action attribute, then any calls made to an ACTION_TILE function will result in the pre-defined action being overwritten. Action expressions can incorporate both the GET_ATTR and GET_TILE functions.

Once the user has modified or selected a tile, information regarding this action is returned to the action expression in the form of a callback. In most cases, callbacks can be generated from every active tile within a dialog box. Callback functions are most often used to update information within a dialog box. This includes issuing error messages, disabling tiles and displaying text in an edit box or list box. The following AutoLISP expressions define three callback functions that incorporate the MODE_TILE function to enable or disable the tiles:

```
(action_tile
    "current_layer"
    "(progn (mode_tile \"layer_pop_list\" (atoi $value))
        (mode_tile \"create_layer_box\" (atoi $value))
        (mode_tile \"field_layer\" (- 1 (atoi $value))))"
)
  (action_tile
    "select_layer"
    "(progn (mode_tile \"create_layer_box\" (atoi $value))
        (mode_tile \"field_layer\" (atoi $value))
        (mode_tile \"layer_pop_list\" (- 1 (atoi $value))))"
)
```

```
(action_tile
 "create_layer"
 "(progn (mode_tile \"layer_pop_list\" (atoi $value))
    (mode_tile \"field_layer\" (atoi $value))
    (mode_tile \"create_layer_box\" (- 1 (atoi $value))))"
)
```

Figures 7–2, 7–3, and 7–4 show each of the three tiles enabled.

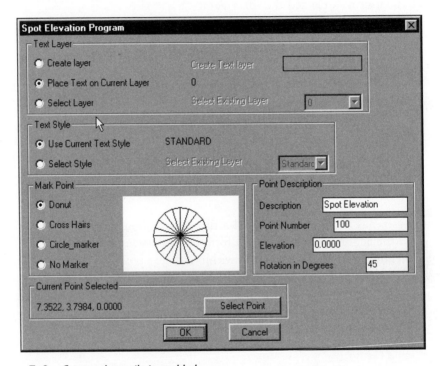

Figure 7–2 *Current_layer tile is enabled*

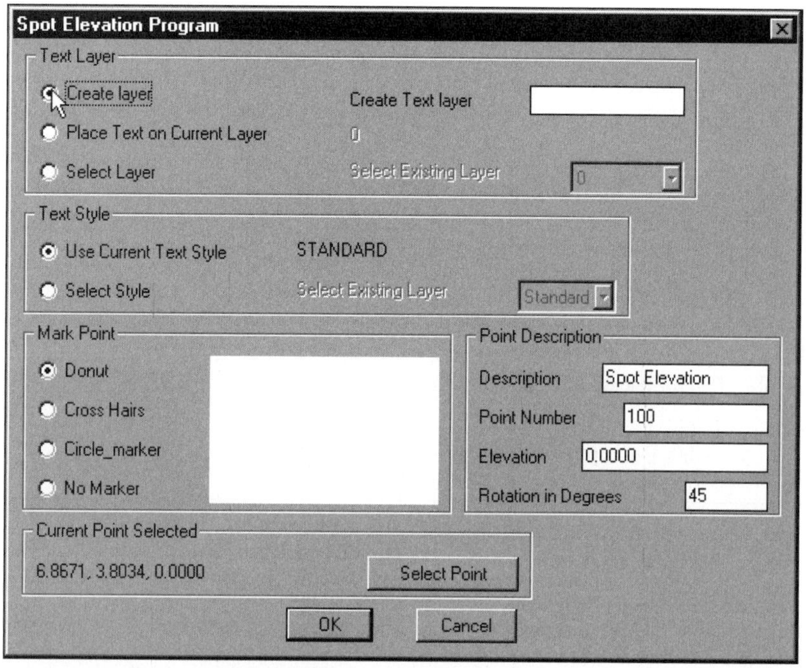

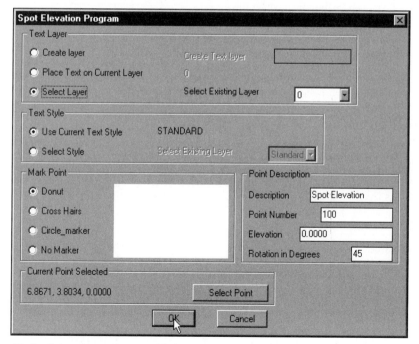

Figure 7–3 *Create_layer tile is enabled*

Figure 7–4 *Select_layer tile is enabled*

In the previous example, the variable *$value* is used in the action expression to retrieve the current state of the tile. That information is then passed to the MODE_TILE function and the appropriate tile is then enabled (disabling the others). Table 7–1 lists the variables that can be used in an action expression. Also note the use of the PROGN function in the above expressions. Because tiles can have only one action expression assigned to them at a time, multiple expressions must be grouped together in a single expression.

Table 7–1   Action Expression Variables

Variable	Description
$Key	The key attribute of the tile that was selected. This variable applies to all actions.
$Value	The string form of the current value of the tile, such as the string from an edit box or a 1 or 0 from a toggle. This variable applies to all actions. If the tile is a list box (or popup list) and no item is selected, the *$value* variable will be nil.
$Data	The application-managed data (if any) that was set just after new_dialog time by means of client_data_tile. This variable applies to all actions, but *$data* has no meaning unless your application has already initialized it by calling client_data_tile.
$Reason	The reason code that indicates which user action triggered the action. This variable indicates why the action occurred; its value is set for any kind of action, but you need to inspect it only when the action is associated with an edit_box, list_box, image_button, or slider tile.

Courtesy of the Visual LISP Developers Guide

Note: The variable names listed in the previous chart are reserved. Their values are returned as read-only.

SETTING THE MODE OF A DIALOG BOX

The initial state or characteristics of a dialog box can be set in the dialog box's definition through DCL attributes. If changes are to be made to a tile's mode during runtime, then the MODE_TILE (mode_tile Key mode) function must be used. The *mode* argument is used to specify the mode setting for the specified tile. Valid settings for this argument are 0 - Enable Tile, 1 - Disable Tile, 2 - Set Focus to Tile, 3 - Select Edit Box Contents, 4 - Flip Image Highlighting On or Off. Once the function has been evaluated, a value of nil is returned. Examples of the MODE_TILE function are shown below.

```
(mode_tile "elevation" 1)          ;The tile elevation is enabled
(action_tile
    "select_layer"
    "(progn (mode_tile \"create_layer_box\" (atoi $value))
```

```
            (mode_tile \"field_layer\" (atoi $value))
            (mode_tile \"layer_pop_list\" (- 1 (atoi $value)))))"
   )
; Mode_tile is embedded in an action expression where the callback
; from the tile select_layer is used to set the state of the tiles
; create_layer, field_layer and layer_pop_list
```

Figures 7–5 and 7–6 show the results of mode settings.

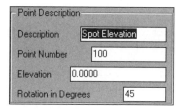

Figure 7–5 *Enabled tiles: mode_tile set to 0, disabled tiles: mode_tile set to 1*

Figure 7–6 *Text in textbox highlighted: mode_tile set to 3*

 Note: When Set Focus is applied to a tile, then that tile becomes active.

LIST BOXES AND POPUP LISTS

User options always enhance the versatility of an application. Chapter 6 introduced five DCL tiles that are specifically designed to provide dialog boxes with this capability. Three of these tiles (radio buttons, buttons and toggles) are considered static (the number of choices will not change). If a developer wishes to provide the user with options that reflect changes made from runtime to runtime, then popup lists or list boxes must be used. Placing a list box or popup list in a dialog box's definition is only the beginning in incorporating these tiles. These lists must be populated before the dialog box is displayed and that requires making several adaptations to the AutoLISP program that calls the dialog box. To populate a list box or popup list, the AutoLISP program must first start a list, then append the entries to that list, and finally close the list.

Starting a List

The first step in populating a list box or popup list is to initialize (start the process) the list used by these tiles. In AutoLISP this is accomplished by the START_LIST (start_list key [operation [index]]) function. This function specifies to the PDB (programmable dialog box) facility the tile that is being manipulated and the type of operation that is to be performed. The operations this function can perform are: 1 - Change Selected List Contents, 2 - Append new List Entry, and 3 - Delete old List and Create New List. The function has one required argument: the *key* associated with the tile being processed. If the *operation* argument is not supplied, then by default the function deletes the old list associated with the specified tile and then creates a new one. If the *index* argument is to be used, then the *operation* argument must also be provided. The *index* argument must be used if one or more entries in an existing list are being modified. Entries contained within a list are numbered, newly created with zero. Once the function has been evaluated, the name of the started list is returned.

```
(start_list "layer_pop_list")
```

Caution: Once a list has been started, do not use the SET_TILE function until the list process has been completed.

If an index argument is supplied and an operation other than 1 is specified, then the index argument is ignored.

Adding Items to a List

Once a list has been started, then entries can be added with the ADD_LIST (add_list string) function. This function either adds a string to the current list or replaces the item specified by the *index* argument of the START_LIST function. The function has only one argument associated with it, and that is a string representing the entry to be applied. When multiple items are to be added to a list, then multiple calls must be made to the ADD_LIST function. If the ADD_LIST function is placed in a loop, a list of the layer names contained within a drawing can be generated. For example:

```
(start_list "layer_pop_list")
(setq Layer_list  '()
    list_item         (tblnext "layer" T)
    index_number 0
)
(while list_item
  (if (= (getvar "clayer") (cdr (assoc 2 list_item)))
    (setq CurrentLayer index_number
```

```
      index_number (1+ index_number)
  )
 )
 (setq    Layer_list (append Layer_list (list (cdr (assoc 2
  list_item))))
    list_item (tblnext "layer")
 )
)
(mapcar 'add_list Layer_list)
(end_list)
```

Ending a List

Once information has been appended to a list, then that list must be closed. This is accomplished with the END_LIST (end_list) function. This function has no argument associated with it and, once evaluated, returns a value of nil.

IMAGES

The processes used to incorporate an image into a dialog box are similar to the processes used to populate a list in either a list box or popup list. The creation of the image must be started, the image must then be defined, and finally the creation process must be stopped. Just as for the list box and popup list, the processes are defined in the AutoLISP program calling the dialog box containing the image.

Starting an Image

The first step in displaying an image in a dialog box is to start the creation process itself. This is accomplished by the START_IMAGE (start_image key) function. Like its list-processing counterpart START_LIST, this function, when supplied with a *key* (associated with either an image or image_button tile), begins the actual creation process. Once the function has been evaluated, it returns either the specified key (if the function was successful) or nil (if it was not).

 Caution: Once an image has been started, do not use the SET_TILE function until the image process has been completed.

Filling an Image Tile

One way that an image can be produced is by constructing the image out of solid (two-dimensional only) regions. In AutoLISP, using the FILL_IMAGE (fill_image x1 y1 width height color) function produces a solid region. This function creates a solid rectangle at a specified point, length, width and color. The first two arguments of this function specify the X and Y starting coordinates of the solid rectangle. In an image tile, the origin is always located at the upper left corner of the tile with the positive Y-axis running top to bottom and the positive X-axis running left to right. The next two arguments, *width* and *height*, specify the width and height of the fill area (in pixels) relative to the *x1* and *y1*arguments. The final argument determines the color of the fill area. Its value can be either an AutoCAD color number or one of the logical color numbers shown in Table 7–2. Once the function has been evaluated, an integer representing the specified color is returned.

Table 7–2 Symbolic Names for Color Attributes

Color Number	ADI Mnemonic	Description
-2	BGLCOLOR	Current background of the AutoCAD graphics screen
-15	DBGLCOLOR	Current dialog box background color
-16	DFGLCOLOR	Current dialog box foreground color (text)
-18	LINELCOLOR	Current dialog box line color

```
(start_image "test")
(fill_image 0 0 (dimx_tile "test") (dimy_tile "test") 0)
(end_image)
```

 Note: The coordinates of the lower right corner of the image tile can be obtained through DIMX_TILE (dimx_tile key) and DIMY_TILE (dimy_tile key). These functions return the maximum allowed X and Y coordinates (lower right corner) for the specified image tile.

Constructing a Line in an Image Tile

Solid regions are not the only means available for constructing an image in a dialog box; vectors can also be employed. In AutoLISP, a vector can be constructed in an image tile through the VECTOR_IMAGE (vector_image x1 y1 x2 y2 color) function. This function, when supplied with a starting and ending point, displays a vector in the color specified. The first and second arguments for this function determine the vector's starting X and Y coordinates. The third and fourth arguments of the function determine the vector's ending X and Y coordinates. The final argument specifies the color of the vector. Like the color argument for the FILL_TILE function, this argument can be supplied as either an AutoCAD color number or one of the logical color numbers. Once the function has been evaluated, an integer representing the specified color is returned.

```
(start_image "test")
(vector_image 40 8 50 8 -12)
```

```
(vector_image 45 8 45 16 -12)
(vector_image 0 20 100 20 1)
(vector_image 40 24 50 24 1)
(vector_image 45 24 45 32 1)
(end_image)
```

Inserting a Slide into an Image Tile

Images displayed in a dialog box can also be constructed using AutoCAD slides. A slide is screen capture of the AutoCAD graphics editor saved in a special format that can be viewed later from within AutoCAD. In AutoCAD, slides are generated with the MSLIDE command and viewed with the VSLIDE command. However, in AutoLISP a slide is displayed in a dialog box with the SLIDE_IMAGE (slide_image x1 y1 width height sldname) function. The function has five arguments that must be specified before the slide can be displayed in the dialog box: the X offset, Y offset, width of the slide, height of the slide, and the slide name. The X and Y offsets (*x1* and *y1*) are positive integers representing the distance in pixels that the slide will be displayed from the upper left corner. The *width* and *height* arguments represent the dimensions in pixels for the slide to be displayed. The last argument supplied to this function is the name of either the slide or the slide library containing the slide. If a slide is specified, then the name is entered as a text string including the slide's path if the slide resides in a location other than one of the AutoCAD search libraries (path/slide name or slide name). If the slide is part of a slide library, then the library name along with the slide name must be specified. If the slide library is not located in one of the AutoCAD search libraries, then the location, name of the slide library, and the slide name (path/library (slide name)) must be supplied. Once the function has been evaluated, the name of the slide inserted in the image tile is returned.

For "Example.sld," a slide not located in a slide library, "/Program files/ACAD2000/My programs/Slides/Example.sld" is a valid slide name.

For "Slide_program_library (Example.sld)," a slide located in a slide library, "/Program files/ACAD2000/My programs/Slides/Slide_program_library (Example.sld)" is a valid slide name.

```
(start_image "test")
   (setq x (dimx_tile "test")
      y (dimy_tile "test")
   )
   (slide_image 0 0 x y "/temp/donut.sld")
(end_image)
```

Closing an Image

Once an image has been opened with the START_IMAGE function and its attributes defined, then the image must be closed before it can be displayed in a dialog box or any calls are made to the SET_TILE function. In AutoLISP this is accomplished much the same way it is for the list-processing functions, with the END_IMAGE (end_image) function. The END_IMAGE function is identical to its list-processing counterpart. First, the function does not accept any arguments. Second, once the function has been evaluated, it returns a value of nil.

APPLICATION - DETERMINING SPOT ELEVATIONS

Chapters 6 and 7 have introduced the reader to the concepts of constructing and managing programmable dialog boxes using DCL and AutoLISP. Many of these concepts are illustrated in the following application. When this application is launched, it allows the user to label spot elevations contained on a civil engineering contour drawing. When the program is first activated, the user is presented with a dialog box where all but three options, Select Point, OK, and Cancel, are disabled (see Figure 7–7). This is accomplished by the MODE_TILE function. The MODE_TILE AutoLISP function is embedded in an IF statement that tests the condition of the variable *elevation*. If the variable is nil, then the tiles are disabled and the user is only allowed to select a point or cancel out of the program. If the user selects the Select Point tile, the dialog box is dismissed and AutoLISP reestablishes control. Once a point has been selected, the dialog box is reissued with the remaining tiles enabled (see Figure 7–8). Dismissing and reissuing the dialog box is accomplished by enclosing the expressions that control the dialog box and point selection portion of the application in a loop. The program remains trapped within the loop until the user either selects a point and selects OK or the program is canceled.

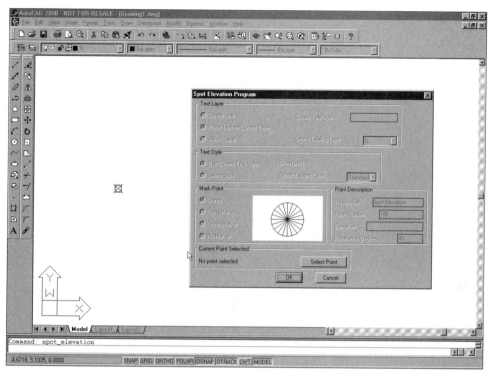

Figure 7–7 *Spot elevation program with all tiles disabled*

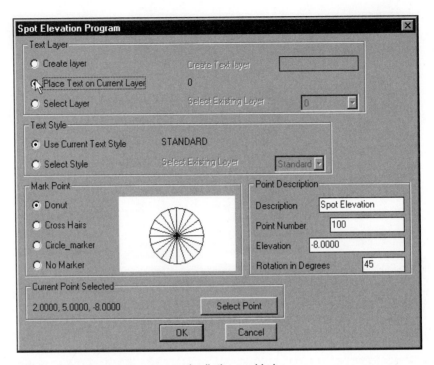

Figure 7–8 *Spot elevation program with all tiles enabled*

Once a point has been selected, the user is able to determine the layer where the elevation label will be placed as well as the type of marker to use. The types of markers that the user can choose from are displayed with an `image` tile. The initial image displayed by the `image` tile is defined in the beginning of the application. However, action expressions are also assigned to the tiles `donut_marker`, `cross_marker`, `circle_marker` and `no_marker` toggles so that, when one of these tiles is selected, the image is changed to the corresponding slide representing the type of marker that was chosen (see Figures 7–9, 7–10, 7–11, and 7–12). Once the user has selected the type of marker to insert and the layer on which to place the elevation label, then the program uses the advanced list-processing functions to generate the spot elevation label and marker.

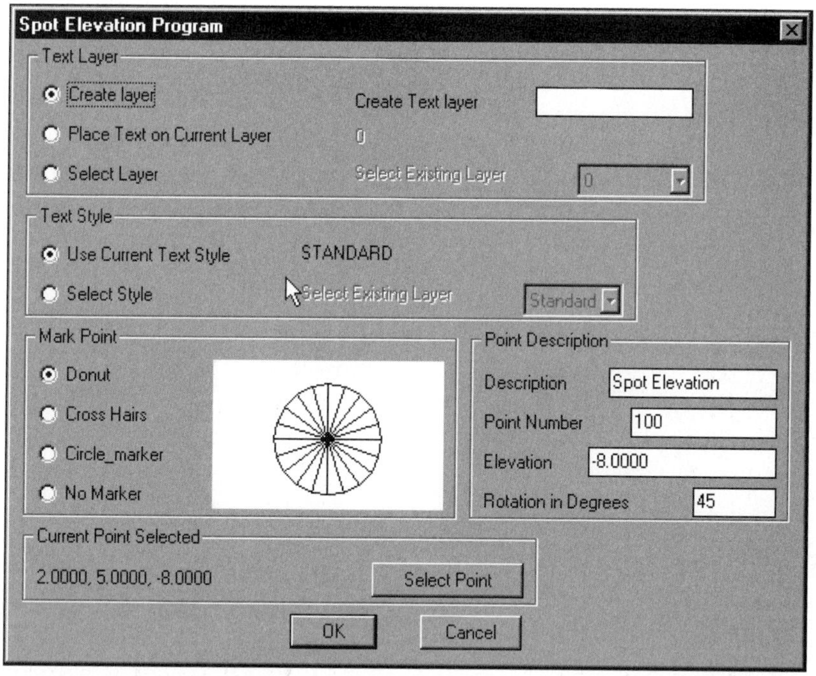

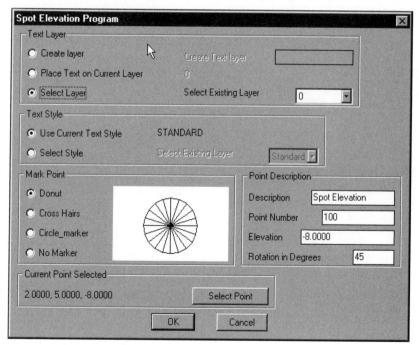

Figure 7–9 *Donut_marker slide*

Figure 7–10 *Cross_marker slide*

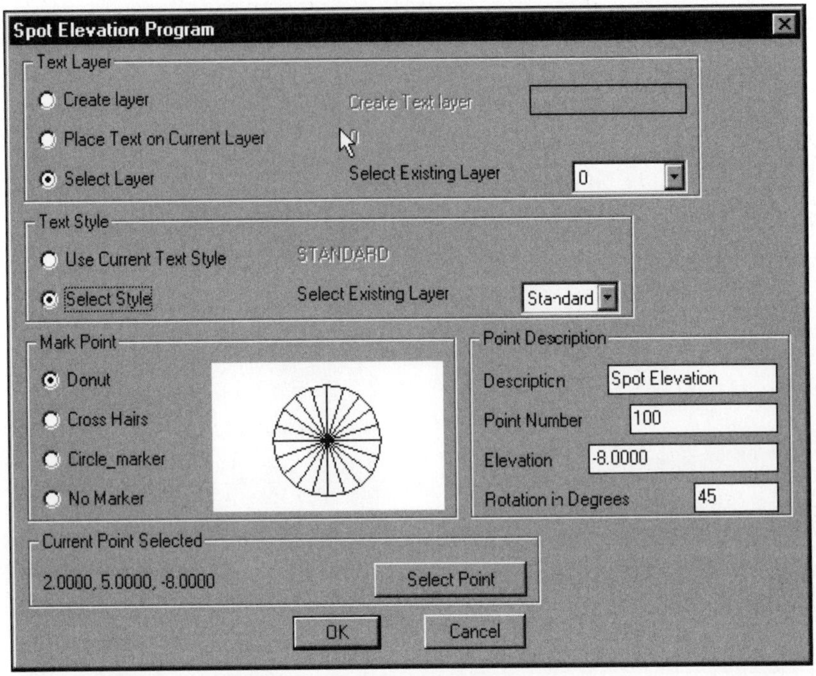

Figure 7–11 *Circle_marker slide*

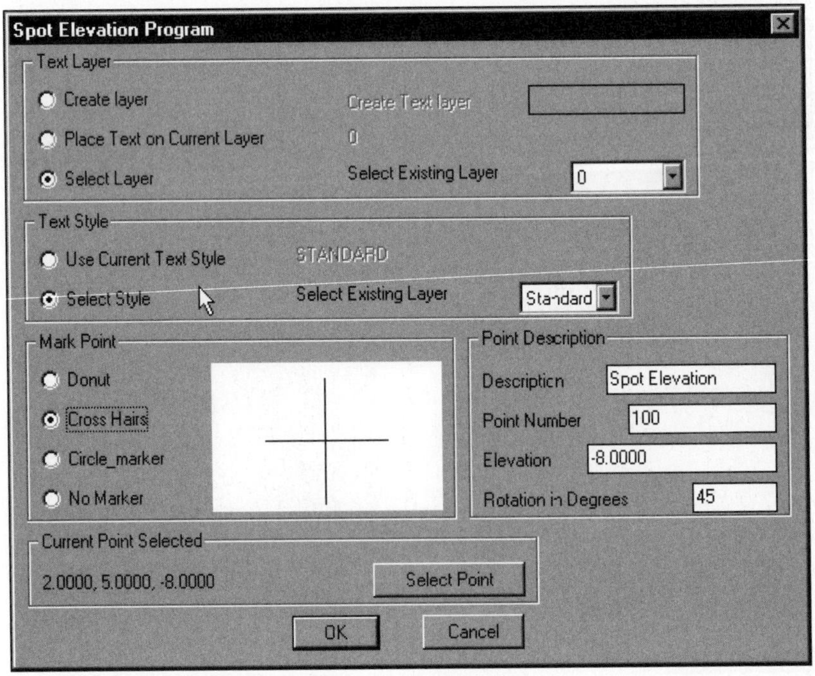

Figure 7–12 *No_marker slide*

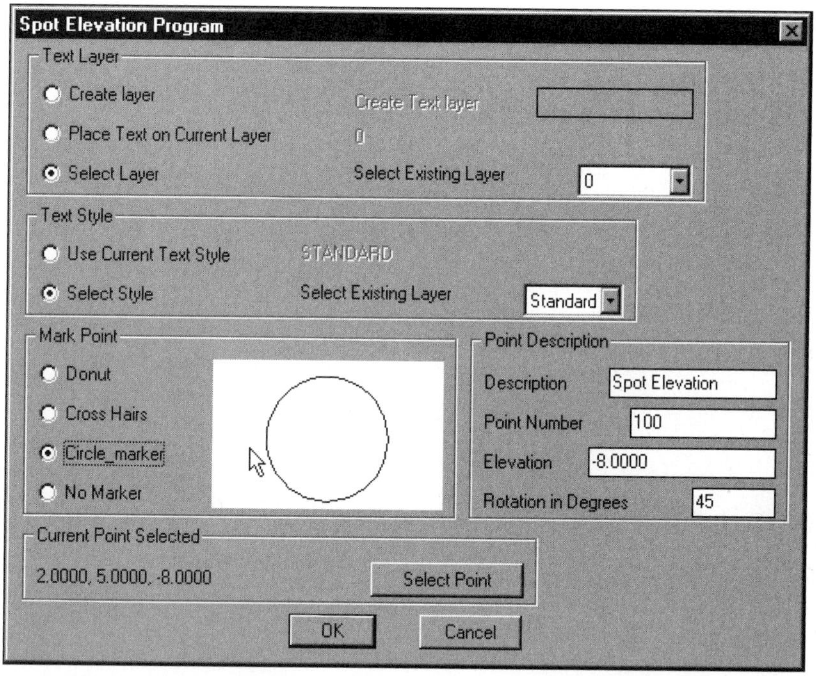

Figure 7–13 *Circle_marker slide*

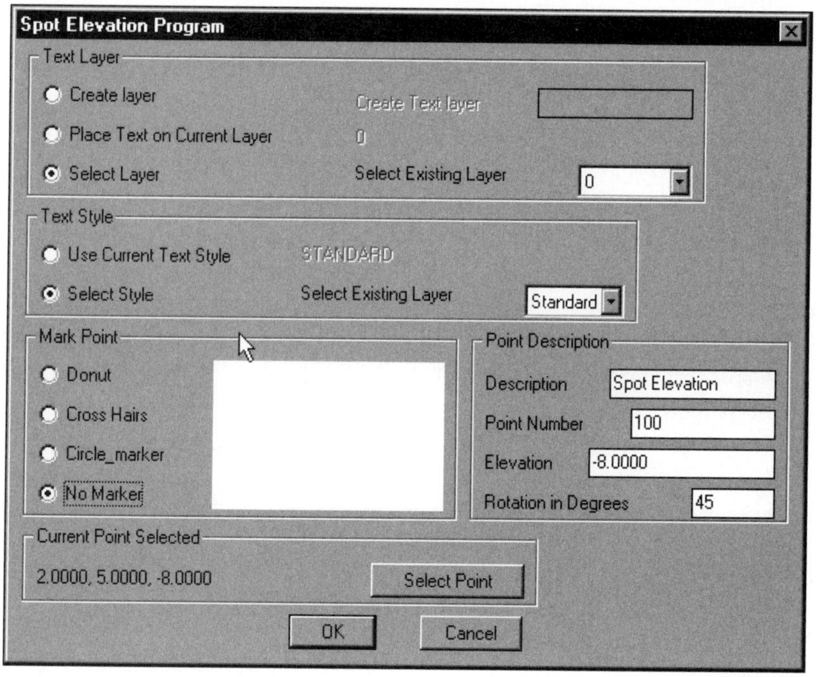

Figure 7–14 *No_marker slide*

```
;;;*******************************************************************
;;;
;;;Program Name: Spot_elevation.lsp
;;;
;;;Program Purpose: Label spot elevations on an AutoCAD drawing
;;;
;;;Programmed By: Kevin Standiford
;;;
;;;Date: 12/22/99
;;;
;;;*******************************************************************
(defun c:spot_elevation (/ result z_pt field_position elevation pt)
 (setq cnt 1)
 (setq result 4)
 (while (> result 1)
  (setq dcl_id4 (load_dialog "C:/WINDOWS/Desktop/Visual_Lisp/Ch6/LSP/
  spot.dcl"))
  (if      (not (new_dialog "spot" dcl_id4))
   (exit)
  )
  (start_list "layer_pop_list")
  (setq layer_list '()
    next_list (tblnext "layer" T)
    index 0
  )
  (while next_list
   (if (= (getvar "clayer") (cdr (assoc 2 next_list)))
    (setq new_index index
       index (1+ index)
    )
   )
    (setq layer_list (append layer_list (list (cdr (assoc 2 next_list)))))
      next_list (tblnext "layer")
   )
  )
  (mapcar 'add_list layer_list)
  (end_list)
  (set_tile "layer_pop_list" (ITOA new_index))
  (start_list "style_pop_list")
  (setq new_style_list    '()
    new_style    (tblnext "style" T)
```

```
    index_style   0
 )
 (while new_style
  (if (= (getvar "textstyle") (cdr (assoc 2 new_style)))
  (setq new_index index_style
      index_style (1+ index_style)
  )
  )
   (setq new_style_list (append new_style_list (list (cdr (assoc 2
new_style)))))
     new_style (tblnext "style")
 )
)
 (mapcar 'add_list new_style_list)
 (end_list)
 (set_tile "point_number" "100")
 (set_tile "description" "Spot Elevation")
 (set_tile "rotation" "45")
 (set_tile "field_layer" (getvar "clayer"))
 (set_tile "field_style" (getvar "cmlstyle"))
(if     (/= field_position nil)
  (set_tile "field_position" field_position)
  (set_tile "field_position" "No point selected")
)
(if     (/= elevation nil)
  (set_tile "elevation" (rtos elevation))
 (progn
 (mode_tile "elevation" 1)
 (mode_tile "rotation" 1)
 (mode_tile "description" 1)
 (mode_tile "point_number" 1)
 (mode_tile "c1" 1)
 (mode_tile "c2" 1)
 (mode_tile "c3" 1)
 (mode_tile "field_style" 1)
 (mode_tile "field_layer" 1)
 )
)
 (set_tile "style_pop_list" (ITOA CL_s))
 (set_tile "current_layer" "1")
 (set_tile "current_style" "1")
```

```
(set_tile "donut_marker" "1")
(mode_tile "layer_pop_list" 1)
(mode_tile "create_layer_box" 1)
(mode_tile "style_pop_list" 1)
(action_tile
  "current_layer"
  "(progn (mode_tile \"layer_pop_list\" (atoi $value))
     (mode_tile \"create_layer_box\" (atoi $value))
     (mode_tile \"field_layer\" (- 1 (atoi $value))))"
)
(action_tile
  "select_layer"
  "(progn (mode_tile \"create_layer_box\" (atoi $value))
     (mode_tile \"field_layer\" (atoi $value))
     (mode_tile \"layer_pop_list\" (- 1 (atoi $value))))"
)
(action_tile
  "create_layer"
  "(progn (mode_tile \"layer_pop_list\" (atoi $value))
     (mode_tile \"field_layer\" (atoi $value))
     (mode_tile \"create_layer_box\" (- 1 (atoi $value))))"
)
(action_tile
  "current_style"
  "(progn (mode_tile \"style_pop_list\" (atoi $value))
     (mode_tile \"field_style\" (- 1 (atoi $value))))"
)
(action_tile
  "select_style"
  "(progn (mode_tile \"field_style\" (atoi $value))
     (mode_tile \"style_pop_list\" (- 1 (atoi $value))))"
)
(start_image "test")
(setq x (dimx_tile "test")
   y (dimy_tile "test")
)
(slide_image 0 0 x y "/temp/donut.sld")
(end_image)
(action_tile
  "donut_marker"
  "(progn (start_image \"test\")
```

```
      (fill_image 0 0 x y 0)
      (slide_image 0 0 x y \"/temp/donut.sld\")
      (end_image))"
  )
  (action_tile
    "cross_marker"
    "(progn (start_image \"test\")
      (fill_image 0 0 x y 0)
      (slide_image 0 0 x y \"/temp/cross.sld\")
      (end_image))"
  )
  (action_tile
    "circle_marker"
    "(progn (start_image \"test\")
      (fill_image 0 0 x y 0)
      (slide_image 0 0 x y \"/temp/circle.sld\")
      (end_image))"
  )
  (action_tile
    "no_marker"
    "(progn (start_image \"test\")
      (fill_image 0 0 x y 0)
      (end_image))"
  )
  (action_tile "accept" "(get_info)(done_dialog 1)")
  (action_tile "s_point" "(done_dialog 4)")
  (action_tile "cancel" "(setq result nil) (done_dialog 0)")
  (setq result (start_dialog))
(if      (and (= result 1) (= pt nil))
  (progn
  (alert "No point Specified")
  (setq result 2
      elevation    nil
      field_position
      nil
  )
  )
)
(if      (= result 4)
  (progn
  (setq pt (getpoint "\nPick insertion point : "))
```

```lisp
        (setq field_position
            (strcat (rtos (car pt))
                    ", "
                    (rtos (cadr pt))
                    ", "
                    (rtos (setq z_pt (caddr pt)))
            )
            elevation   z_pt
        )
      )
     )
    )
   (princ)
  )
 (defun get_info   ()
   (setq      description      (get_tile "description")
      elevation        (get_tile "elevation")
      rotation         (get_tile "rotation")
      point_number     (get_tile "point_number")
      no_marker        (get_tile "no_marker")
      cross_marker     (get_tile "cross_marker")
      donut_marker     (get_tile "donut_marker")
      circle_marker    (get_tile "circle_marker")
      style_pop        (nth (atoi (get_tile "style_pop_list")) ll_s)
      select_style     (get_tile "select_style")
      current_style    (get_tile "current_style")
      layer_pop        (nth (atoi (get_tile "layer_pop_list")) ll)
      create_layer_box (get_tile "create_layer_box")
      select_layer     (get_tile "select_layer")
      current_layer    (get_tile "current_layer")
      create_layer     (get_tile "create_layer")
   )
   (if (/= pt nil)
    (progn
     (setq pt (list (car pt) (cadr pt) (atof elevation)))
     (if (= no_marker "1")
     (no_marker_function)
     )
     (if (= circle_marker "1")
     (circle_marker_function)
     )
```

```
    (if (= donut_marker "1")
    (donut_marker_function)
    )
    (if (= cross_marker "1")
    (cross_marker_function)
    )
  )
 )


)
(defun no_marker_function ()
 (setq     object_1 nil
    object_2 nil
 )
 (make_object_function)
)
(defun circle_marker_function ()
 (setq     object_1 (LIST (CONS 0 "CIRCLE")
                 (CONS 10 pt)
                 (CONS 40 0.5)
                 (CONS 8 "pt_layer")
             )
    object_2 nil
 )
 (make_object_function)
)
(defun donut_marker_function ()
 (setq     object_1
     (list (cons 0 "LWPOLYLINE")
        (cons 100 "AcDbEntity")
        (cons 67 0)
        (cons 410 "Model")
        (cons 8 "0")
        (cons 100 "AcDbPolyline")
        (cons 90 2)
        (cons 70 1)
        (cons 43 0.5)
        (cons 38 0.0)
        (cons 39 0.0)
        (cons 10 pt)
        (cons 40 0.5)
```

```
                    (cons 41 0.5)
                    (cons 42 1.0)
                    (cons 10 (list (+ 0.5 (car pt)) (cadr pt) (caddr pt)))
                    (cons 40 0.5)
                    (cons 41 0.5)
                    (cons 42 1.0)
                )
            object_2 nil
    )
    (make_object_function)
)
(defun cross_marker_function ()
    (setq     object_1 (list
                    (cons 0 "line")
                    (CONS 10 (list (- (car pt) 0.5) (cadr pt) (caddr pt)))
                    (cons 11 (list (+ (car pt) 0.5) (cadr pt) (caddr pt)))
                    (cons 8 "pt_layer")
                )
        object_2 (LIST
                    (CONS 0 "line")
                    (CONS 10 (list (car pt) (- (cadr pt) 0.5) (caddr pt)))
                    (CONS 11 (list (car pt) (+ (cadr pt) 0.5) (caddr pt)))
                    (CONS 8 "pt_layer")
                )
    )
    (make_object_function)
)
(defun make_object_function ()
    (setvar "cmdecho" 0)
    (setvar "cmddia" 0)
    (if (= current_layer "1")
        (setq layer (getvar "clayer"))
    )
    (if (= create_layer "1")
        (setq layer create_layer_box)
    )
    (if (= select_layer "1")
        (setq layer layer_pop)
    )
    (if (= current_style "1")
        (setq style (getvar "cmlstyle"))
```

```
)
(if (= select_style "1")
  (setq style style_pop)
)
(ENTMAKE (list (cons 0 "block")
               (CONS 2 "point_block")
               (cons 10 pt)
               (cons 70 64)
     )
)
(ENTMAKE object_1)
(ENTMAKE object_2)
(ENTMAKE
 (list (cons 0 "TEXT")
    (cons 100 "AcDbEntity")
    (cons 67 0)
    (cons 410 "Model")
    (cons 8 (strcat layer "_ptn"))
    (cons 100 "AcDbText")
    (cons 10
          (list (+ (car pt) (* 1.7143 (getvar "textsize")))
                (+ (cadr pt) (* 1.7143 (getvar "textsize")))
                (caddr pt)
          )
    )
    (cons 40 (getvar "textsize"))
    (cons 1 point_number)
    (cons 50 (* (atof rotation) (/ 3.14 180)))
    (cons 41 1.0)
    (cons 51 0)
    (cons 7 style)
    (cons 71 0)
    (cons 72 0)
    (cons 100 "AcDbText")
    (cons 73 0)
 )
)
(ENTMAKE
 (list (cons 0 "TEXT")
    (cons 100 "AcDbEntity")
```

```
(cons 67 0)
(cons 410 "Model")
(cons 8 layer)
(cons 100 "AcDbText")
(cons 10
       (list (+ (car pt) (* 1.7143 (getvar "textsize")))
            (cadr pt)
            (caddr pt)
         )
)
(cons 40 (getvar "textsize"))
(cons 1 elevation)
(cons 50 (* (atof rotation) (/ 3.14 180)))
(cons 41 1.0)
(cons 51 0)
(cons 7 style)
(cons 71 0)
(cons 72 0)
(cons 100 "AcDbText")
(cons 73 0)
 )
)
(ENTMAKE
 (list (cons 0 "TEXT")
     (cons 100 "AcDbEntity")
     (cons 67 0)
     (cons 410 "Model")
     (cons 8 (strcat layer "_des"))
     (cons 100 "AcDbText")
     (cons 10
            (list (+ (car pt) (* 1.7143 (getvar "textsize")))
                (- (cadr pt) (* 1.7143 (getvar "textsize")))
                (caddr pt)
             )
     )
     (cons 40 (getvar "textsize"))
     (cons 1 description)
     (cons 50 (* (atof rotation) (/ 3.14 180)))
     (cons 41 1.0)
     (cons 51 0)
     (cons 7 style)
```

```
        (cons 71 0)
        (cons 72 0)
        (cons 100 "AcDbText")
        (cons 73 0)
     )
   )
   (ENTMAKE (list (cons 0 "endblk")))
   (ENTMAKE (list (cons 0 "INSERT")
                (cons 2 "point_block")
                (cons 10 pt)
        )
   )
   (loc)
 )
(defun loc ()
  (setq
    ent1 (tblsearch "block"
                (cdr (assoc 2 (setq entkev (entget (entlast)))))
        )
  )
  (setq rnam (cdr (assoc -1 entkev)))
  (setq namkev (assoc 2 entkev))
  (setq ent1 (subst (cons 2 "*Unnn") (assoc 2 ent1) ent1))
  (setq ent1 (subst (cons 70 1) (assoc 70 ent1) ent1))
  (entmake ent1)
  (setq ent1 (entget (cdr (assoc -2 ent1))))
  (entmake ent1)
  (while (/= ent1 nil)
    (setq ent1 (entnext (cdr (assoc -1 ent1))))
    (if      (/= ent1 nil)
     (progn
      (setq ent1 (entget ent1))
      (entmake ent1)
     )
   )
  )
  (setq l (entmake '((0 . "endblk"))))
  (setq entkev (subst (cons 2 l) (assoc 2 entkev) entkev))
  (entmod entkev)
  (princ)
)
```

```
//*****************************************************************
//
// Dialog Box Name : Spot.dcl
//
// Purpose : Provide the user interface for the spot elevation
//                  program.
//
//*****************************************************************
spot : dialog {
 label = "Spot Elevation Program";
 //value = "This is an example of the value attribute";
    : boxed_row {
    fixed_width = true;
    children_fixed_width = true;
    label = "Text Layer";
    : column {
    key = "c1";
            : radio_button {
            key = "create_layer";
    label = "Create layer";
    }
    : radio_button {
    key = "current_layer";
    label = "Place Text on Current Layer";
    }
    :radio_button {
    key = "select_layer";
    label = "Select Layer";
    }
    }
    : column {
    : edit_box {
    label = "Create Text layer";
    key = "create_layer_box";
    edit_width = 15;
    }
    : text {
    key = "field_layer";
    }
    : popup_list {
            label = "Select Existing Layer";
```

```
key = "layer_pop_list";
width = 35;
     }
  }
  }
 : boxed_row {
fixed_width = true;
children_fixed_width = true;
label = "Text Style";
: column {
key = "c2";
: radio_button {
key = "current_style";
label = "Use Current Text Style";
}
:radio_button {
key = "select_style";
label = "Select Style";
}
}
: column {
: text {
key = "field_style";
}
: popup_list {
       label = "Select Existing Layer";
key = "style_pop_list";
width = 35;
     }
  }
  }
 : row {
 : boxed_row {
fixed_width = true;
children_fixed_width = true;
label = "Mark Point";
: column {
key = "c3";
       : radio_button {
       key = "donut_marker";
```

```
label = "Donut";
}
: radio_button {
key = "cross_marker";
label = "Cross Hairs";
}
:radio_button {
key = "circle_marker";
label = "Circle_marker";
}
:radio_button {
key = "no_marker";
label = "No Marker";
}
}
: image {
key = "test";
color = 0;
width = 25;
height = 5;
}
}
: boxed_column {
label = "Point Description";
: edit_box {
label = "Description";
key = "description";
}
: edit_box {
label = "Point Number";
key = "point_number";
}
: edit_box {
label = "Elevation";
key = "elevation";
}
: edit_box {
label = "Rotation in Degrees";
key = "rotation";
}
}
```

```
    }
    : boxed_row {
     label = "Current Point Selected";
     fixed_width = true;
     :text {
      key = "field_position";
      width = 35;
     }
     : retirement_button {
      label = "Select Point";
      key = "s_point";
     }
    }
   is_default = true;
   ok_cancel;
}
```

SUMMARY

Before a dialog box can be used by an AutoLISP application, several steps must be completed within the application that employs the dialog box. The dialog box definition must be loaded into memory, assigned an identification number, activated, its tiles initialized, images loaded, any action associated with the dialog box specified, verification of the data entered and the dialog session ended. Loading a dialog box into memory is accomplished by the LOAD_DIALOG function. In addition to loading the dialog box into memory, this function also assigns the dialog box an identification number. The identification number assigned by the LOAD_DIALOG function is used whenever subsequent function calls are made to the dialog box. After a dialog box has been loaded into memory and an identification number is assigned to it, the dialog box must be initialized. This is accomplished by the NEW_DIALOG function. Once the dialog box has been initialized, the dialog box becomes active and will remain active until either an action expression or callback function executes a DONE_DIALOG function. Active dialog boxes can be forced to display with the START_DIALOG function. At this point the dialog box does not have any arguments assigned to it; however, if the dialog box is activated, the dialog box will return the optional status argument that was assigned by the DONE_DIALOG function.

When a dialog box is displayed, all control is shifted from the AutoCAD graphics screen to the dialog box. The dialog box remains in control until the dialog box has been dismissed by the DONE_DIALOG function. When a DONE_DIALOG expression is used, it must be contained within either an action expression or a callback function. An action expression associates an AutoLISP expression with a specific tile defined in the dialog box. The AutoLISP expression defines the action that is to be taken once the tile has been selected. Information relating to how the user has selected or modified a tile's content is known as a callback function. Because dialog boxes can be nested within other dialog boxes, sometimes it becomes necessary to terminate all dialog boxes displayed with the TERM_DIALOG function. The TERM_DIALOG function terminates all current dialog boxes, active or not. Once a dialog box has been terminated, it is often a good idea to unload that dialog box's definition from memory to prevent possible conflicts with other dialog boxes. This is accomplished by the UNLOAD_DIALOG function.

Setting the initial value of a tile contained within a dialog box can be accomplished through either the DCL attribute value or the AutoLISP function SET_TILE. If the value is set using the DCL attribute value, then the value set is considered static, meaning that the value cannot be changed. When the value assigned to a tile needs to change from runtime to runtime, then the AutoLISP SET_TILE function must be used.

Once a dialog box has been terminated, the information entered by the user must be extracted before it can be used by the AutoLISP application that called the dialog box. In AutoLISP this is accomplished by the GET_TILE function.

To provide the user with options that change from runtime to runtime, then a popup list or list box must be employed. The tiles associated with the features are defined in the DCL source file through the list_box and popup_list tiles. Defining the tiles in the DCL source file is only the beginning; the list associated with these tiles must be populated. To populate a list box or popup list, the AutoLISP program must first start a list, then append the entries to that list, and finally close the list.

The way images are incorporated into dialog boxes is similar to the way they are used in popup lists. The creation of the image must be started, the image is defined, and then the creation process is stopped. The processes used to accomplish this are done in the AutoLISP program that calls the dialog box.

REVIEW QUESTIONS

1. Define the following terms
   Action Expression
   Callback Function
   Static
   Dynamic

2. What are the steps required to use a dialog box in an AutoLISP application?

3. Why is it sometimes necessary to terminate all active and non-active dialog boxes?

4. How are dialog boxes terminated in AutoLISP?

5. What AutoLISP function is used to force a dialog box to display?

6. True or False: When a dialog box is displayed, control is split between the AutoCAD graphics screen and the dialog box. (If the answer is false, explain why.)

7. What is the importance of the identification number assigned to the dialog box's definition?

8. Why is it sometimes necessary to unload a dialog box's definition from memory?

9. True or False: Calls made to the SET_TILE function can be done before calls made to the NEW_DIALOG function. (If the answer is false, explain why.)

10. List the steps used to incorporate a popup list box into a dialog box.

11. List the steps used to incorporate images into a dialog box.

12. What is the purpose of a callback function?

13. What is the purpose of an action expression?

14. How are action expressions defined?

15. List the two methods used to set the value of a tile in a dialog box. What are the advantages and disadvantages of each?

16. True or False: The DONE_DIALOG function can be called from the AutoCAD command prompt. (If the answer is false, explain why.)

17. True or False: All dialog control functions can be called from the AutoCAD command prompt. (If the answer is false, explain why.)

18. What is an AutoCAD slide?

19. What two methods can be used to disable a tile in a dialog box? What are their advantages and disadvantages?

20. How is a dialog box reissued once it has been dismissed?

CHAPTER 8

Introduction to Visual LISP

OBJECTIVES

Upon completion of this chapter the reader will be able to do the following:

- Distinguish between AutoLISP and Visual LISP
- Define the terms IDE and API
- Launch the Visual LISP IDE from the AutoCAD command prompt or by selecting it from the proper AutoCAD menu option
- Identify the main components of the Visual LISP IDE
- Identify the nine main menu categories in Visual LISP and describe the purpose of each
- Identify the five main categories of toolbars in Visual LISP and describe the purpose of each
- Describe the difference between the Visual LISP console window and the Visual LISP text editor
- Understand the purpose of the trace output window
- Open, close, and save files in Visual LISP
- Identify the different file types Visual LISP has the ability to maintain
- Understand the purpose of the syntax coloring, text formatting, parentheses matching, and multiple file searching features
- Use the Visual LISP syntax checker to check an AutoLISP application for common errors
- Use bookmarks to navigate the Visual LISP text editor
- Advance the cursor in Visual LISP with the aid of bookmarks
- Use the Visual LISP word complete feature
- Use the Apropos feature to complete a word

- Use the Apropos feature to search the symbol table
- Use the results returned by the Apropos feature
- Understand the importance of the watch and inspection sindows
- Understand what the stack is used for in Visual LISP
- Understand the importance of a Visual LISP project
- Understand the difference between a FAS and a VLX application

KEY WORDS AND AUTOLISP FUNCTIONS

Application Program Interface

Apropos

Bookmarks

Console Window

Error Stack

FAS

Formatter

Inspection Window

Integrated Development Environment

Multiple File Search

Parenthesis Matching

PRJ

Stack

Symbol Services

Syntax Checking

Syntax Coloring

Text Editor

Text Formatting

Trace Output Window

Visual LISP Project

Visual LISP Text Editor

VLIDE

VLISP

VLX

Watch Window

INTRODUCTION TO VISUAL LISP

By now it should be clear why Autodesk chose AutoLISP as its initial *A*pplication *P*rogramming *I*nterface (API) to be used for the customization of AutoCAD. As demonstrated in the previous chapters, AutoLISP is a powerful programming language that is well suited for the AutoCAD design environment. It allows developers to create programs that are targeted to solve a wide range of engineering and architectural design problems, tasks that could not be accomplished using ordinary AutoCAD commands. Moreover, with very little effort, a skilled user can develop short routines that alleviate some of the tedious drafting tasks during the various phases of any project. However, with all of its powerful assets and mind-boggling possibilities, AutoLISP still falls short in some areas. In the visual age of computers and computer programming, AutoLISP programs are still developed using many of the techniques pioneered by earlier versions of the well-established programming languages used today. In AutoLISP, programs are created with an independent word processor, and then tested and debugged using AutoCAD. While at first glance this might not appear to be a very big deal, compared to other programming languages it can be a rather time-consuming process when debugging large complex programs. While the programming language itself has undergone dramatic development by Autodesk since its release, very little work had been done to address some of these issues until the release of Visual LISP in 1999. Visual LISP solves many of these problems by providing the developer with a fully integrated text editor, compiler, debugger, and numerous tools (collectively referred to as an *Integrated Development Environment* or *IDE*) designed to increase productivity. Originally released as an add-on package for AutoCAD Release 14.01 and now fully integrated within AutoCAD 2000, Visual LISP was not designed as a replacement for AutoLISP but rather as the next evolutionary phase of the AutoLISP programming language. While numerous enhancements have been made to Visual LISP, it is essential that developers have a good understanding of AutoLISP before they can effectively use Visual LISP.

ENHANCEMENTS MADE TO AUTOLISP

One of Autodesk's objectives when designing Visual LISP was to help relieve the burden of developing an AutoLISP application by providing the developer with a streamlined IDE, allowing the developer to spend more time writing applications and less time debugging. This required borrowing concepts and features that have been employed by other IDEs for years. To compete with the other languages currently available to customize AutoCAD, the new generation of AutoLISP would have to include syntax checkers, file compilers, source code debuggers, text file editors, AutoLISP formatters, comprehensive inspection and watch tools, context-sensitive help, and a project manager. In addition, Autodesk incorporated the concepts of *Object Oriented Programming* (OOP), and ActiveX and introduced reactors, thereby releasing the most powerful version of AutoLISP ever.

WORKING WITH VISUAL LISP

As mentioned earlier, Visual LISP is a fully integrated development package that allows its user to create, test, and debug AutoLISP applications. The Visual LISP development environment contains its own windows to develop programs. However, the Visual LISP IDE still relies on the AutoCAD engine and cannot be deployed independently of AutoCAD. This is because AutoLISP programs are designed to interact with, and not separately from, AutoCAD.

LAUNCHING THE VISUAL LISP IDE

Visual LISP can be launched from either the AutoCAD command prompt or from the AutoCAD menu. To activate Visual LISP from the AutoCAD command prompt, the AutoCAD command VLIDE (for version 14.01 and 2000) or VLISP (version 2000 only) are used. To launch Visual LISP from the AutoCAD menu, choose Tools|AutoLISP|Visual LISP Editor (see Figure 8–1).

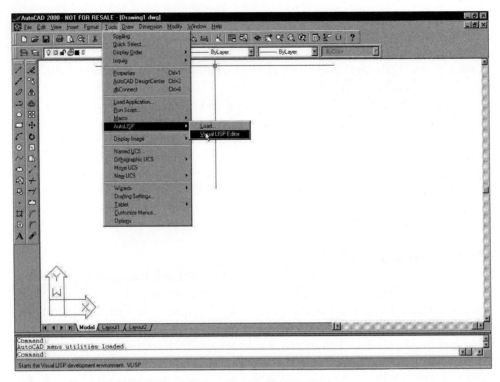

Figure 8–1 *Launching Visual LISP from the AutoCAD menu*

NAVIGATING THE VISUAL LISP IDE MENUS

After Visual LISP has been activated, the Visual LISP user interface is displayed (see Figure 8–2). The interface is made up of menus, toolbars, a main window, a console window, a trace window, text editor, and a status bar.

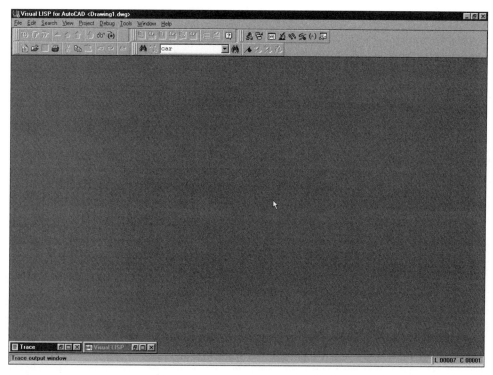

Figure 8–2   *Visual LISP interface*

Each area of the user interface serves a specific purpose. The menu region allows the developer to select Visual LISP commands from a menu system, which is divided into nine main categories: Files, Edit, Search, View, Project, Debug, Windows, and Help. Once a menu item has been highlighted, Visual LISP displays a brief description of that item in the status bar (see Figure 8–3). Table 8–1 provides a brief description of each pull down menu category and Figures 8–4 through 8–12 show the options for each menu.

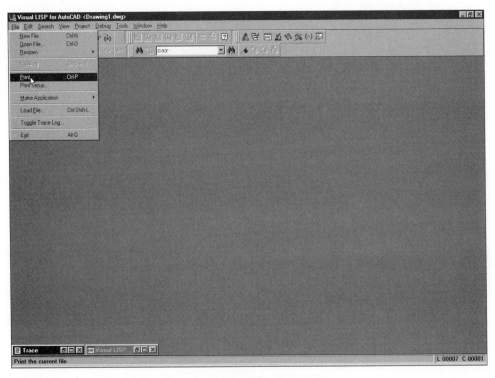

Figure 8–3 *Description of menu item displayed in the status bar*

Table 8–1 Visual LISP Menu Categories Summary

Menu Category	Description
File	Create a new AutoLISP program file for editing, open an existing file, save changes to program files, build Visual LISP application files, and print program files.
Edit	Copy and paste text, undo the last change you made to text (or undo the last command entered in the console window), select text in the Visual LISP editor or console windows, match parentheses in expressions, and redisplay previous commands entered in the console window.
Search	Find and replace text strings, set bookmarks, and navigate among bookmarked text.
View	Find and display the value of variables and symbols in your AutoLISP code.
Project	Work with and compile programs.
Debug	Set and remove breakpoints in your program and step through program execution one expression at a time. Then check the state of variables and the results of expressions.

Menu Category	Description
Tools	Set Visual LISP options for text formatting and various environment options, such as the placement of windows and toolbars.
Windows	Organize the windows currently displayed in your Visual LISP session or activate another Visual LISP or AutoCAD window.
Help	Display online help.

Courtesy of the AutoLISP Developers Guide

Figure 8–4 *Visual LISP File menu*

Figure 8–5 *Visual LISP Edit menu*

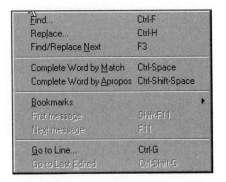

Figure 8–6 *Visual LISP Search menu*

Figure 8–7 *Visual LISP View menu*

Figure 8–8 *Visual LISP Project menu*

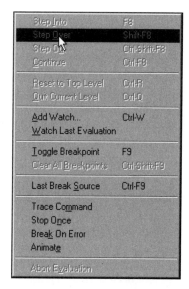

Figure 8–9   *Visual LISP Debug menu*

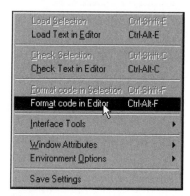

Figure 8–10   *Visual LISP Tools menu*

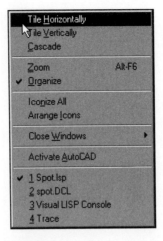

Figure 8–11 *Visual LISP Windows menu*

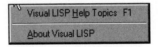

Figure 8–12 *Visual LISP Help menu*

NAVIGATING THE VISUAL LISP IDE TOOLBARS

The toolbar region of the Visual LISP user interface is design to accomplish the same task as the menus by allowing the developer to quickly access frequently used Visual LISP commands. Although many of the Visual LISP commands can be activated through the toolbar buttons, some commands can only be executed from the menus. Toolbars, like menus, are also divided into categories. They are Standard, Search, Tools, Debug, and View (see Figures 8–13 through 8–17).

In Visual LISP, when the user leaves the mouse pointer positioned over a tool button for a few seconds, a tool tip is displayed (see Figure 8–18). Positioning the mouse over a toolbar button also causes a description to display in the status bar (see Figure 8–19). Table 8–2 provides a brief description of each toolbar category.

Table 8–2　Visual LISP Toolbar Categories Summary

Toolbar Category	Description
Standard	New File, Open File, Save File, Print, Cut, Copy, Paste, Undo, Redo, Complete Word
Search	Find, Replace, Find Toolbar String, Toggle Bookmark, Next Bookmark, Previous Bookmark, Clear all Bookmarks
Tools	Load Active Edit Window, Load Section, Check Edit Window, Check Section, Format Edit Window, Format Section, Comment Block, Uncomment Block, Help
Debug	Step Into, Step Over, Step Out, Continue, Quit, Reset, Toggle Break Point, Add Watch, Last Break
View	Activate AutoCAD, Select Window, Visual LISP Console Window, Inspect, Trace, Symbol Service, Apropos, Watch Window

Figure 8–13　*Visual LISP Standard toolbar*

Figure 8–14　*Visual LISP Search toolbar*

Figure 8–15　*Visual LISP Tools toolbar*

Figure 8–16　*Visual LISP Debug toolbar*

Figure 8–17 *Visual LISP View toolbar*

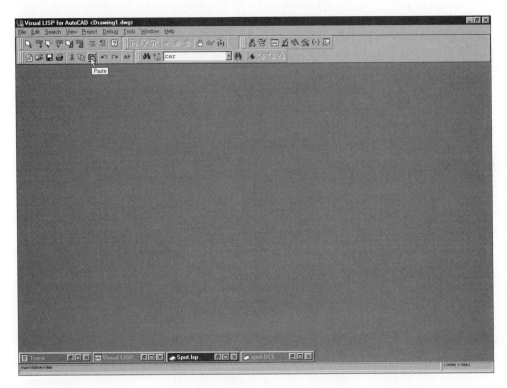

Figure 8–18 *Tool tip*

Figure 8–19   *Description of tool button displayed in the status bar*

THE VISUAL LISP CONSOLE WINDOW

The console window is a separate window that is equipped with scroll bars and buttons to minimize, restore, and close windows (see Figure 8–20). It has many built-in benefits that are designed to aid the developer in the construction of AutoLISP programs. Its primary purpose is to allow the developer to test and evaluate AutoLISP expressions by entering them at the console command prompt much the way they would be entered at the AutoCAD command prompt.

Figure 8–20 *Visual LISP console window*

AutoLISP expressions that are entered at the console command prompt are instantly evaluated and their results displayed (see Figure 8–21). Multi-line AutoLISP expressions can also be entered at the console window's command prompt if CTRL+ENTER are pressed to continue to the next line (see Figure 8–22). This allows for multiple expressions to be evaluated at one time.

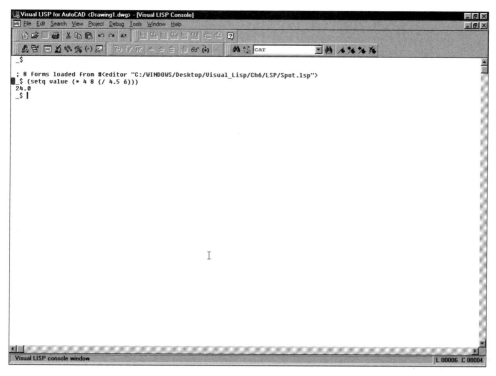

Figure 8–21 *Results from a single expression displayed*

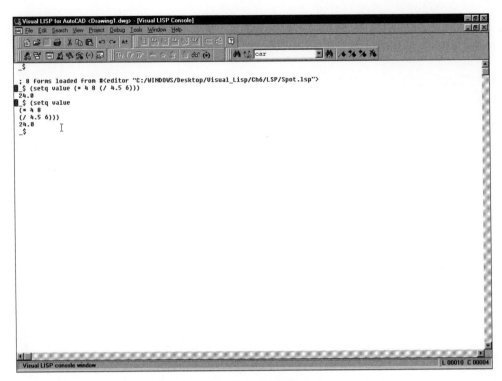

Figure 8–22 *Results from multi-line expression displayed*

To verify the value assigned to a variable at the console window's command prompt, enter the variable name without the exclamation point (see Figure 8–23). To launch AutoLISP programs from the console window's command prompt, enter the user-defined function name as it appears in the DEFUN function (see Figure 8–24).

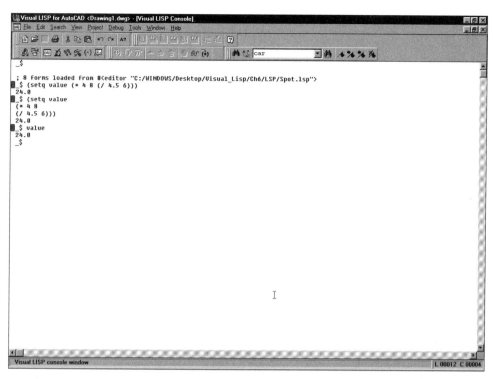

Figure 8–23 *Value of a variable returned*

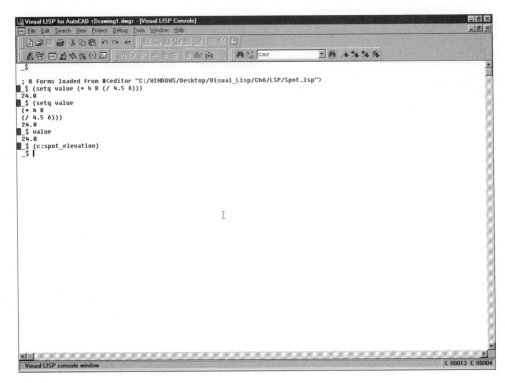

Figure 8–24 *Launching a program*

To retrieve previous expressions and commands from the console history, press TAB (see Figure 8–25). Commands and expressions are toggled in the reverse order from which they were originally entered. However, to switch the direction in which the commands are toggled, press SHIFT+TAB (see Figure 8–26).

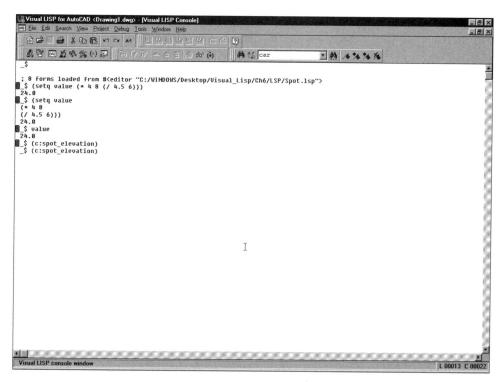

Figure 8–25 *Scrolling console history*

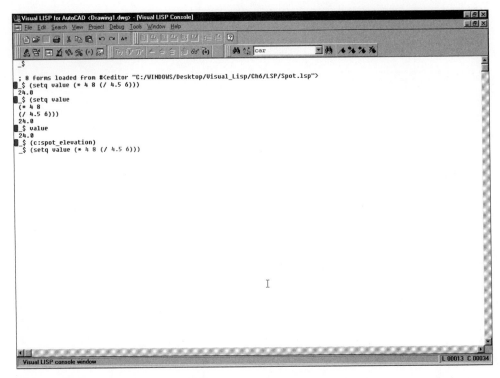

Figure 8–26 *Reversing console history*

Once an expression has been entered at the console window's command prompt, to clear it, press ESC. However, if SHIFT+ESC are pressed, the text is not cleared or evaluated and a new console prompt is displayed (see Figure 8–27).

Figure 8–27 *New console command line started*

The console window is also where Visual LISP displays diagnostic messages returned from the evaluation of AutoLISP programs and expressions (see Figure 8–28). Text can be transferred to and from the console window by standard Windows' OLE commands, Copy, Cut and Paste (see Figure 8–29).

284

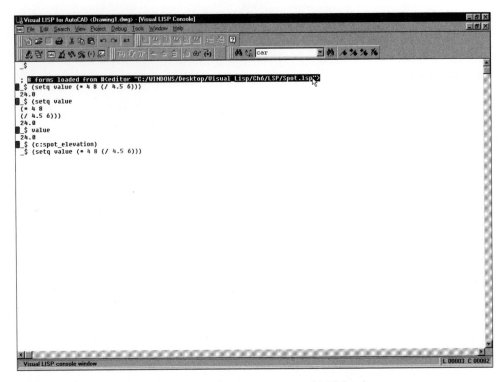

Figure 8–28 *Console window with a diagnostic message highlighted*

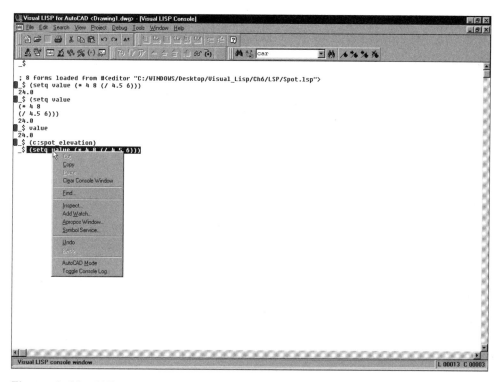

Figure 8–29 *OLE menu*

The latest release of AutoCAD exposes the end user to a multiple document interface environment, also known as MDI. (MDI is covered in more detail in Chapter 11.) This new environment allows the user to open multiple documents in a single AutoCAD session. In AutoCAD, when multiple documents are opened, each document has its own text window; in Visual LISP only one console window is opened and all documents share the same console window. All commands entered at the Visual LISP console window's command prompt affect only the current active AutoCAD document.

 Note: Switching from the Visual LISP console window to the AutoCAD graphics screen will clear any expressions or commands that have not been completely entered.

THE VISUAL LISP TRACE WINDOW

The trace window (also called the trace output window) is specifically designed to display the output of the trace functions. When Visual LISP is first activated, a console window and a trace window are opened. On startup, the trace window displays information regarding the current release of Visual LISP as well as any information concerning problems or errors that Visual LISP may have encountered (see Figure 8–30).

Figure 8–30 *Visual LISP trace window*

THE VISUAL LISP TEXT EDITOR
The crown jewel of the Visual LISP integrated development environment is the text editor (see Figure 8–31). Unlike Notepad or WordPad, the Visual LISP text editor is a collection of sophisticated programming tools integrated into a word-processing application specifically designed for the creation of AutoLISP programs. Some of the major tools integrated into the Visual LISP text editor are syntax coloring, text formatting, parentheses matching, AutoLISP expression execution, and multiple file searching.

Figure 8–31   *Visual LISP text editor*

Syntax Coloring

One of the more difficult aspects of programming is identifying the different components of a computer program. Distinguishing function calls from variable names can be a rather time-consuming task. This has been somewhat alleviated in Visual LISP with the introduction of syntax coloring. By default, Visual LISP displays functions and protected symbols in blue, strings in magenta, integers in green, real numbers in teal, comments in magenta (on a gray background), parentheses in red, and unrecognized items in black. This is a valuable tool for the developer: Different portions of the programming language are easily distinguishable from one another. Misspellings and typing errors are quickly detected. Meanwhile, the color scheme displayed by Visual LISP has no effect on the program. The color scheme used by Visual LISP can be changed or modified to meet the developer's specifications at any time by choosing Tools|Window Attributes|Configure Current (see Figures 8–32 and 8–33).

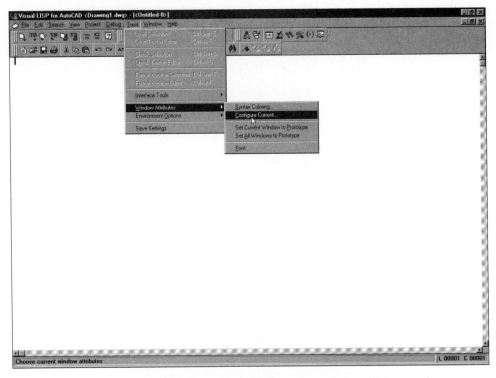

Figure 8–32 *Menu options for changing the default syntax coloring scheme*

Figure 8–33 *Window attributes dialog box*

By default, when the Visual LISP text editor is first activated, it is set for AutoLISP syntax coloring style, but if the developer is writing in a language other than AutoLISP, then the coloring style can also be changed to reflect the language used. To change the coloring style in Visual LISP, choose Tools|Window Attributes|Syntax Coloring (see Figures 8–34 and 8–35).

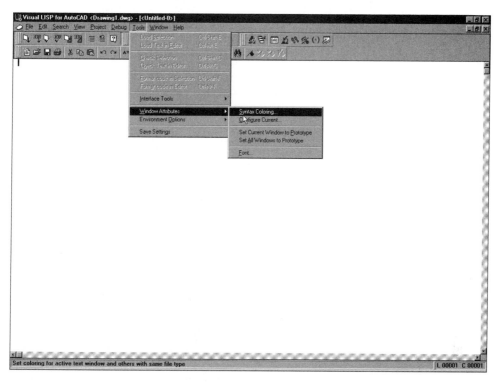

Figure 8–34 *Menu options for changing default syntax coloring*

Figure 8–35 *Color Style dialog box*

Note: The console window treats all text entered as AutoLISP code

Text Formatting

Recall from Chapter 1 that text formatting and spacing are solely for cosmetic purposes and, like syntax coloring, they have no effect on the performance of the program. It simply makes the source code easier to read and follow. In Visual LISP, formatting is performed automatically as the code is entered in the text editor. The entire program or just a portion can be reformatted at any time by invoking the formatter. This feature is useful for reformatting text that has been copied and pasted from another source or reformatting AutoLISP programs that were created outside Visual LISP. To reformat a portion of a program, the section to be formatted is first selected, and then the formatter is invoked by either pressing CTRL+SHIFT+F, selecting the ▇ (located on the Tools toolbar), or selecting the Format Code in Selection option from the Tools menu (see Figure 8–36). The entire source code file can be reformatted by either pressing CTRL+ALT+F, selecting the ▇ (located on the Tools toolbar), or selecting the Format Code in Editor option from the Tools menu (see Figure 8–37).

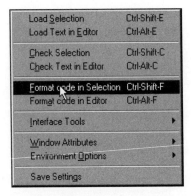

Figure 8–36 *Activating the formatter through the menu to reformat a portion of an AutoLISP program*

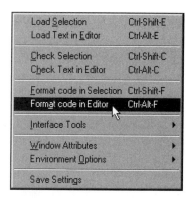

Figure 8–37 *Activating the formatter through the menu to reformat an entire AutoLISP program*

The Visual LISP formatter follows the rules associated with four different format styles: PLane, WIde, NArrow and COlumn.

- The PLane style places all arguments on the same line separated by a mere space. This style is applied when the last character of an expression does not exceed the right margin, the value of the approximate line length environment option is greater than the expression printing length, or the expression is free of embedded comments with newline characters.

- The WIde style arranges the arguments so that the first argument is contained on the same line as the function. Any remaining arguments associated with the function are placed in a column directly below the first argument. This style is applied when the PLane style cannot be employed and the first element is a symbol whose length is less than the Maximum Wide Style Car Length environment option.

- The NArrow style places the first argument on the line following the function with all remaining arguments arranged in a column positioned below the first argument. This style is applied when the PLane and WIde styles cannot be used. This style is also applied to all PROGN expressions.

- The COlumn style formats the code so that all elements are placed in a column. This style is used for quoted lists and all CONS expressions. The formatter chooses the correct style to apply according to rules established by the format options dialog box (see Figure 8–38). This dialog box can be activated by choosing Tools|Environment Options|Visual LISP Format Options (see Figure 8–39).

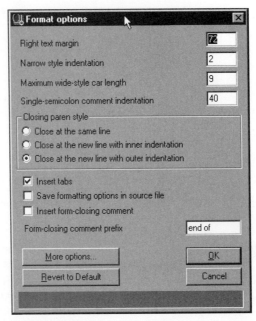

Figure 8–38 *Format Options dialog box*

Figure 8–39 *Selecting Visual LISP Format Options from the menu*

Note: The formatter can format only valid AutoLISP expressions and comments. If an invalid expression is encountered, then the formatter will issue an error message. If the formatter encounters unbalanced parentheses, then an error message is issued alerting the developer of this condition. In addition to the error message, the developer is given the opportunity to either fix the error condition or allow the formatter to automatically fix the error. If the developer chooses to allow the formatter to automatically correct the error, then it is very unlikely that the formatter will place the missing parenthesis in the correct location. For this reason, it is recommended that this option not be selected if this error should occur (see Figure 8–40).

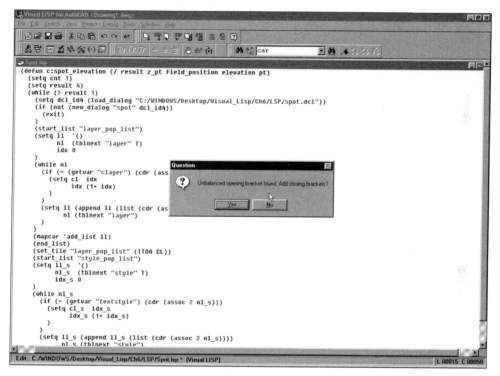

Figure 8–40 *Format error*

Parenthesis Matching

One of the hardest aspects of writing an AutoLISP program is keeping track of the parentheses used to group expressions and functions together. The parenthesis matching command in Visual LISP facilitates this task. Position the cursor in front of an opening parenthesis and press CTRL+] to have Visual LISP automatically find the closing parenthesis that goes with that expression. Reverse the process by placing the cursor just past the closing parenthesis and press CTRL+[to locate its corresponding opening parenthesis.

Executing AutoLISP Expressions

An AutoLISP program, or a portion of an AutoLISP program, can be made available to the command interpreter from the text window and then executed from the console window 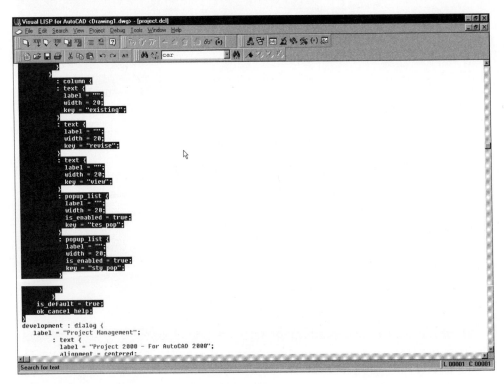 toolbar button (from the Tools toolbar) to run the entire program or by selecting the toolbar button (from the Tools toolbars). If the entire program is made available to the command interpreter, the application can be executed by entering the DEFUN expression defining the program at the console window (see Figure 8–41). If a portion of the program is made available to the command interpreter, the expression(s) are evaluated and the result(s) displayed in the console window (see Figure 8–42).

Figure 8–41 *Loading a program from the text editor*

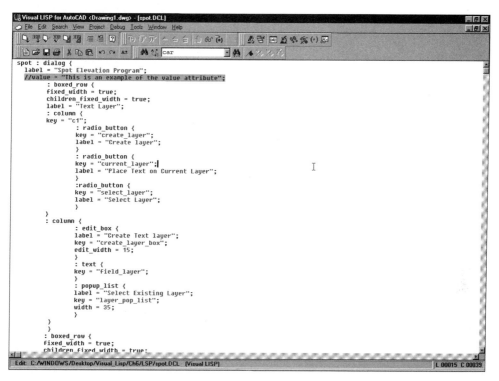

Figure 8–42 *Executing a portion of a program*

Multiple File Searching

Often programmers will build a library of specific functions that can be used as building blocks when applications are created. As the programmer develops more applications and their libraries become more densely populated, it becomes more difficult to locate a particular function. Also, as different versions of AutoLISP become available, more functions are added, and often older functions are either deleted or modified, making it necessary to update applications that contain obsolete functions. The task of searching a library for a particular function that contains a particular expression or searching a program for an obsolete function can be time consuming and tedious. Autodesk has addressed this by adding an extremely powerful Find command to the Visual LISP IDE. Unlike Find commands in other applications, the Find command in Visual LISP allows the developer to search either a portion of the current file, the entire file, a specified project, or a directory for a key word or AutoLISP expression. Activate the Find command by either pressing CTRL+F or choosing the Find option from the Search menu (see Figures 8–43 and 8–44).

Visual LISP for AutoCAD <Drawing1.dwg> - [spot.DCL]

File Edit Search View Project Debug Tools Window Help

```
                Find                Ctrl-F
                Replace             Ctrl-H
                Find/Replace Next   F3

                Complete Word by Match     Ctrl-Space
                Complete Word by Apropos   Ctrl-Shift-Space

                Bookmarks                            ▶
                First message       Shift-F11
                Next message        F11

                Go to Line...       Ctrl-G
                Go to Last Edited   Ctrl-Shift-G
```

```
spot : di
  label =
  //valu                        lue attribute";

                : radio_button {
                key = "create_layer";
                label = "Create layer";
                }
                : radio_button {
                key = "current_layer";|
                label = "Place Text on Current Layer";
                }
                :radio_button {
                key = "select_layer";
                label = "Select Layer";
                }
        }
        : column {
                : edit_box {
                label = "Create Text layer";
                key = "create_layer_box";
                edit_width = 15;
                }
                : text {
                key = "field_layer";
                }
                : popup_list {
                label = "Select Existing Layer";
                key = "layer_pop_list";
                width = 35;
                }
        }
        }
        : boxed_row {
        fixed_width = true;
        children_fixed_width = true;
```

Search for text   L 00015 C 00039

Figure 8–43 *Accessing the Find command through the Search menu*

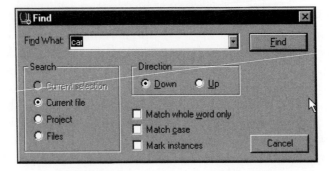

Figure 8–44 *Find command's dialog box*

Checking AutoLISP Code for Syntax

Before an application is deployed for testing, the developer can check the application for common errors while still working in the Visual LISP text editor. To do this the developer can either select the Check Editor Window toolbar button ▨ or the

Check Select toolbar button 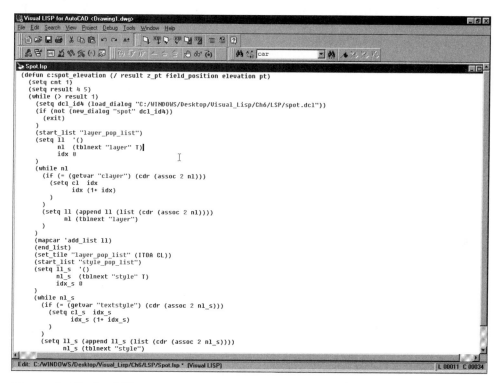 located on the Tools toolbar. These commands can also be activated from the keyboard by pressing either CTRL+SHIFT+C (to check only a portion of a program) or CTRL+ALT+C (to check the entire program). The common errors that the syntax check searches for are incorrect number of arguments supplied to a function, invalid variable names passed to a function, and incorrect syntax for special form functions. The checker does not check for runtime errors, for example, supplying a real number to a function that requires integer input. If an error is detected, then a new build window is constructed containing the error message and the expression where the error occurred. The portion of the AutoLISP program below contains an error in the expression (setq result 4 5) (see Figure 8–45). When the syntax checker evaluates this program, the error is detected and the expression containing the error is displayed in a new build window (see Figure 8–46).

Figure 8–45 *AutoLISP program before the syntax checker is activated*

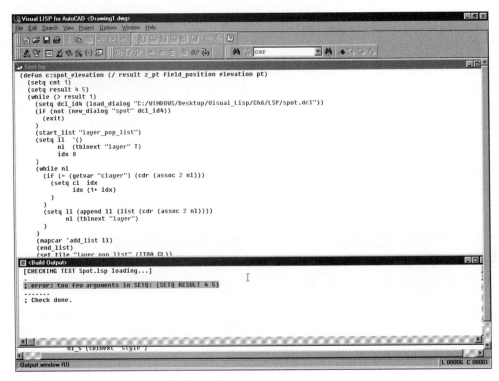

Figure 8–46 *Build window displaying the error*

OPENING, CLOSING, CREATING, AND SAVING FILES

The Visual LISP IDE has the ability to open, edit, create, and save four types of source code files: AutoLISP, DCL, SQL (*S*tructured *Q*uery *L*anguage, SQL was developed by IBM in the 1970s for creating, updating and querying relational database management systems), and C/C++. While each file type is different in its structure and the functions they use to create applications, the process for opening, saving, creating (a new file) and closing is the same. To open an existing file in Visual LISP, the Open command must be used (see Figure 8–47). Activate the Open command by either pressing CTRL+O, selecting the Open File menu option from the Files menu (see Figure 8–48), or selecting the Open File 📄 toolbar button from the Standard toolbar.

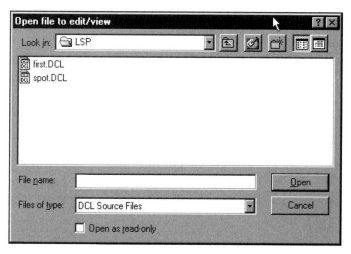

Figure 8–47 *Visual LISP Open command dialog box*

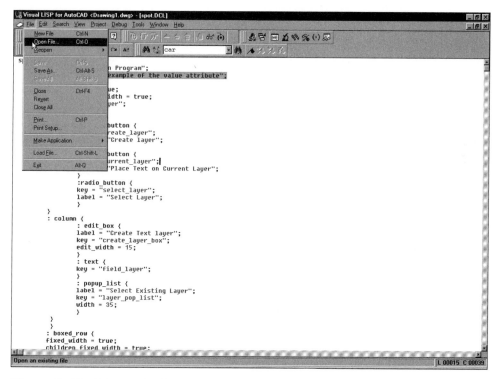

Figure 8–48 *Selecting the Open File command from the File menu*

To create a new file, the New command must be used. Upon executing this command, Visual LISP starts a new text editor window for the new source code. Activate the command by either pressing CTRL+N, selecting the New File menu option from the File menu (see Figure 8–49), or selecting the New File 📄 toolbar button from the Standard toolbar.

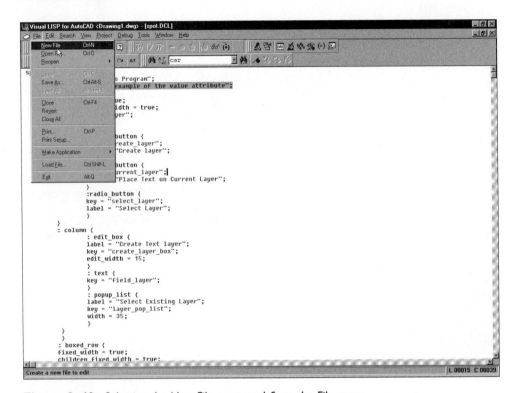

Figure 8–49 *Selecting the New File command from the File menu*

Regardless of the file type created in Visual LISP, the act of saving a file is the same. A file can be saved through either the Save or Save-As command. The Save command automatically saves a file without prompting the developer for the file name if the file has been saved at least once since it was created, while the Save-As command always prompts the developer for the file name (see Figure 8–50). Execute both the Save and Save-As commands by selecting them from the File menu, (see Figure 8–51) or by selecting the Save File 💾 toolbar button from the Standard toolbar.

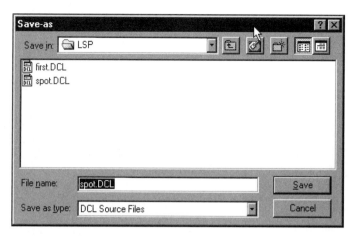

Figure 8–50 *Visual LISP Save-As command dialog box*

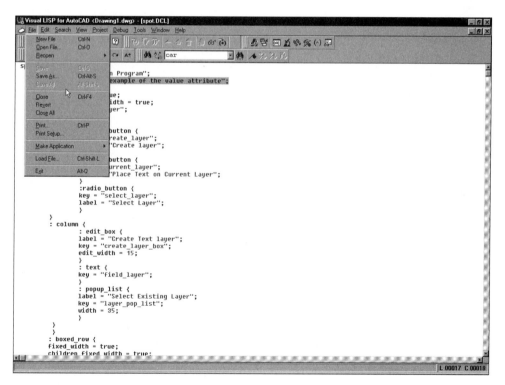

Figure 8–51 *Selecting the Save and Save-As commands from the File menu*

Finally, in Visual LISP a file can be closed by either selecting the Close menu option from the File menu (see Figure 8–52) or by selecting the Close Window toolbar button located in the upper right corner of the text editor window (see Figure 8–53).

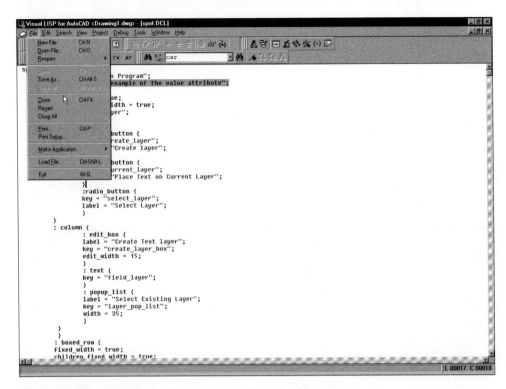

Figure 8–52 *Selecting the Close command from the File menu*

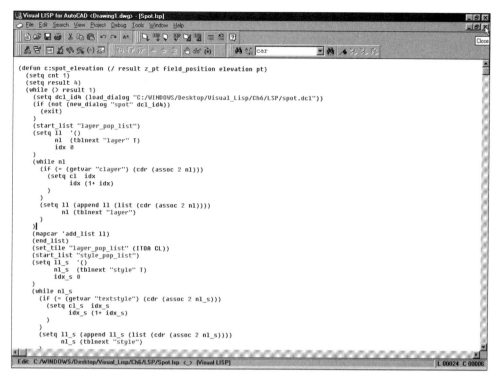

Figure 8–53 *Close toolbar button*

DEVELOPING APPLICATIONS WITH VISUAL LISP

Among the many features previously mentioned, Visual LISP offers the developer several tools that are specifically designed for the creation, testing, and implementation of AutoLISP applications. Using these tools can speed up the development of an AutoLISP application, allowing the developer to concentrate on the actual source code.

MARKING SECTIONS OF THE SOURCE CODE WITH BOOKMARKS

Anyone who has worked with the World Wide Web is familiar with the concept of bookmarks. Bookmarks are non-printable anchors that allow the developer to review a particular section of an application without having to scroll through all the source code. Visual LISP currently allows the developer to set up to 32 bookmarks in an application's source code file. Each application maintains its own set of bookmarks. If the developer tries to set more than 32 bookmarks in an application, then Visual LISP responds by deleting the oldest bookmark.

Setting, Deleting, and Navigating with Bookmarks

Before a bookmark can be set in Visual LISP, the cursor must be positioned at the location where the bookmark is to be placed. Once the cursor has been positioned, a bookmark can be inserted by either pressing ALT+PERIOD, selecting the Toggle Bookmarks option from the Search menu (see Figure 8–54), or by selecting the 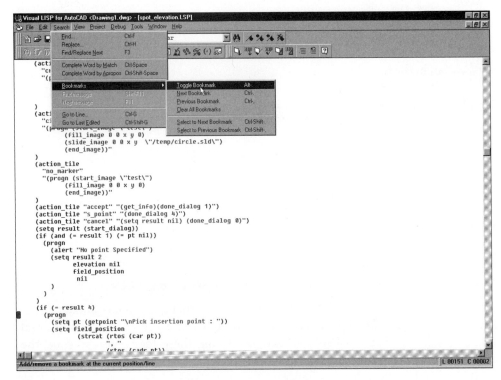 toolbar button from the Search toolbar. To delete a bookmark, the cursor must be positioned on the same line as the bookmark, and the same menu options that were used to place the bookmark are selected again.

Figure 8–54 *Setting a bookmark from the Visual LISP menu*

Once a bookmark has been placed in an application, the location associated with that bookmark can be recalled by pressing either CTRL+PERIOD or CTRL+COMMA. These keyboard combinations also allow the developer to navigate back and forth between bookmarks. If more than one bookmark is placed, then pressing CTRL+COMMA advances the cursor forward to the next bookmark. If more than one bookmark is inserted and CTRL+PERIOD are pressed, then the cursor is moved to the previous bookmark.

Selecting Code between Two Bookmarks

Bookmarks also allow the developer to highlight sections of code listed between two bookmarks. To highlight a section of code positioned between two bookmarks, the developer would press either CTRL+SHIFT+COMMA to select the code between the current location and the next bookmark, or CTRL+SHIFT+PERIOD to select the code between the current location and the previous bookmark.

ADVANCING THE CURSOR WITHOUT USING BOOKMARKS

Visual LISP provides the developer with two tools, independent of bookmarks, that allow the developer to reposition the cursor to a predetermined location. These options are Go to Line and the Go to Last Edited. The Go to Line option allows the developer to move the cursor to a specified location in the application. This feature can be activated either by selecting the Go to Line option from the Search menu or by pressing CTRL+G. Once the option has been activated, a dialog box appears where the developer enters the line number representing the new location of the cursor (see Figure 8–55).

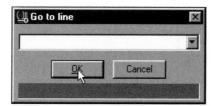

Figure 8–55 *Go to Line dialog box*

The other option, Go to Last Edited, allows the developer to return to the location of the last line edited. This option can be called by either pressing CTRL+SHIFT+G or selecting the Go to East Edited option from the Search menu (see Figure 8–56).

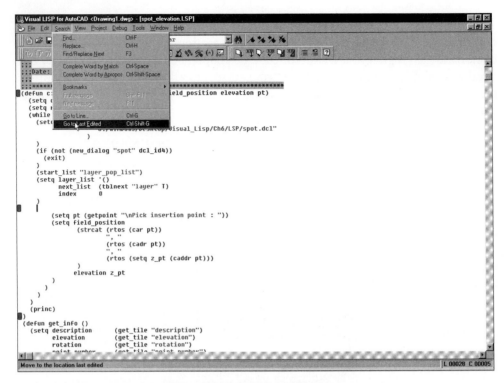

Figure 8–56 *Selecting Go to Last Edited from the Search menu*

USING VISUAL LISP TO COMPLETE A WORD BY MATCHING

Visual Lisp has the ability to complete a word in either the console window or the text editor if CTRL+SPACEBAR is pressed. This feature scans the current window for a previously entered word that matches the current portion. If the matching feature does not return the correct word, then repeat the process by pressing CTRL SPACEBAR again, each time scanning from the previous match. If a match is not found, then the feature does nothing. The Complete Word by Match feature is not case sensitive.

THE APROPOS FEATURE AND WORKING WITH SYMBOLS

Visual LISP keeps a listing of all symbols used by AutoLISP as well as user-defined symbols contained in an application in the form of a symbol table. The developer can search the symbol table from either the text editor or the console window by using the Apropos feature. The results returned by the Apropos feature can then be used in a variety of ways.

Searching the Symbol Table

To search the symbol table for a particular symbol name, the Apropos function must first be launched. This can be accomplished by pressing CTRL+SHIFT+A. Once the feature has been activated, a dialog box appears that allows the developer to define the parameters for the search (see Figure 8–57).

Figure 8–57 *Apropos search options*

The Match by Prefix Option

When this option is selected, the Apropos search engine matches the symbol name entered with the first character of the symbol name in the symbols table. When this option is not selected, the Apropos search engine is permitted to match the symbol name with any character of the symbol names in the symbol table. To illustrate this point, checking the Match by Prefix option and searching for the symbol name "set" returns the following list:

```
SET
SET1
SETCFG
SETENV
SETFUNHELP
SETQ
SETURL
SETVAR
SETVIEW
SET_TILE
```

When the Match by Prefix option is not selected, Apropos returns the following list:

```
ACET-INI-SET
ACRX_CMD_USEPICKSET
APPSET
C:MVSETUP
C:QLATTACHSET
C:QLDETACHSET
defun-q-list-set
PICKSET
SET
SET1
SETCFG
SETENV
SETFUNHELP
SETQ
SETURL
SETVAR
SETVIEW
SET_TILE
SSSETFIRST
VL-BB-SET
VL-DOC-SET
```

The WCMATCH Option

When WCMATCH option is selected, the Apropos search engine permits the use of the wild-card character (*). When the option is not selected, the Apropos search engine treats a wild-card character as if it is actually part of a symbol name. This can be illustrated by performing a search for the wild-card character asterisk. When the WCWATCH option is selected, Apropos returns a list containing 856 symbol names. When the option is not selected, Apropos returns a list of only eleven symbol names.

The Lowercase Symbols Option

When this option is selected, any symbol names that are copied to the clipboard are converted to lowercase. If the option is not selected, then all symbol names are copied to the clipboard in all caps.

Using Flags and Filter in a Search

The Apropos feature provides the developer with two additional options, flags and filters, for narrowing the developer's search. The filter flag option, when selected, displays a dialog box that enables the developer to instruct Apropos to consider only symbols that match specific flags. The filters option also allows the developer to specify filters to narrow the Apropos search.

RESULTS OBTAINED FROM APROPOS

Once the developer has defined the Apropos feature's search requirements and selected OK, then Apropos begins its search. If Apropos is unsuccessful in locating a symbol that matches the developer's search criteria, then Apropos displays an error message in the bottom of its dialog box. If the search is successful, then Apropos displays a list of the symbols matching the developers description in a new dialog box (see Figure 8–58).

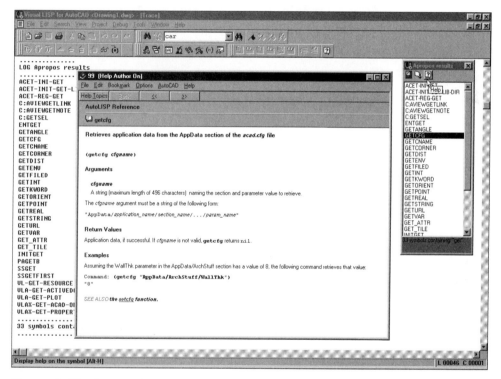

Figure 8–58 *Apropos Results dialog box*

The Apropos Results dialog box has three options, Apropos Options, Copy to Trace/ Log, and Help. Apropos Options, when selected, returns the Apropos Options dialog box (see Figure 8–59).

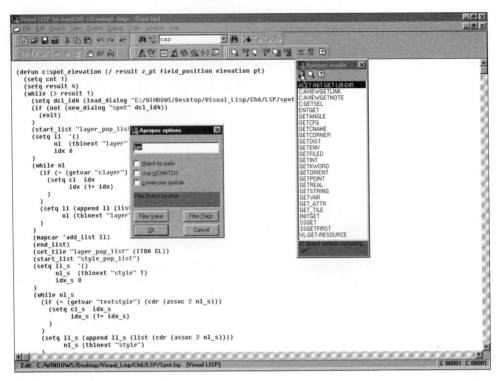

Figure 8–59 *Apropos Options dialog box recalled*

The Copy to Trace/Log option allows the developer to copy the result to the Visual LISP trace window. Once the list has been copied to the trace window, the developer can copy either all or a portion of the list using Windows' Copy command. If the developer has activated trace logging, then the list is also copied to the trace log file (see Figure 8–60).

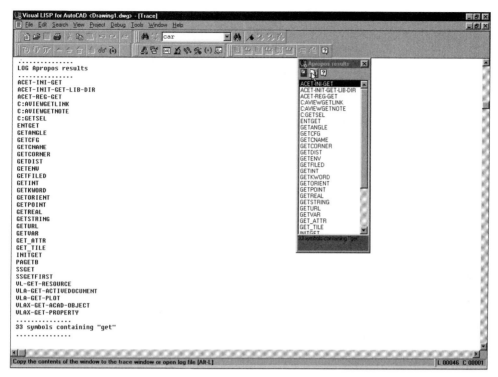

Figure 8–60 *Apropos Copy to Trace/Log option*

If the developer highlights a symbol name and then selects the Help option, Apropos responds by activating the Visual LISP online help for the entry selected (see Figure 8–61).

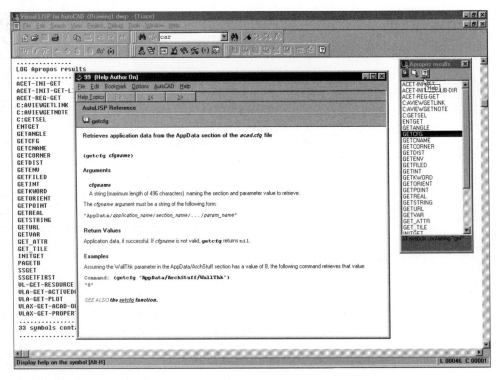

Figure 8–61 *Apropos Help option*

Note: If the specified search criteria cause over 1,000 matches to be returned, then Apropos will be unable to display all matching symbol names in the results dialog box. The entire list can be viewed by copying the results to the trace window using the Copy to Trace/Log option.

Results displayed in the Apropos Results dialog box can also be processed using the Apropos shortcut menu. The Apropos shortcut menu allows the developer to copy a symbol name to the clipboard or to an inspection window, to print the symbol name to the console window, to launch the Visual LISP symbol service, to copy the symbol name to the *OBJ* system variable, or to add a symbol name to the watch window (see Figure 8–62). The Apropos shortcut menu is activated by selecting a symbol name and then performing a right mouse click.

Note: Once a symbol name has been copied to the console window using the Apropos shortcut menu, the symbol name can be recopied to the console window prompt by highlighting the symbol name (in the console window) and then pressing ENTER (see Figure 8–63).

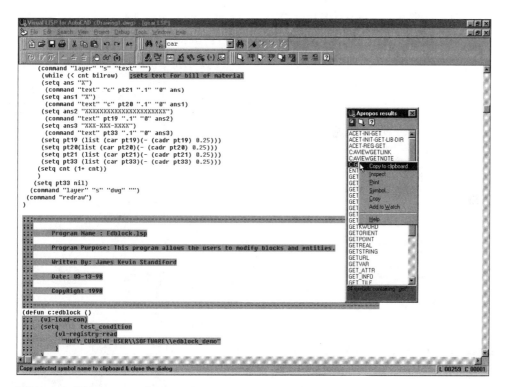

Figure 8–62 *Apropos shortcut menu*

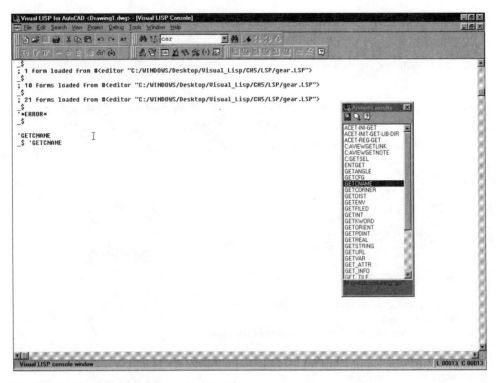

Figure 8–63 *Apropos Copy to Console option*

USING APROPOS TO COMPLETE A WORD

In Visual LISP, word completion can also be accomplished using the Apropos feature. When word completion is done using Apropos, any suggestions provided by Visual LISP for word completion are based on the symbol table and not on previously typed information. The Apropos word completion feature can be activated in either the text or console window by pressing CTRL+SHIFT+SPACEBAR (see Figure 8–64). If Apropos is unable to locate a match for the incomplete word, then the Apropos Options dialog box is displayed. If more than 15 matches are found by Apropos, then Visual LISP displays the Apropos Results dialog box.

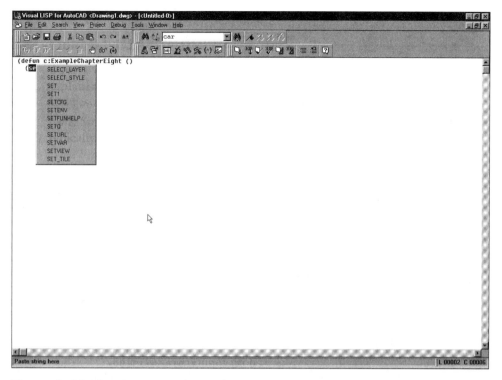

Figure 8–64 *Completing a word using Apropos*

WATCH WINDOW AND INSPECTION WINDOW

Among the various tools provided in the Visual LISP IDE, one of the most useful is the watch window (see Figure 8–65). The watch window allows the developer to monitor the value of a variable during a program's execution. Each line in the watch window represents a variable name and its current value. Updates are performed automatically with each step of an interactive session. When a program is executed, the watch window is updated after each expression is evaluated. To activate the Visual LISP watch window, the developer starts by highlighting a variable and then pressing CTRL+SHIFT+W. If the watch window is activated without a variable being first selected, then Visual LISP responds by displaying the Add Watch dialog box (see Figure 8–66). The Add Watch dialog box allows the developer to enter the name of the variable that is to be monitored.

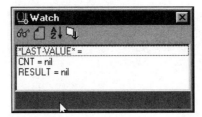

Figure 8–65 *Visual LISP watch window*

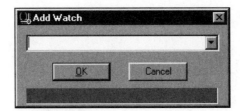

Figure 8–66 *Add Watch dialog box*

Once a variable has been added to the watch window, its current value and the name of the variable associated with that value can be copied to the Visual LISP trace window by selecting the ▧ toolbar button. The contents of the watch window can be resorted in alphabetical order at any time by selecting the ▧ toolbar button. Additional variable names can be added to the watch window by selecting the ▧ toolbar button. To clear the contents of the watch window, the ▧ toolbar button must be selected.

The Visual LISP watch window displays the current value of a selected variable; however, if a closer examination of a variable's content is needed, then the Visual LISP inspection window can be used (see Figure 8–67).

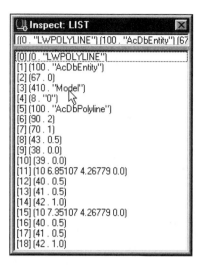

Figure 8–67 *Visual LISP inspection window*

The Visual LISP inspection tool allows the developer to navigate, view, and modify both AutoLISP and AutoCAD objects. For each item that the developer wishes to inspect, Visual LISP opens a separate inspection window. Activate the inspection window either by double-clicking a variable name listed in the watch window or by highlighting a variable name and then pressing CTRL+SHIFT+I.

All inspection windows contain three main components: a caption, an object line, and the object element list. The caption displays the entity type of the object being inspected, while the object line displays a printed representation of the selected object. The element list reveals the individual components of the selected object. All entries displayed in the object element list follow the same basic format, name then content. When an element name is enclosed in square brackets, that element may be modified by performing the Modify command from the shortcut menu. If the element name is enclosed in curly brackets, then that element cannot be changed. All entries contained within the object element list can be expanded even further by double-clicking on the element name. When Visual LISP expands an element contained within a list, a separate inspection window is opened containing the expanded information (see Figures 8–68 and 8–69).

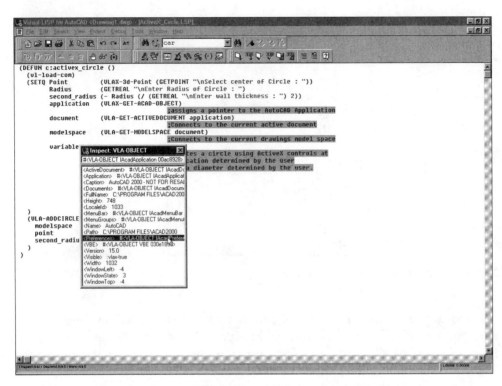

Figure 8–68 *Element selected in object element list*

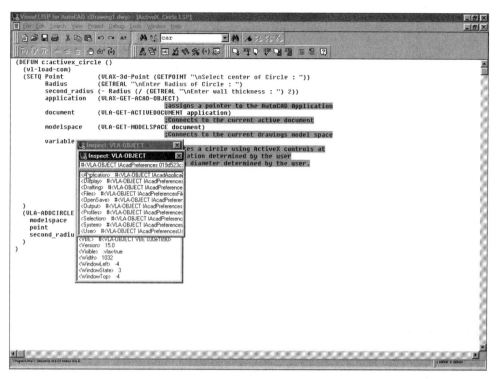

Figure 8–69 *Expanded element window*

THE VISUAL LISP TRACE STACK

Because AutoLISP has the ability to evaluate nested expressions and functions, it is necessary for the program to remember its way out of nested expressions and functions. AutoLISP accomplishes this by storing a historical record of the execution of expressions and functions in a stack by an application (see Figure 8–70). When an AutoLISP application is executed, AutoLISP places an entry in the stack. When the function calls nested functions, additional entries are added to the stack. Entries are placed inside the stack only when AutoLISP needs to remember a way out. The stack is a useful tool when a program is locked in a suspended state or an error causes the program to stop.

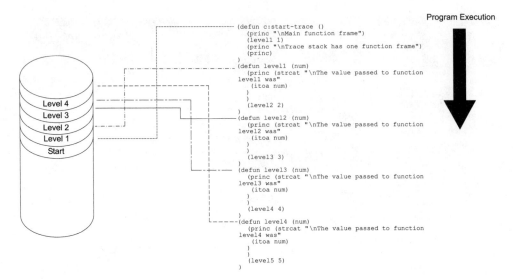

Figure 8–70 *Stack*

Viewing the Trace Stack

When an application is suspended at a breakpoint, the trace stack window may be activated by selecting the Trace Stack option from the View menu. When this function is activated, it displays a list of stack elements (see Figure 8–71). The stack elements represent a line-item history as to where the application is currently located. The trace stack window in Figure 8–71 indicates that the application is currently suspended in function *level6*. Visual LISP has five types of elements that can appear in the trace stack. Table 8–3 lists and describes each of these elements.

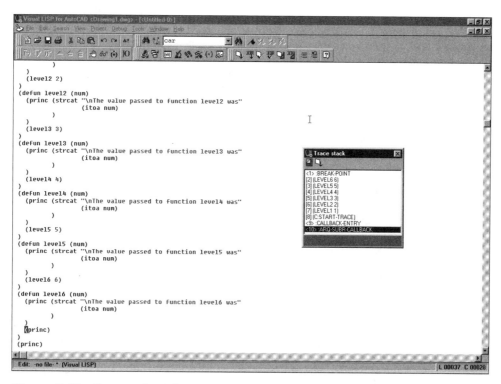

Figure 8–7I   *Trace stack window*

Table 8–3   Trace Stack Elements

Element	Description
Function Call Frame	Shows a function call. Each function call frame displays the name of the function, the function's arguments, and the function's level within the trace stack.
Keyword Frame	These frames appear at the top and bottom of the stack and are displayed in the form: level : keyword - {optional-data}. The *keyword* portion of the element indicates the type of frame displayed. The *optional-data* portion of the frame displays additional information regarding the state of the program.
Top Form	Indicates an action that was initiated from the console window by the typing of an expression. The call to a function by the loading of a file or a selection made within the Visual LISP editor window.
Lambda Form	Indicates a lambda function was called within the program.
Special Form	Indicates that either a FOREACH or REPEAT function was called within the program.

In the following AutoLISP program, the reader can get a better understanding of how the trace stack window operates. In the program, a function call is made from the main function to the *level1* function. Within that function, another function call is made to the function *level2*. This arrangement is continued throughout the remaining functions, producing an application that contains six levels of nested functions:

```
(defun c:start-trace ()
   (princ "\nMain function frame")
   (level1 1)
   (princ "\nTrace stack has one function frame")
   (princ)
)
(defun level1 (num)
   (princ (strcat "\nThe value passed to function level1 was"
            (itoa num)
        )
    )
   (level2 2)
)
(defun level2 (num)
   (princ (strcat "\nThe value passed to function level2 was"
            (itoa num)
        )
    )
   (level3 3)
)
(defun level3 (num)
   (princ (strcat "\nThe value passed to function level3 was"
            (itoa num)
        )
     )
   (level4 4)
)
(defun level4 (num)
   (princ (strcat "\nThe value passed to function level4 was"
            (itoa num)
        )
     )
   (level5 5)
)
```

```
(defun level5 (num)
   (princ (strcat "\nThe value passed to function level5 was"
             (itoa num)
     )
  )
   (level6 6)
 )
(defun level6 (num)
   (princ (strcat "\nThe value passed to function level6 was"
             (itoa num)
     )
  )
   (princ)
 )
(princ)
```

When the application is loaded into Visual LISP and executed, the program produces the following results:

Command: **start-trace** (ENTER)

Main function frame
The value passed to function level1 was1
The value passed to function level2 was2
The value passed to function level3 was3
The value passed to function level4 was4
The value passed to function level5 was5
The value passed to function level6 was6
Trace stack has one function frame

Command:

As each function is called, a new element is added to the stack. Once the last nested function has completed executing, the program can use the stack history to backtrack to the main function. By inserting a breakpoint 🖑 at the beginning of the last PRINC function in level6, the trace stack can be viewed (see Figure 8–71).

Viewing the Error Trace Stack

When an error occurs in an application, the contents of the trace stack are flushed (see Figure 8–72). To retain a copy of the trace stack when a program crashes, the developer can either select the Break on Error option from the Debug menu or activate the error trace feature from the View menu.

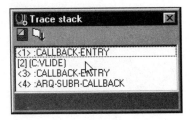

Figure 8–72 *Trace Stack window*

When the Break on Error debugging tool is selected, the trace stack retains its entries after the application has crashed (see Figure 8–73). To view the trace stack as it appeared at the time the error occurred without setting the break on error option and rerunning the application, the developer can launch the error trace feature. The error trace feature is an exact copy of the trace stack at the time of the error (see Figure 8–74).

Figure 8–73 *Trace stack with Break on Error*

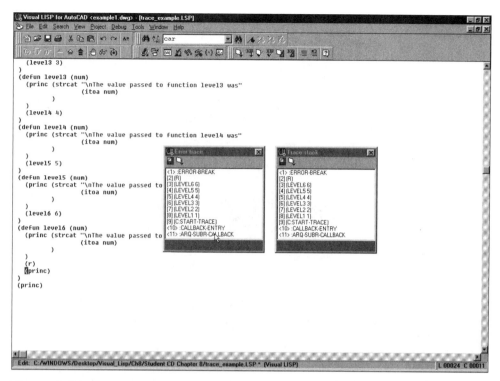

Figure 8–74   *Error trace window*

SYMBOL SERVICE

The Visual LISP symbol service provides the developer with quick access to several of the Visual LISP debugging options available for symbols. The Symbol Service dialog box contains a toolbar, a name field, a value field, and symbol flags (see Figure 8–75). The toolbars allow the developer to add the symbol to the watch window, display the value of the variable in an inspection window, show the defined functions, and display the online Visual LISP help for the specified symbol.

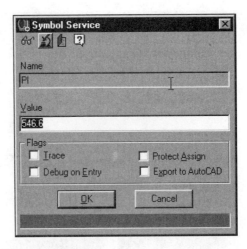

Figure 8–75 *Symbol Service dialog box*

The symbol flags associated with the symbol service provide the developer direct access to the specified symbol flags options. These flag options are Trace (Tr), Protect Assign (Pa), Debug on Entry (De), and Export to AutoCAD (Ea).

- When the Trace flag is selected, tracing is activated when the specified user defined symbol is used in an expression as a function.

- When the Protect Assign flag is selected, protected symbols may be redefined.

- The Debug on Entry flag automatically sets a breakpoint at each occurrence of the specified symbol.

- The Export to AutoCAD flag makes the function associated with the specified symbol name an external subroutine.

BROWSING THE AUTOCAD DATABASE

Chapters 4 and 5 introduced the concepts of entity data, extended entity data, tables, and dictionaries as well as how to view and manipulate them by constructing one or more AutoLISP expressions. In Visual LISP this can be accomplished by selecting one of the four AutoCAD database browsers.

Viewing All Objects

Visual LISP provides a browser for viewing the entity data of all objects contained within an AutoCAD drawing. Selecting View|Browse Drawing Database|Browse all Entities launches the browser (see Figure 8–76).

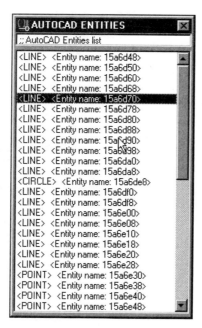

Figure 8–76 *Entities inspection window*

The browser displays only the name and type of the parent entities. To view a nested entity or display a detailed listing of an entity, double-click the parent name of the entity to view (see Figure 8–77).

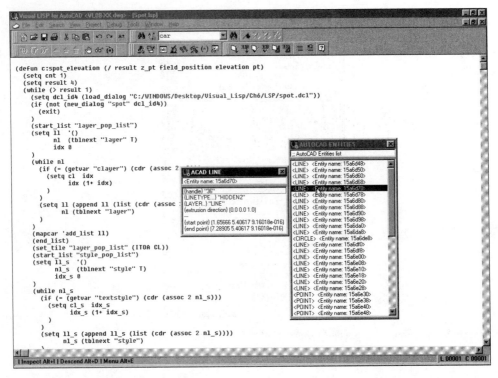

Figure 8–77 *Line inspection window*

Viewing a Drawing's Tables

Visual LISP provides a browser for viewing the drawing tables of an AutoCAD drawing. Selecting View|Browse Drawing Database|Browse Tables launches the browser (see Figure 8–78).

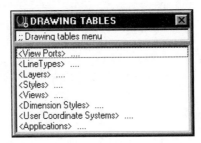

Figure 8–78 *Drawing tables inspection window*

The browser displays only the table names associated with the current AutoCAD drawing. Double-clicking a table name launches an inspection window where the names of the entities associated with the selected table are revealed in a separate inspection window (see Figure 8–79). To inspect further the entity names contained within an inspection window, double-click on each entity name (see Figure 8–80).

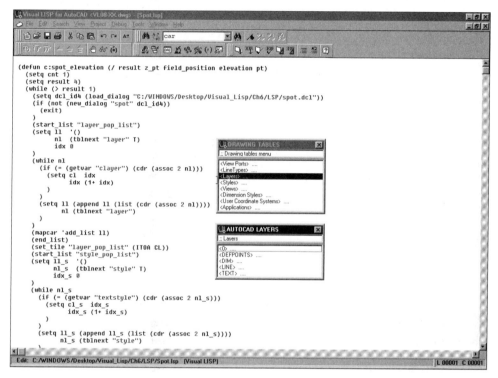

Figure 8–79 *Inspection window containing the entity names associated with a table*

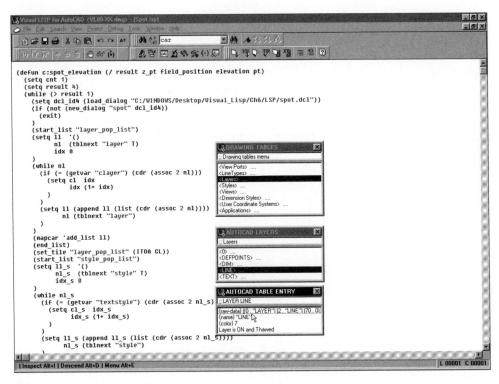

Figure 8–80 *Examining an entity*

Viewing Block Information

Visual LISP provides a browser for viewing the blocks contained within an AutoCAD drawing. Selecting View|Browse Drawing Database|Browse Blocks launches the browser (see Figure 8–81).

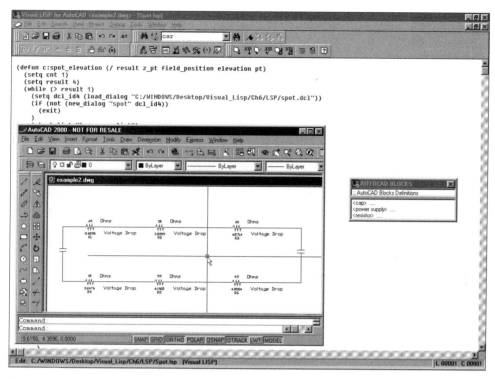

Figure 8–81 *Block browser*

The browser displays only the block name. To obtain information regarding the block names displayed in the browser, double-click an entry in the browser's list (see Figure 8–82).

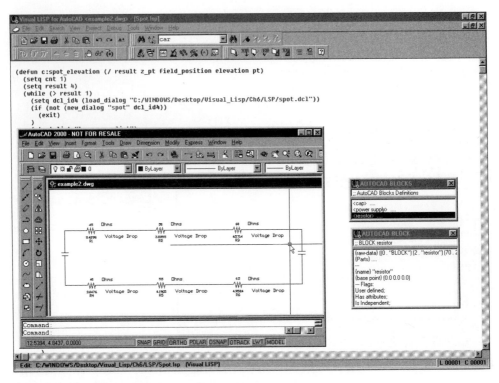

Figure 8–82 *Block inspection window*

Viewing Selection Sets

Visual LISP provides a browser for viewing a selection set. Selecting View|Browse Drawing Database|Browse Selection launches the browser (see Figure 8–83). When the function is activated, Visual LISP responds by prompting the developer to define a selection set. Once the selection set has been defined, Visual LISP displays the selected entities in an inspection window.

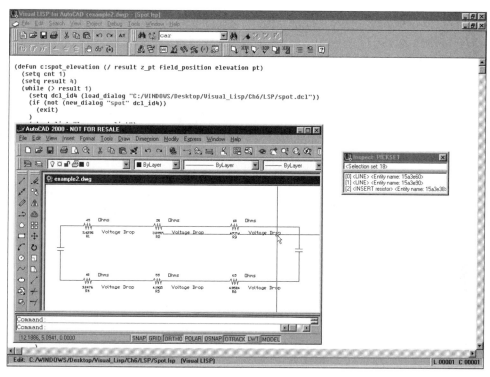

Figure 8–83   *Selection set browser*

Viewing Xdata

Visual LISP provides a browser for viewing a list of applications currently registered as containing extended entity data in an AutoCAD drawing. Selecting View|Browse Drawing Database|Inquire Extended Entity Data launches the browser (see Figure 8–84). Once the browser is activated, a list of registered applications for the current drawing is displayed. From this list the developer may select the name of an application containing extended entity data to view. After the developer has made a selection, the extended entity data may be viewed by selecting the AutoCAD object containing the extended entity data from the AutoCAD drawing and then launching the Browse Selection feature. If an application name is specified before the Browse Selection feature is launched, Visual LISP includes that object's extended entity data in the browse selection inspection window.

Figure 8–84 *List of registered applications displayed*

PREVIEWING A PORTION OF A DIALOG BOX AND THE ENTIRE DIALOG BOX

When a dialog box is being developed for an AutoLISP application, it is often necessary for the programmer to preview the dialog box as it will be presented to the end user. This helps ensure that the dialog box is aesthetically pleasing in appearance and ergonomically designed. In Visual LISP this can be accomplished by the text editor's interface tools. These tools allow the developer to preview either a portion of a dialog box by selecting Tools|Interface|Preview DCL in Selection (for a portion of a DCL source file) or Preview DCL in Editor (see Figures 8–85, 8–86, and 8–87).

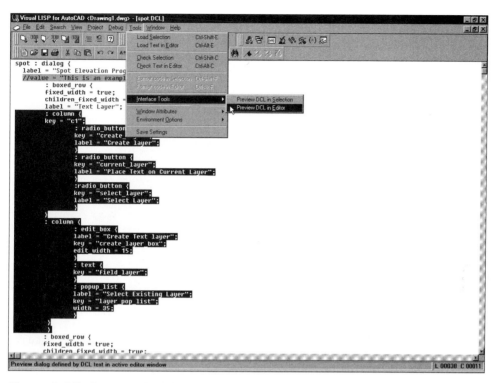

Figure 8–85 *Activating the DCL preview tools from the Tools menu.*

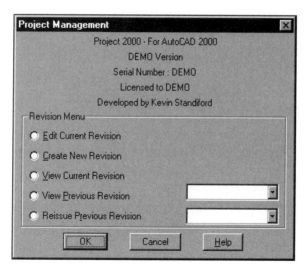

Figure 8–86 *Previewing a portion of a DCL file*

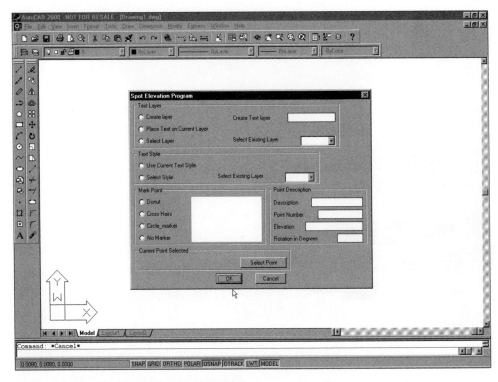

Figure 8–87 *Previewing the entire DCL file*

USING THE ANIMATE DEBUGGING FUNCTION

The animate feature allows the developer to watch as Visual LISP steps through each line of code when executing a program. When the feature is activated and the program is executed, Visual LISP highlights each expression in the text editor as the expressions are being evaluated (see Figure 8-88). Any watch windows activated at the time the program is executed are continuously updated.

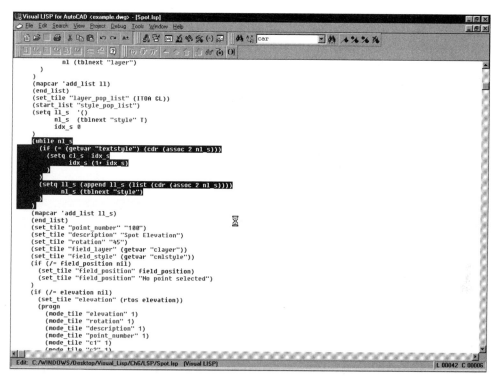

Figure 8–88 *Highlighting shows expression being evaluated*

VISUAL LISP PROJECTS

The creation of custom applications for AutoCAD 2000 can consist of programs that require only a few lines of code or applications that incorporate other source code files, as well as dialog boxes. When an application consists of a single source code file, then maintaining that application is not very complicated. However, when an application consists of several AutoLISP programs and one or more dialog boxes, then maintaining the application can be somewhat tricky. To help alleviate this burden, Autodesk has incorporated the concept of a project into Visual LISP. A project is nothing more than a list of AutoLISP source code files that are associated with a particular application and a set of rules as to how those files are to be compiled. Incorporating the concept of a project into the development of larger and more complex AutoLISP applications offers numerous benefits to the developer. The developer can access source code files associated with a project with a single click. A search can be limited to a particular project for an expression or a group of expressions. A search can be performed for expressions contained in one of the project files. Visual LISP can automatically check and recompile AutoLISP source code files that have been modified within a project. Finally, compiled applications can be optimized by defining links to corresponding source code files.

CREATING A PROJECT

The first step in setting up a project is to create a project definition file (.prj) using the New Project menu option (see Figure 8–89). A project definition file contains all the locations and file names of the source codes that make up a project. When the command is first executed, the developer is presented with the New Project dialog box (see Figure 8–90). This dialog box allows the developer to create the actual project files containing the project parameters.

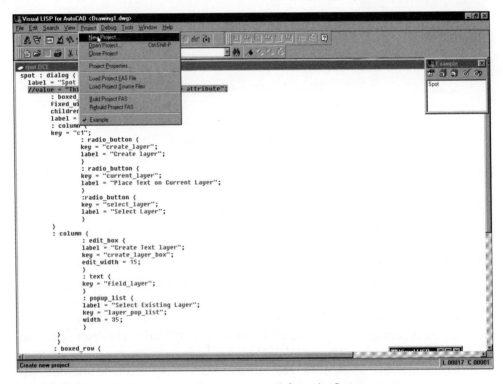

Figure 8–89 *Activating the New Project command from the Project menu*

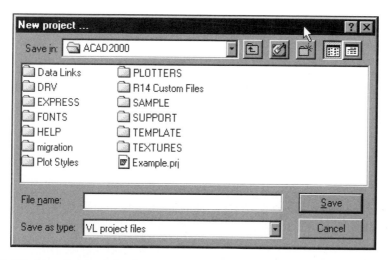

Figure 8–90 *New Project dialog box*

Once the project has been created, the developer is given the Project Properties dialog box (see Figure 8–91). The Project Properties dialog box has two tabs, Project Files and Build Options. It is in the Project Properties dialog box that the actual parameters for the project are set (AutoLISP source files to include and the compiler configurations).

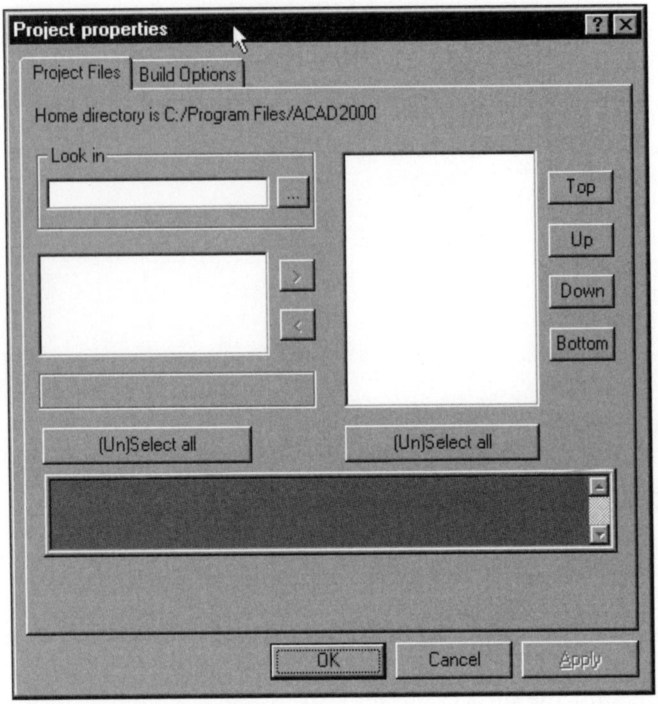

Figure 8–91 *Project Properties dialog box*

Selecting the Project Files

The Project Files tab is used to specify the AutoLISP source code files that are to be included in the project. AutoLISP source code files are added to a project by first selecting the location of the files, using the ▣ located in the Look In drop-down list. Once a location has been selected, all source code files residing in that location are displayed on the left side of the dialog box (see Figure 8–92). To add a source code file to the project, the developer highlights the name of the file(s) to add and then presses the ▣ button. When a file has been added to the project, its name will appear on the right side of the Project Properties dialog box (see Figure 8–93).

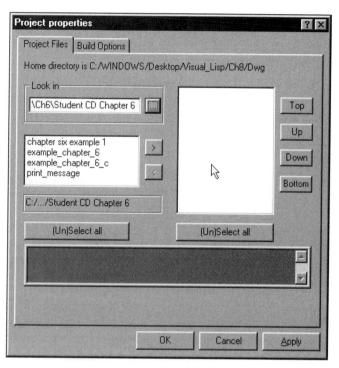

Figure 8–92 *Source code files displayed on the left side*

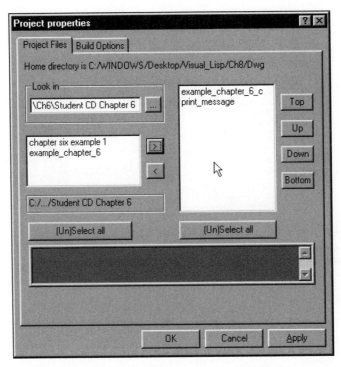

Figure 8–93 *Files added to the project displayed on the right side*

Specifying the Build Options

The Build Options tab allows the developer to specify numerous options regarding how the project's source code is to be compiled (see Figure 8–94). These options have an effect on the efficiency of the compiled program(s).

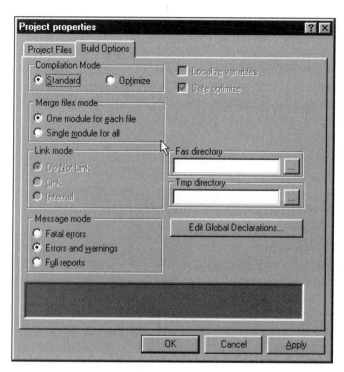

Figure 8–94 *Build Options tab*

Compilation Mode

The Compilation Mode allows the developer to determine how the source code files are to be compiled, as either standard or optimized binary files. When the Standard option is selected, all symbol names that are associated with either functions or global variables are retained. Preserving the application names in the application allows the program to reference functions associated with other files.

When a program is executed that has been compiled in standard mode, Visual LISP uses a table to determine the assigned memory location of symbols. When the Optimize option is selected, the Visual LISP compiler discards all symbol names, allowing the program, when executed, to go directly to the memory location containing the symbol's value. Discarding the symbol names contained within an application reduces the size of the application. This produces a program that executes much faster than programs compiled in standard mode. However, optimized mode is not always suitable for each application and in some situations, optimizing can create problems that would otherwise not exist in non-optimized code. This is due in part to the fact that LISP allows the developer to create applications that have the ability to either create or modify functions at runtime.

When an application is compiled in optimized mode, the Visual LISP compiler analyzes the code as it is compiled, looking for possible situations that may cause problems. When the compiler is finished, it generates a report outlining all sections that are of possible concern. While the compiler is able to recognize the vast majority of the situations that may cause problems when the optimized mode is selected, there are some situations that the compiler cannot detect. These situation are interaction with ObjectARX applications, evaluation of dynamic-built code using EVAL, APPLY, MAPCAR and/or LOAD, using SET for dynamically supplied variables, and dynamic action strings in ACTION_TILE and NEW_DIALOG.

Merge Files Mode

This Merge Files Mode instructs the compiler to either merge all source code files into a single FAS file or create separate FAS files for each source code file in a project. If the source code files are combined into a single FAS file, the application will run much faster. Because FAS files do not allow for debugging, it is recommended that the developer create a separate FAS file for each source code file if the program has not been completely debugged. Doing so allows the developer to load each file one at a time.

 Note: FAS files are compiled AutoLISP applications that execute and load much faster than standard AutoLISP source code files. Because these files are compiled, their content appears illegible when opened in either a word processing application or the Visual LISP editor, thereby protecting the developer's source code (see Figure 8–95). Even though FAS files are compiled AutoLISP source code files, they are not standalone AutoCAD applications. A standalone AutoCAD application file is a VLX file. These files can be created from FAS files through the Visual LISP Make Application wizard.

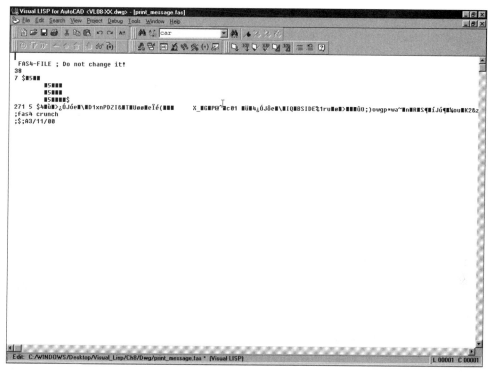

Figure 8–95 *FAS file opened in the Visual LISP text editor*

Edit Global Declarations

This option creates a global declaration file for compatibility with the first release of Visual LISP (AutoCAD Release 14.01).

FAS Directory

This option allows the developer to specify the directory in which to place compiled AutoLISP files. If this option is left blank, then Visual LISP places all compiled files in the same directory as the project definition file.

Tmp Directory

This option allows the developer to specify the location of all project temporary files.

Link Mode

The Link Mode is available only if the optimized compilation mode has been selected. This option allows the developer to specify how the Visual LISP compiler will optimize function calls within an application by providing the developer with three choices: Do Not Link, Link, and Internal.

- If the Do Not Link option is selected, Visual LISP stores the address of the symbol name defining the function.

- The Link option allows the compiler to directly address function definitions and any calls to the referenced functions.

- The Internal option, when selected, removes all function names and directly links function calls.

 Note: Applications that redefine or create functions should not be directly linked to other applications through the link mode. All symbols exported to AutoCAD retain their symbol names once the application has been compiled, regardless of the link mode selected.

Localize Variables

When this check box is selected, the compiler removes all local symbol names and directly links the address where the variables are stored.

Safe Optimize

This check box provides the developer with a margin of safety by instructing the Visual LISP compiler to refuse optimizing the source code if there is a chance that doing so could produce an incorrect FAS file.

Message Mode

The Message Mode options allow the developer to specify the amount of information passed by the Visual LISP compiler to the trace window when a file is compiled.

- If the Fatal Errors option is selected, the compiler reports only errors that cause the compiler to terminate abruptly.

- The Errors and Warnings option instructs the compiler to report any warnings or errors encountered during compilation.

- If the Full Report option is selected, the compiler provides the developer with a report indicating all errors, warnings, and statistics regarding the compilation of the application.

An example of a full report is provided below for the AutoLISP application spot.lsp.

```
; Make example_1; Date: 3/15/00; Time:8:04 PM;
; Make one module for each file
; Make standard:
; Make message level: 2


;;; CREATING PDB...
[Analyzing file "C:/WINDOWS/Desktop/Visual_Lisp/Ch8/Projects/Spot.lsp"]
........


;;; COMPILING source files...
[COMPILING C:/WINDOWS/Desktop/Visual_Lisp/Ch8/Projects/Spot.lsp]
;;C:SPOT_ELEVATION
; === STATISTICS: C:SPOT_ELEVATION
;                    external globals are (CL CL_S CNT DCL_ID4 IDX IDX_S LL
  LL_S NL NL_S T X Y)
;                    bound globals are (ELEVATION FIELD_POSITION PT RESULT
  Z_PT)
;                    not linked function calls (/= 1+ = > ACTION_TILE ALERT
  APPEND ASSOC CADDR CADR CAR CDR DIMX_TILE DIMY_TILE END_IMAGE END_LIST
  EXIT GETPOINT GETVAR ITOA LIST LOAD_DIALOG MAPCAR MODE_TILE NEW_DIALOG
  NOT PRINC RTOS SET_TILE SLIDE_IMAGE START_DIALOG START_IMAGE
  START_LIST STRCAT TBLNEXT)
;;GET_INFO
; === STATISTICS: GET_INFO
;                    external globals are (CIRCLE_MARKER CREATE_LAYER
  CREATE_LAYER_BOX CROSS_MARKER CURRENT_LAYER CURRENT_STYLE DESCRIPTION
  DONUT_MARKER ELEVATION LAYER_POP LL LL_S NO_MARKER POINT_NUMBER PT
  ROTATION SELECT_LAYER SELECT_STYLE STYLE_POP)
;                    not linked function calls (/= = ATOF ATOI CADR CAR
  CIRCLE_MARKER_FUNCTION CROSS_MARKER_FUNCTION DONUT_MARKER_FUNCTION
  GET_TILE LIST NO_MARKER_FUNCTION NTH)
;;NO_MARKER_FUNCTION
; === STATISTICS: NO_MARKER_FUNCTION
;                    external globals are (OBJECT_1 OBJECT_2)
```

```
;                 not linked function calls (MAKE_OBJECT_FUNCTION)
;;CIRCLE_MARKER_FUNCTION
;  === STATISTICS: CIRCLE_MARKER_FUNCTION
;                 external globals are (OBJECT_1 OBJECT_2 PT)
;                 not linked function calls (CONS LIST
  MAKE_OBJECT_FUNCTION)
;;DONUT_MARKER_FUNCTION
;  === STATISTICS: DONUT_MARKER_FUNCTION
;                 external globals are (OBJECT_1 OBJECT_2 PT)
;                 not linked function calls (+ CADDR CADR CAR CONS LIST
  MAKE_OBJECT_FUNCTION)
;;CROSS_MARKER_FUNCTION
;  === STATISTICS: CROSS_MARKER_FUNCTION
;                 external globals are (OBJECT_1 OBJECT_2 PT)
;                 not linked function calls (+ - CADDR CADR CAR CONS
  LIST MAKE_OBJECT_FUNCTION)
;;MAKE_OBJECT_FUNCTION
;  === STATISTICS: MAKE_OBJECT_FUNCTION
;                 external globals are (CREATE_LAYER CREATE_LAYER_BOX
  CURRENT_LAYER CURRENT_STYLE DESCRIPTION ELEVATION LAYER LAYER_POP
  OBJECT_1 OBJECT_2 POINT_NUMBER PT ROTATION SELECT_LAYER SELECT_STYLE
  STYLE STYLE_POP)
;                 not linked function calls (* + - / = ATOF CADDR CADR
  CAR CONS ENTMAKE GETVAR LIST LOC SETVAR STRCAT)
;;LOC
;  === STATISTICS: LOC
;                 external globals are (ENT1 ENTKEV L NAMKEV RNAM)
;                 not linked function calls (/= ASSOC CDR CONS ENTGET
  ENTLAST ENTMAKE ENTMOD ENTNEXT PRINC SUBST TBLSEARCH)
;  === STATISTICS: "C:/WINDOWS/Desktop/Visual_Lisp/Ch8/Projects/Spot.lsp"
;                 not dropped function names (C:SPOT_ELEVATION
  CIRCLE_MARKER_FUNCTION CROSS_MARKER_FUNCTION DONUT_MARKER_FUNCTION
  GET_INFO LOC MAKE_OBJECT_FUNCTION NO_MARKER_FUNCTION)
;                 were exported to AutoCAD (C:SPOT_ELEVATION)
"C:/WINDOWS/Desktop/Visual_Lisp/Ch8/Projects/Spot.ob"
```

```
[FASDUMPING object format -> "C:/WINDOWS/Desktop/Visual_Lisp/Ch8/
    Projects/Spot.fas"]
; Make complete.
```

OPENING AN EXISTING PROJECT

Once the parameters of a project have been set and a project file created, the developer can recall that project at any time by selecting the Open Project option from the Project menu. When a project is either created or an existing project is reloaded, Visual LISP displays a project dialog box. The dialog box contains five toolbar buttons across the top and a list of the source code files defined in the project (see Figure 8–96). The toolbar allows the developer to edit the project's properties, load the FAS file into memory, load the source code into memory, built the project FAS file, and rebuild the project FAS file. The project dialog box also allows the developer to load source code file into the Visual LISP editor by double-clicking on a file name contained within the project's dialog box or by highlighting the file name and then launching the project dialog box shortcut menu (single right mouse click).

Figure 8–96 *A project dialog box*

CREATING A PROJECT FAS FILE

After a project has been defined, the developer can generate a compiled version of the application by either selecting the Build Project FAS File option from the Project menu or the Build the Project FAS toolbar button from the project dialog box. Creating a FAS file of the project can be used to help the developer determine how the application will perform once compiled.

CREATING A STANDALONE APPLICATION

Once an application has been developed and debugged, the developer can create a compiled standalone version of the application by selecting the Make Application Wizard option from the File menu. The wizard contains a series of dialog boxes that are designed to guide the developer through the creation process. The first dialog box displayed by the wizard allows the developer to specify the mode in which the wizard operates (see Figure 8–97). If the Simple option is selected, the wizard only prompts the developer to specify the AutoLISP FAS files or project file to use in creating the application. If the developer selects the Expert mode, then the wizard asks the developer to include DCL, VBA, and additional AutoLISP source code files in the application as well as change some of the default compiler settings.

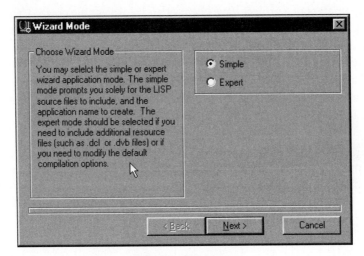

Figure 8–97 *Visual LISP Make Application wizard*

SUMMARY

Because LISP is a powerful programming language well suited for the graphical environment, Autodesk chose to structure their initial Application Development Interface (AutoLISP) based on this language. Even though Autodesk has continued supporting and enhancing the AutoLISP programming language, it was not until the release of Visual LISP that the techniques used in developing AutoLISP applications changed to match those used by other current languages. In reality, Visual LISP is not a new programming language, but instead the next evolution of the AutoLISP programming language. Features incorporated in the new version of AutoLISP (Visual LISP) include a text editor, compiler, debugger, Object Oriented Programming, ActiveX, and numerous other development tools. The initial release of Visual LISP was an add-on package for AutoCAD Release 14.01; however, with AutoCAD 2000 this changed and Visual LISP became a part of the actual AutoCAD engine.

Because Visual LISP is a fully integrated development environment, it contains its own windows, separate from the windows used in AutoCAD for developing applications. To launch Visual LISP from the AutoCAD command prompt, the VLIDE or VLISP is used. Once this command has been executed, the Visual LISP user interface is displayed. The Visual LISP interface is comprised of menus, toolbars, a main window, console window, trace window, text editor and a status bar. The menu region allows the developer to select Visual LISP commands from a menu system. The toolbars provide the developer with a shortcut to many of the commands used in Visual LISP. The Visual LISP console window is designed to allow the developer to test and evaluate AutoLISP expressions by entering them at a console prompt much the way they would be entered at the AutoCAD command prompt. The trace window (also known as the trace output window) is specifically designed to display the output of the trace function. The main window of the Visual LISP IDE is the text editor. Unlike WordPad, the Visual LISP text editor is a sophisticated collection of programming tools specifically designed for the creation of AutoLISP programs. Some of the major tools integrated into the Visual LISP text editor are syntax coloring, text formatting, parentheses matching , AutoLISP expression execution, and multiple file searching. Other features in Visual LISP IDE include bookmarking and text selection, word matching, Apropos, watch windows, inspection windows, trace stack viewing, and AutoCAD database browsing.

The Visual LISP IDE also allows the developer to create and maintain projects. A project is nothing more than a collection of source code files and a set of rules on how these files are to be compiled. Incorporating the concepts of a project into the development of larger and more complex AutoLISP applications offers numerous benefits to the developer: The developer can access source code files associated with a project with a single click. A search can be limited to a particular project for an expression or a group of expressions. A search can be performed for expressions contained in one of the project files. Visual LISP can automatically check and recompile AutoLISP source code files that have been modified within a project. Finally, compiled applications can be optimized if links are defined to corresponding source code files.

REVIEW QUESTIONS

1. Define the following terms:
   IDE
   OOP
   API

2. True or False: Visual LISP is a separate language from AutoLISP. (If the answer is false, explain why.)

3. List five enhancements that have been made to AutoLISP programming with the introduction of Visual LISP.

4. What AutoCAD commands can be used to launch the Visual LISP IDE?

5. What are the seven areas that make up the Visual LISP interface? What is the purpose of each area?

6. True or False: The Visual LISP toolbars can be divided into nine areas. (If the answer is false, explain why.)

7. True or False: The Visual LISP toolbars execute different commands from the Visual LISP menus. (If the answer is false, explain why.)

8. What key sequence can be used to format text in the text editor window?

9. What key sequence can be used to format a portion of text in the text editor window?

10. Why is text formatting important?

11. What is the purpose of the syntax-coloring feature of Visual LISP?

12. What are the four different styles Visual LISP uses to format text? Give a brief explanation of each.

13. What is the purpose of the parentheses-matching feature in Visual LISP?

14. What key sequence can be used to find an expression's closing parenthesis?

15. True or False: To display the value of a variable in the Visual LISP console window, an exclamation mark must be placed in front of the variable name. (If the answer is false, explain why.)

16. True or False: Multiple file searching cannot be performed in Visual LISP. (If the answer is false, explain why.)

17. What is the purpose of bookmarks?

18. What key sequence can be used to advance the cursor forward when bookmarks are not used?

19. What key sequence can be used to complete a word in Visual LISP?

20. What is the purpose of Apropos?

21. How are words completed using Apropos?

22. What is the purpose of the Visual LISP watch window?

23. What is the purpose of the inspection window?

24. What is a stack, and how is it used in Visual LISP?

25. What is an error trace stack?

26. True or False: Visual LISP does not supply a means of browsing the AutoCAD database. (If the answer is false, explain why.)

27. What are projects used for in Visual LISP? Why are they important?

28. Define the following terms: FAS, VLX, and PRJ.

29. Why is it important to compile an AutoLISP application once it has been debugged?

30. What is safe optimizing?

Object Oriented Programming and ActiveX

OBJECTIVES

Upon completion of this chapter the reader will be able to:

- Understand the difference between spaghetti programming and structured programming
- Understand the difference between event-driven programming and object oriented programming
- Understand what a method is and how it is called
- Define the following terms: Class, Object, Data, Property, Method, Encapsulation, Polymorphism, Inheritance, Array, Multidimensional Array, SafeArray
- Understand the difference between OLE and ActiveX
- Use the Visual LISP inspection tool to view the properties of an object
- Use the AutoLISP function VLAX-NAME->VLA-OBJECT to convert entity names extracted through the AutoLISP ENTXXX functions to VLA-Objects
- Use AutoLISP to view the contents of an array
- Use AutoLISP to retrieve the upper and lower boundaries of a SafeArray
- Use AutoLISP to convert a SafeArray to a Visual Basic variant
- Use ActiveX in an AutoLISP application to access an object's properties
- Dump a list of properties associated with an object to the text screen using the AutoLISP function VLA-DUMP-OBJECT
- Use the VL-LOAD-COM function to load the supporting code of ActiveX controls into memory

- Establish a connection to the AutoCAD object model using the AutoLISP function VLAX-GET-ACAD-OBJECT

- Use the VLA_GET-ACTIVEDOCUMENT function to establish a connection to the current drawing

- Obtain access to graphic entities in an AutoCAD drawing using the VLA-GET-MODELSPACE function

- Obtain access to non-graphic entities in an AutoCAD drawing using the VLA-GET-PAPERSPACE function

- Use the VLAX-FOR function to evaluate a series of expressions for each object contained in a collection

- Understand the importance of releasing all references to an object once an application has completed evaluating that object

- Establish a connection between AutoLISP and an external application using VLAX-GET-OBJECT, VLAX-CREATE-OBJECT, or VLAX-GET-OR-CREATE-OBJECT

KEY WORDS AND AUTOLISP FUNCTIONS

ActiveX	**VLA-GET-PAPERSPACE**
Array	**VLAX-3D-POINT**
AutoCAD Object Model	**VLAX-GET-ACAD-OBJECT**
Child Object	**VLAX-GET-PROPERTY**
Class	**VLAX-INVOKE-METHOD**
Collection	**VLAX-MAKE-SAFEARRAY**
COM	**VLAX-MAKE-VARIANT**
Compound Object Modeling	**VLAXNAME->VLA-OBJECT**
Data	**VLAX-PUT-PROPERTY**
Encapsulation	**VLAX-SAFEARRAY->LIST**
Event-Driven Programming	**VLAX-SAFEARRAY-FILL**
Inheritance	**VLAX-SAFEARRAY-GET-ELEMENT**
Inspect Tool	**VLAX-SAFEARRAY-GET-L-BOUNDS**
Method	**VLAX-SAFEARRAY-GET-U-BOUNDS**
Object	**VLAX-SAFEARRAY-PUT-ELEMENT**
Object Linking and Embedding	**VLAX-VARIANT-CHANGE-TYPE**
Object Oriented Programming	**VLAX-VARIANT-TYPE**
Parent Object	**VLAX-VARIANT-VALUE**
Polymorphism	**VLAX-VBARRAY**
Properties	**VLAX-VBBOOLEAN**
SafeArray	**VLAX-VBDOUBLE**

Spaghetti Programming	**VLAX-VBEMPTY**
Structured Programming	**VLAX-VBINTEGER**
Variant	**VLAX-VBLONG**
Visual Basic	**VLAX-VBNULL**
Visual Basic Application	**VLAX-VBOBJECT**
VLA-ADDCIRCLE	**VLAX-VBSINGLE**
VLA-DUMP-OBJECT	**VLAX-VBSTRING**
VLA-GET-ACTIVEDOCUMENT	**VL-LOAD-COM**
VLA-GET-MODELSPACE	**Wrapper Functions**

INTRODUCTION TO OBJECT ORIENTED PROGRAMMING

When computers were first developed, programmers constructed applications using a technique known as *spaghetti programming*. This technique produced applications that were unreliable, difficult to maintain, and almost impossible to follow. It wasn't until the 1970s that computer scientists made a major breakthrough in computer programming by introducing the concept of *structured programming*. The technique was developed as a means of simplifying computer programs so that they were easier to read and maintain. It allowed developers to produce applications that were more reliable and stable than those produced using the earlier method. Structured programming does this by assuming that all programs are interpreted by the computer, starting at the beginning of the application and proceeding until it reaches the end. Statements and expressions are executed one after the other following the order in which they are presented.

Structured programming allowed the developer to create applications in a modular fashion, meaning that the program is broken up into manageable portions. Each portion was defined as a subroutine that can be called by the main program. Structured programming also provided the developer with an avenue for making decisions based upon information that has been tested against a specific condition. Finally, structured programming introduced the concept of loops. Although structured programming helped transform the art of computer programming from an unsure, unreliable set of expressions strung together on a whim into the industry it is today, the technique was still lacking. In order to incorporate new features into an application, more subroutines had to be added and/or serious modifications made to the existing application. In most cases the developer was required to completely rewrite an application.

The next big advancement in programming came with the introduction of *event-driven programming*. Event-driven programming is the basis for the Graphical User Interface. A program remains idle until the operator performs some action that triggers an event, such as the moving of the mouse, or the pressing of a button.

Event-driven programming was employed in Chapters 6 and 7 in the development and management of programmable dialog boxes. The introduction of event-driven programming not only extended programming possibilities over previous methods of structured programming, it also made it possible for programmers to develop applications that were user friendly. However, event-driven programming still incorporates the techniques used in structured programming. This becomes apparent if the sample program (spot_elevation) featured at the end of Chapter 7 is re-examined. In the spot_elevation program, the user is presented with a dialog box to select the point to label, the layer to place the label on, as well as the type of marker, if any, to identify the point labeled. The dialog box and controlling program remain idle until the user makes these selections. When the user makes a selection in the dialog box, areas of the dialog box corresponding to the selection automatically update, indicating a selection has been made. In reality, the dialog box is either calling subroutines or assigned expressions to complete the task assigned by the user.

While event-driven programming has made its fair share of contributions to the advancement of computer programming, it was and is still not enough. Another push was needed before software could evolve at an astonishing rate. Even with event-driven programming, the developer or vendor was responsible for every aspect of an application. In other words, buying generic programming components was almost impossible.

It wasn't until computer scientists and developers started reinventing the way problems were addressed that the next major development in computer programming was made. When the developers stopped thinking about the processes involved in completing a task and started thinking about the actual data generated, the concept of object oriented programming (OOP) was born.

In *object oriented programming*, the data becomes the principal items in the program, instead of the process itself. OOP technology is based on the concepts of classes, objects, methods, and properties. Once the concept of object oriented programming was introduced, new languages started evolving (Java) and existing languages (C, Basic, Pascal, Visual LISP, etc.) were modified to include this new technology. Object oriented programming allows the developer to reuse applications that have been previously developed instead of reinventing the wheel every time. Before OOP can be incorporated into an AutoLISP application, the developer must first understand some of the basic concepts associated with OOP. These concepts are the backbone for incorporating ActiveX controls into AutoLISP.

CLASSES, OBJECTS, METHODS, AND PROPERTIES

In OOP, a *class* is a template used to create an object. The class defines the properties and methods that will be common to all objects within it. A class is a group of related objects that share common properties and methods. An *object* is an individual member of a class, or an instance of that class. Objects consist of two primary parts, the properties (also called data) and methods. The *methods* are the procedures or functions used to manipulate the data associated with an object. The information regarding the characteristics of the object is called the *properties*. These may include such information as the height, width, length, weight, and spatial location of the object in a graphical database.

ENCAPSULATION, POLYMORPHISM, AND INHERITANCE

All OOP languages use a procedure known as *encapsulation* to ensure the integrity of the object's properties from unauthorized or even inappropriate access. This is handled by letting the object control how the data is accessed. To accomplish this, the data is encapsulated by the object and therefore direct access to the data is not permitted without going through the object methods supplied. When a program treats different objects as if they were identical, and each object reacts as it was originally intended, then this is called *polymorphism*. If it weren't for polymorphism, then writing generic programs that are capable of handling a variety of classes would be difficult, if not impossible, to accomplish. When an object is created from a particular class, then that object inherits the methods and properties of that class. In OOP, an object can be created from other objects and therefore can inherit the methods and properties of the original object's class ,as well as any special methods and properties that may be associated with the original object (also called a parent). Inheritance is the factor in OOP that allows for program reusability. Existing objects and classes can be copied to create new objects and classes without the developer having to start from scratch. There are two types of inheritance associated with OOP: implementation inheritance and interface inheritance. In *implementation inheritance*, when a program activates a method that has been inherited, then it is the parent's methods that are executed and not the child's. With *interface inheritance*, the definitions of the methods are inherited and therefore the child must actually provide the means of processing the program's request.

INTRODUCTION TO ACTIVEX

From the introduction of OOP emerged a new technology, ActiveX, to better cope with some of the issues surrounding the creation of complex applications. ActiveX, which was introduced by Microsoft Corporation, was originally called OLE or Object Linking and Embedding Version 1. It was intended as a means of creating compound documents. In a compound document, elements created by other applications are either linked or embedded in a single document. The reasoning behind the creation of OLE 1 was to allow the user to focus more on the data contained within a document and not on the applications used to create the data. This was the first step in creating a technology that would allow various software packages and/or components of packages to share their services with other applications.

OLE version 1 soon underwent a refinement process. This process produced a new version of OLE where the version number as well as the acronym's meaning were dropped. Also, a new technology was introduced: COM (compound object modeling) which provided a means of establishing interaction between software libraries, applications, system software, etc. COM does this by defining a standard by which applications can communicate with one another.

In 1996, Microsoft introduced another new technology, ActiveX, using COM as its base. The technology, originally intended for Internet-based applications, soon began to emerge in other applications. Today, ActiveX is used to refer to all applications that use COM-based technologies.

OBJECT ORIENTED PROGRAMMING AND AUTOCAD

Object oriented programming is not a new concept to the AutoCAD user and developer. The primary focus of AutoCAD since its release has been objects. To illustrate this point, consider what happens when a line is placed on a layer. Depending upon the line's current property settings, the line can inherit characteristics from the layer (color and linetype). If an object is copied in AutoCAD, then the replica will have the same characteristics as the original.

ACTIVEX AND AUTOCAD

The introduction of ActiveX has opened up new possibilities that could not have been achieved through traditional programming techniques and philosophies. In 1998, Autodesk decided to make these possibilities available to the AutoLISP developer by including ActiveX in the Visual LISP IDE. This inclusion has opened the doors to unlimited potential for the AutoLISP programming language, in addition to providing the developer with a means of manipulating AutoCAD drawing objects faster than with traditional AutoLISP expressions. Without the inclusion of ActiveX controls in Visual LISP, reactors would not be possible.

UNDERSTANDING HOW ACTIVEX WORKS WITH AUTOCAD AND VISUAL LISP

ActiveX technology is centered on OOP. It makes extensive use of objects, methods, and properties. In OOP, not only are entities treated as objects, but the drawing and the AutoCAD application are viewed as objects as well. This viewpoint taken by ActiveX and OOP regarding AutoCAD is known as the AutoCAD object model. The AutoCAD object model is arranged in a tree structure format, where the AutoCAD application is considered to be the root. Access to AutoCAD objects contained within a drawing is only gained by following the appropriate branches of the object model (see Figure 9–1).

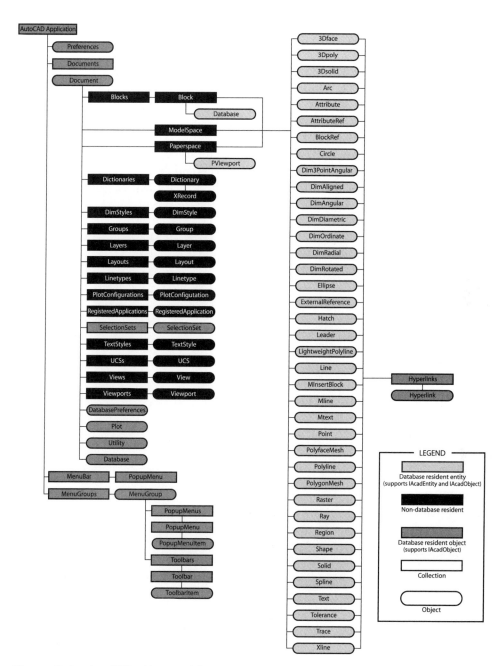

Figure 9–1 *AutoCAD object model*

A hatch pattern on a drawing can only be accessed by first going through the AutoCAD application object, to the document object, to the modelspace object, and finally to the hatch object. All objects contained within the AutoCAD object model contain at least one property. Before a property can be manipulated, its exact name must be known. All objects have at least one method associated with them. Some methods are typical to almost all AutoCAD objects, while others are applicable only to a specific object. Methods associated with AutoCAD objects are treated like AutoLISP functions. The *ActiveX and VBA Reference* manual can be used to determine which properties are available for specific AutoCAD objects and the methods used to modify them. The reference is available through either the Visual LISP or AutoCAD Help menus (see Figures 9–2, 9–3, and 9–4).

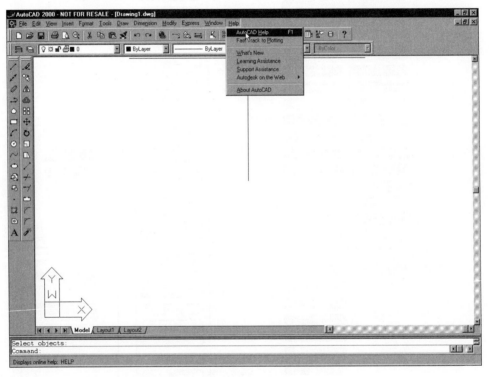

Figure 9–2 *Accessing the help files in AutoCAD*

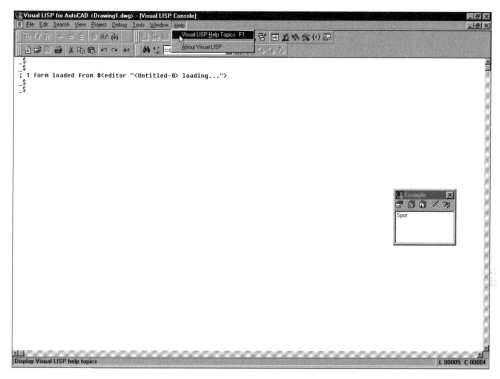

Figure 9–3 *Accessing the help files in Visual LISP*

Figure 9–4 *ActiveX and VBA Reference*

The reference can also be accessed by executing either the acad_vlr.hlp, acad_vlt.hlp, or acad_vlq.hlp file located in the ACAD2000 Help directory (see Figure 9–5).

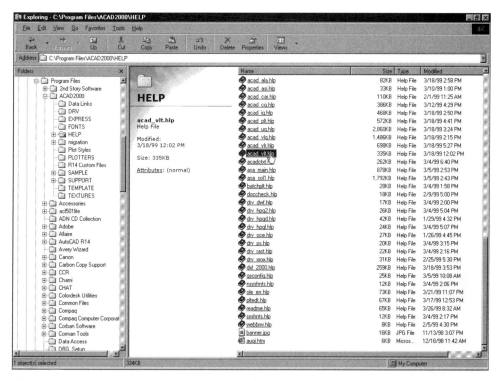

Figure 9–5 *Accessing the acad_vlt.hlp file in the ACAD2000\help directory*

The *ActiveX and VBA Reference* was originally set up for the Visual Basic programming language. Before it can be used by an AutoLISP programmer, a few key concepts must be understood. Autodesk has added numerous functions to the AutoLISP programming language for accessing ActiveX controls. These functions start with VLA-, VLAX-, and VLR-. The VLA- functions activate ActiveX support in AutoLISP programs by calling methods (VLA-) and retrieving (VLA-GET) and updating (VLA-PUT) property values. The VLAX- functions provide more general ActiveX functions such as utility and data conversion functions, dictionary handling functions, and curve measurement functions. The VLAX- functions also allow the developer to access objects from other applications as well as AutoCAD. To access the methods associated with an object created by an application other than AutoCAD, the VLAX-INVOKE-METHOD function should be used. To access and update the properties of the object just initialized, the VLAX-GET-PROPERTY and VLAX-PUT-PROPERTY functions should be used. The VLR- functions are used to provide support for AutoCAD reactors. (Reactors are covered in more detail in Chapter 10.)

Once the method's name has been obtained, the Visual LISP function used to call that method can easily be determined by adding the prefix VLA- to the method name. To add a circle to a drawing, the *addcircle* method must be used. To call the *addcircle* method from within an AutoLISP program, the VLA-ADDCIRCLE function must be employed. Determining which Visual LISP functions should be used to access which ActiveX methods is one thing; however, matching the supporting arguments to those functions is another. In Visual Basic, the object that the operation affects is listed first, followed by the method that is to be employed. The object and the method in Visual Basic are separated by a period, and any supporting arguments (radius, center point, length, height, etc.) are normally placed in parentheses. When this is converted to Visual LISP, the object becomes the first argument that is supplied, followed by any remaining arguments that the method may require. For example, the *addcircle* (RetVal = object.AddCircle (Center, Radius)) method when converted from Visual Basic to AutoLISP becomes (SETQ variable (VLA-ADDCIRCLE modelSpace (VLAX-3d-Point (GETPOINT "\nSelect center of Circle : ")) (GETREAL "\nEnter Radius of Circle : "))) (see Figure 9–6).

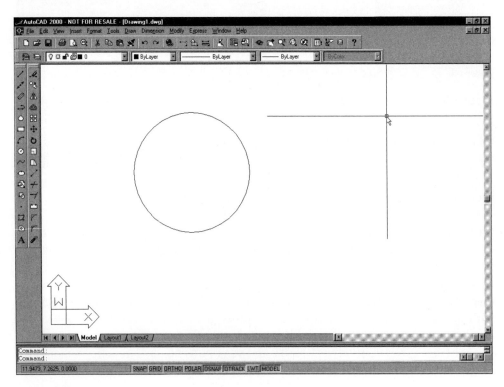

Figure 9–6 *Circle created using ActiveX controls*

Finally, all objects in the AutoCAD object model are grouped into collections. A *collection* is a grouping of all related AutoCAD objects into categories. The modelspace collection consists of all graphic entities contained within a drawing's model space.

 Note: When an AutoCAD object is created through ActiveX functions, the result will be that Visual LISP stores the object as a VLA object type.

USING ACTIVEX IN CONJUNCTION WITH AUTOLISP

Before ActiveX controls can be added to an AutoLISP application, the supporting code must be loaded into memory. To load ActiveX controls into memory, the VL-LOAD-COM (VL-LOAD-COM) function must be called before any other function calls are made to ActiveX controls. All function calls made to ActiveX controls before the supporting source code is loaded will result in an error. Once the supporting source code has been loaded into memory, a connection to the AutoCAD object model can be established by the VLAX-GET-ACAD-OBJECT (VLAX-GET-ACAD-OBJECT) function. Because the AutoCAD application is the root of the AutoCAD object model, a connection must be established to the AutoCAD application before access to objects contained within a drawing can be made. If the AutoLISP function VLAX-GET-ACAD-OBJECT is successful in evaluation, then the function returns a pointer to the AutoCAD application in the form of a VLA-Object (Visual LISP ActiveX Object).

Once a pointer has been established, a connection can be made to the current drawing. In AutoLISP this is accomplished by the VLA-GET-ACTIVEDOCUMENT (vla_get_activeDocument vla-Object type) function. This function retrieves the current document and returns it as a VLA-Object type. The function does require one argument, the VLA-Object type representing the AutoCAD application. After the document has been obtained, then access to non-graphic entities (objects) can be obtained through name-like properties such as Layer, Linetype, Style, etc.

Before graphic objects can be modified, the space (paper or model) where the object resides must be obtained. In Visual LISP, this is done by using the VLA-GET-MODELSPACE or VLA-GET-PAPERSPACE function. Both functions require one argument, the VLA-Object type representing the current document. To load the ActiveX controls and access the model space of the current drawing, the following expressions would be used:

```
(VL-LOAD-COM)
(SETQ application (VLAX-GET-ACAD-OBJECT)
      document (VLA-GET-ACTIVEDOCUMENT application)
      modelspace (VLA-GET-MODELSPACE document)
)
```

To access the paper space:

```
(VL-LOAD-COM)
(SETQ application (VLAX-GET-ACAD-OBJECT)
        document (VLA-GET-ACTIVEDOCUMENT application)
        paperspace (VLA-GET-PAPERSPACE document)
)
```

Once the object space has been obtained, then new objects can be created using ActiveX controls. To create the circle from the previous example, the following expressions would be added in model space:

```
(DEFUN c:activex_circle    ()
(SETQ       application (VLAX-GET-ACAD-OBJECT)
                            ;assigns a pointer to the AutoCAD application
document       (VLA-GET-ACTIVEDOCUMENT application)
                  ;Connects to the current active document
modelspace    (VLA-GET-MODELSPACE document)
                            ;Connects to the current drawings model space
variable      (VLA-ADDCIRCLE
                      ;Creates a circle using ActiveX controls at
              ;a location determined by the user
                  ;and a diameter determined by the user.
                modelSpace
(VLAX-3d-Point (GETPOINT "\nSelect center of Circle : "))
(GETREAL "\nEnter Radius of Circle : "))
)
)
```

To create the circle in paper space:

```
(DEFUN c:activex_circle    ()
(SETQ       application (VLAX-GET-ACAD-OBJECT)
                            ;assigns a pointer to the AutoCAD application
document       (VLA-GET-ACTIVEDOCUMENT application)
                  ;Connects to the current active document
modelspace    (VLA-GET-PAPERSPACE document)
                            ;Connects to the current drawings Paper space
variable      (VLA-ADDCIRCLE
                      ;Creates a circle using ActiveX controls at
              ;a location determined by the user
                  ;and a diameter determined by the user.
                PaperSpace
```

```
(VLAX-3d-Point (GETPOINT "\nSelect center of Circle : "))
(GETREAL "\nEnter Radius of Circle : "))
)
)
```

 Caution: When an object is supplied to an ActiveX function, that object must be supplied as a VLA-Object. Entity name extraction using the ENTGET function cannot be used with ActiveX functions unless they are converted using the VLAX-NAME->VLA-OBJECT function. Repeated calls to VLA-GET-ACAD-OBJECT, VLA-GET-ACTIVEDOCUMENT, VLA-GET-MODELSPACE, or VLA-GET-PAPERSPACE have a negative impact on the performance of AutoCAD and should be avoided. These functions should be called once and their returned values set to a variable.

VIEWING THE PROPERTIES OF AN OBJECT WITH THE INSPECTION TOOL

Once the supporting code for ActiveX controls has been loaded into memory and a connection has been established to the AutoCAD application and the current AutoCAD drawing, then other objects within the drawing can be accessed. As mentioned earlier, the properties associated with an object can be found by using the *ActiveX and Visual LISP Reference Guide*. Properties of an object can also be found by using the Visual LISP inspection tool (see Figure 9–7).

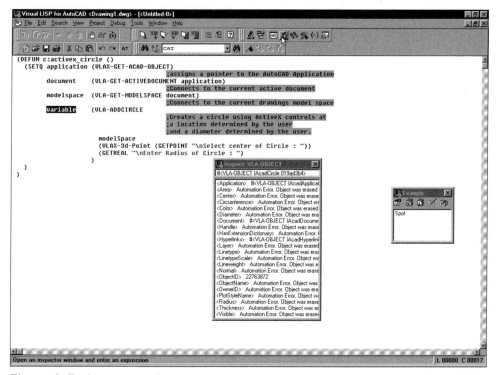

Figure 9–7 *Inspection tool*

The text contained between the less than (<) and greater than (>) signs is the property name. The text to the right of the property name is the data associated with that property. Properties that are identified as #<VLA-OBJECT…> are the exceptions to this rule. The #<VLA-Object refers to other ActiveX objects. To view the properties associated with an ActiveX object contained within the Inspect dialog box, either double-click the object name (see Figure 9–8), or right-click the property name and select Inspect (see Figures 9–9 and 9–10).

Figure 9–8 *Properties of ActiveX object*

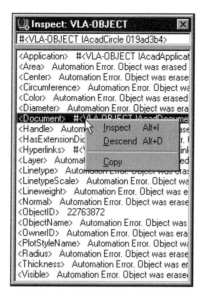

Figure 9–9 *Right-click on ActiveX object*

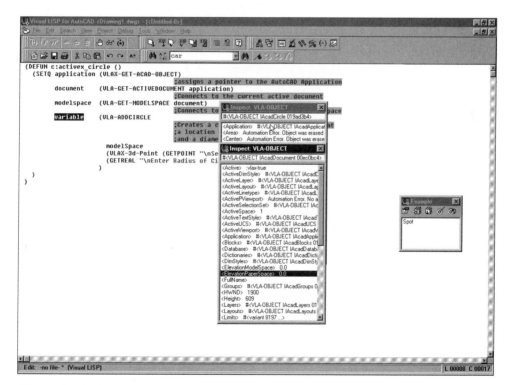

Figure 9–10 *Properties of ActiveX object*

WORKING WITH VISUAL BASIC DATA TYPES

Because the ActiveX reference guide is set up for Visual Basic programming, the developer must first become accustomed to a few key concepts used in Visual Basic before ActiveX controls can effectively be used in an AutoLISP program. These concepts include variants, single-precision floating-point number, double-precision floating-point number, long integer, arrays and SafeArrays.

Working with Variants

Unlike AutoLISP, variables in Visual Basic must be defined as a specific data type before they can be used. This allows the program to reserve the correct amount of memory for the variable. Valid Visual Basic data types assigned to variables are shown in the table below. If a variable is to be used in a program where the data type might change, then that variable can be defined as a *variant*. A variant is a variable that can contain different data types. Many of the variables used to pass information to an ActiveX control are in the form of a variant. This offers one advantage over using specifically defined variables: it allows for ActiveX controls that have been written in any language to understand the data type passed to it. In AutoCAD 2000, several AutoLISP functions have been added to use variants when integrating ActiveX controls into a program. These functions are VLAX-MAKE-VARIANT, VLAX-VARIANT-TYPE, VLAX-VARIANT-VALUE, and VLAX-VARIANT-CHANGE-TYPE.

The VLAX-MAKE-VARIANT (vlax-make-variant [value] [type]) function is used to create variants. The function has two optional arguments associated with it, *value* and *type*. The *value* argument assigns a value to the variant, while the *type* argument assigns the value data type. If the *value* argument is omitted, then the variant is created as an empty type (vlax-vbempty). If the type is not specified, then the variant is created using a default data type based on the value that was assigned. Table 9–1 lists valid data types that can be assigned to variants.

Table 9–1　Data Types and Visual LISP Functions Used to Create Them

Data Type	Description	AutoLISP Function used to Create Data Type	Variant Type
Integer	Contains an integer that can range from −32,768 to 32,767.	VLAX-VBINTEGER	2
Long integer	Contains an integer that can range from −2,147,483,648 to 2,147,483,647.	VLAX-VBLONG	3

Data Type	Description	AutoLISP Function used to Create Data Type	Variant Type
Single-precision floating-point number	Contains a real number that can range from -3.40 x 10^{38} to -1.40 x 10^{-45} for negative values and 1.4 x 10^{-45} to 3.40 x 10^{38} for positive values.	VLAX-VBSINGLE	4
Double-precision floating-point number	Contains a real number that can range from -1.80 x 10^{308} to -4.94 x 10^{-324} for negative values and 4.94 x 10^{-324} to 1.80 x 10^{308} for positive values.	VLAX-VBDOUBLE	5
String	Contains text that can range in length from 0 characters to approximately 2 billion characters.	VLAX-VBSTRING	8
Object	Contains an address for a specific object.	VLAX-VBOBJECT	9
Boolean	Contains the Boolean value of either true or false.	VLAX-VBBOOLEAN (vlax-true or vlax-false)	11

Courtesy of the AutoLISP Developers Guide

 Note: The VLAX-VBEMPTY, VLAX-VBNULL, and VLAX-VBARRAY functions can also be used to specify the data type of a variant. The VLAX-VBEMPTY (returns the value 0) function creates an un-initialized variant using default values. The VLAX-VBNULL (returns the value 1) function creates a variant that does not contain valid data and the VLAX-VBARRAY (returns the value (8192)) function creates a variant used to contain an array.

Once the VLAX-MAKE-VARIANT function has been evaluated, an integer value representing the variant that was created is returned. To create a variant that is assigned the integer value 3, the following expressions can be used:

```
_$ (SETQ variant_example (VLAX-MAKE-VARIANT 3))
#<variant 3 3>
_$ (SETQ variant_example (VLAX-MAKE-VARIANT 3 VLAX-VBLONG))
#<variant 3 3>
```

Note: In the first example the variant type is not specified, but because an integer value is assigned to the variant, the long integer type is automatically assigned to the variant. By default, the VLAX-MAKE-VARIANT function automatically assigns a long integer data type to all variants that have integers assigned to them. It is recommended that when a variant is created or assigned, a numeric data type should also be specified.

Once a variant has been declared, the developer can check the assigned data type by using the AutoLISP function VLAX-VARIANT-TYPE (vlax-variant-type var). This function, when supplied with a variable, returns an integer value representing the variant's type. To check the data type assigned to the variant created in the previous example, the following expression would be used:

```
_$ (VLAX-VARIANT-TYPE example_variant)
3                                        ; Returned Value
```

To extract the value of a variant, the VLAX-VARIANT-VALUE (vlax-variant-value var) function must be used. This function, when supplied with a variant, returns the assigned value. To extract the value assigned to the variant declared in the previous example, the following expression would be used:

```
(SETQ variant_value (VLAX-VARIANT-VALUE example_variant))
3                                       ; Value returned in the form of an
                                        ; integer. If the value assigned to the
                                        ; variant had been text then the returned
                                        ; value would be in the form of text.
```

A variant data type can be changed at any time using the VLAX-VARIANT-CHANGE-TYPE (vlax-variant-change-type var type) function. This function changes the variant to a specified type. To change the data type of the variant in the previous example, the following expressions would be used:

```
_$ (SETQ example_variant
       (VLAX-VARIANT-CHANGE-TYPE
     example_variant
     VLAX-VBDOUBLE
     )
)
#<variant 5 5>                          ; Variant returned
```

Arrays

Visual Basic provides a means of assigning more than one value to a variable, known as an *array*. An array is a collection of data assigned to a single variable with all values sharing the same data type. Unlike AutoLISP, an array in Visual Basic must first be defined and then an upper limit specified. In other words, the number of elements that populate the array must be given when the array is first defined. To create an array consisting of 25 integers in Visual Basic, the following statement would be used:

```
DIM example_array (1 to 25) as Integer
```

In this example, the DIM function is used to declare the variable example_array while the argument (1 to 25) specifies the maximum number of entities (25) that will populate the array as well as the starting index number (1) used to assign and recall values assigned to the array. Finally, the as Integer function determines the data type of the array. Arrays can be constructed that contain multiple sets of lists. This type of array is known as a *multidimensional array*. In Visual LISP, whenever an array is used to pass information to an ActiveX control, the array must be in the form of a *SafeArray*. A SafeArray is one where values cannot be assigned outside the specified limits or boundaries. This eliminates the possibility of a data exception error occurring. To define a SafeArray in Visual LISP, the VLAX-MAKE-SAFEARRAY must be used. To populate the array, the VLAX-SAFEARRAY-PUT-ELEMENT and VLAX-SAFEARRAY-FILL functions are used.

Creating the Array

In AutoLISP, an array is created through the VLAX-MAKE-SAFEARRAY (vlax-make-safearray type '(1-bound . u-bound) ['(1-bound . u-bound)...)] function. This function has two required arguments, *type* and *'(1-bound . u-bound)*. The *type* argument determines the data type of the information stored in the array. The data type can be an integer, long integer, single-precision floating-point number, double-precision floating-point number, string, object, boolean or variant. The second argument is the upper limit and starting index number (lower limit) of the array. The argument is supplied to the function in the form of a dotted pair and must either follow a single quote (') or be supplied by the CONS function. The first number in the dotted pair represents the lower limit of the array, while the second number represents the upper. To create a single dimensional array where the lower limit is set at 0 and the upper is set at 10, the following expression would be used:

```
_$ (setq example_array (vlax-make-safearray vlax-vbstring (cons 0 10)))
#<safearray...>
```

Or the following expression could be used:

```
_$ (setq example_array (vlax-make-safearray vlax-vbstring '( 0 . 10)))
#<safearray...>
```

In this example, the lower limit of the array is set to 0 and the upper is set to 10. This produces an array that can contain 11 elements (example_array 0, example_array 1, example_array 2, etc.). To create a multidimensional array, more than one upper and lower boundary would be specified. For example:

```
(setq example_array_1 (vlax-make-safearray
                vlax-vbstring
                (cons 0 1)
                (cons 2 3)
                (cons 4 5)
                )
```

The first example creates a three-dimensional array where the first dimension starts at 0 and ends at 1, the second dimension starts at 2 and ends at 3, and the third dimension starts at 4 and ends at 5.

```
        example_array_2 (vlax-make-safearray
                vlax-vbstring
                (cons 0 1)
                (cons 2 3)
                )
```

The second example creates a two-dimensional array where the first dimension starts at 0 and ends at 1, and the second dimension starts at 2 and ends at 3.

```
        example_array_3 (vlax-make-safearray
                vlax-vbstring
                (cons 0 1)
                (cons 2 3)
                (cons 4 7)
                (cons 8 10)
                )
    )
```

The last example creates a four-dimensional array where the first dimension starts at 0 and ends at 1, the second dimension starts at 2 and ends at 3, the third dimension starts at 4 and ends at 7, and the last dimension starts at 8 and ends at 10.

Populating the Array

Once an array has been created, it can then be populated with data. This is done through either the VLAX-SAFEARRAY-PUT-ELEMENT (vlax-safearray-put-element var index... value) or VLAX-SAFEARRAY-FILL (vlax-safearray-fill var 'element-values) function. While either function can be used to populate an array, the methods that these functions use to achieve the end result are quite different. The VLAX-SAFEARRAY-FILL function requires that all elements of the array be furnished at one time. If a multidimensional array is populated with this function, then all dimensions must be furnished. To populate an array that starts at 0 and contains four elements, the following expressions would be used:

```
_$ (setq example (vlax-make-safearray vlax-vbstring (cons 0 3)))
#<safearray...>
_$ (vlax-safearray-fill example '("1" "2" "3" "4"))
#<safearray...>
```

When an array is to be populated one element at a time, then the VLAX-SAFEARRAY-PUT-ELEMENT function must be used. This function can also be used to change the value of a single element in an array. The first argument associated with this function is the name of the SafeArray where the element belongs. This is followed by the index number of the element and finally the value. To assign a new value to the second element in the previous array, the following expression would be used:

```
(VLAX-SAFEARRAY-PUT-ELEMENT example 1 "New Value")
```

When a multidimensional array is populated or a value reassigned, then the starting index number of the dimension as well as the element number must be specified. For example:

```
_$ (setq example_1 (vlax-make-safearray vlax-vbString '(1 . 2) '(1 . 2)
))
#<safearray...>                  ; Returned value
_$ (vlax-safearray-put-element example_1 1 1 "a")
"a"                              ; Returned Value
_$ (vlax-safearray-put-element example_1 1 2 "b")
"b"                              ; Returned Value
_$
```

Viewing the Contents of a SafeArray

Once a value has been assigned to an element in an array, that value may be recalled through either the VLAX-SAFEARRAY-GET-ELEMENT (vlax-safearray-get-element var element...) or VLAX-SAFEARRAY->LIST (vlax-safearray->list var) function. While the VLAX-SAFEARRAY-GET-ELEMENT function retrieves the value of one element in the array, the VLAX-SAFEARRAY->LIST function returns the values of all elements. AutoLISP also provides a means of retrieving the upper and lower boundaries of a SafeArray. This is achieved through either the VLAX-SAFEARRAY-GET-L-BOUNDS (vlax-safearray-get-l-bound var dim) or the VLAX-SAFEARRAY-GET-U-BOUNDS (vlax-safearray-get-u-bound var dim) function.

Using SafeArrays in ActiveX Controls

Before ActiveX will accept a SafeArray as an argument, the SafeArray must be passed in the form of a variant through the VLAX-MAKE-VARIANT (vlax-make-variant [value] [type]) function. Once the array has been defined and populated, the array must be converted to a variant or ActiveX will reject the array. When an ActiveX function requires a three-element array of doubles, AutoLISP provides a shortcut function. The VLAX-3D-POINT function (vlax-3D-point list) or (vlax-3D-point x y [z]) creates a SafeArray of three doubles and then converts the SafeArray to a variant. This function is used primarily with ActiveX controls that require a point. The point argument can be furnished in the form of a list containing all three points or each point can be furnished individually. For example:

```
_$ (VLAX-3D-POINT (getpoint "\nSelect point : "))
#<variant 8197 ...>
_$ (VLAX-3D-POINT '(2.0 4.0))
#<variant 8197 ...>
_$ (VLAX-3D-POINT 2.4 5.6)
#<variant 8197 ...>
_$
```

In the first example, the point list is furnished to the VLAX-3D-POINT function using the GETPOINT function, while in the second example the points are supplied in a list. In the third example the elements of the point are passed to the VLAX-3D-POINT function as numbers instead of a list. In the second and third examples, the third element is omitted. When this occurs, AutoLISP automatically assigns a value of zero to the third element.

USING ACTIVEX TO ACCESS AN OBJECT'S PROPERTIES

Using the Inspect tool to view an object's properties is not the only means available to the AutoLISP developer. Autodesk has also provided numerous functions that can access an object's properties as well as change them using ActiveX.

Typically, the functions used to retrieve an object's properties start with the VLA-GET-prefix and conclude with the property name being extracted (vla-get-property object). These functions all follow the same basic format, function name followed by the *object* that contains the property to be extracted. This holds true for the functions designed to change the value of an object's property. These functions all start with the VLA-PUT- prefix and conclude with the name of the property to change. These functions also follow a basic format, function name followed by the *object* that the property belongs to and its value (vla-put-property object new-value).

Finally, the properties associated with an object can be listed using the VLA-DUMP-OBJECT (vlax-dump-object obj [T]) function. This function, when supplied with an object, dumps a list of the properties and their current values to the text screen. To list the properties associated with the model space object in the current drawing, the following expression would be used:

```
_$ (SETQ application     (VLAX-GET-ACAD-OBJECT)
                           ;assigns a pointer to the AutoCAD
                           ; Application
        document        (VLA-GET-ACTIVEDOCUMENT application)
                           ; Connects to the current active document
        modelspace      (VLA-GET-MODELSPACE document)
)
#<VLA-OBJECT IAcadModelSpace 019acda4>
_$ (VLAX-DUMP-OBJECT modelspace)
; IAcadModelSpace: A special Block object containing all model space
  entities
; Property values:
;    Application (RO) = #<VLA-OBJECT IAcadApplication 00ac8928>
;    Count (RO) = 2
;    Document (RO) = #<VLA-OBJECT IAcadDocument 00ec0bc4>
;    Handle (RO) = "1F"
;    HasExtensionDictionary (RO) = 0
;    IsLayout (RO) = -1
;    IsXRef (RO) = 0
;    Layout (RO) = #<VLA-OBJECT IAcadLayout 019a9044>
;    Name = "*Model_Space"
;    ObjectID (RO) = 22760696
;    ObjectName (RO) = "AcDbBlockTableRecord"
;    Origin = (0.0 0.0 0.0)
```

```
;   OwnerID (RO) = 22760456
;   XRefDatabase (RO) = AutoCAD.Application: No database
T
_$
```

 Note: In this example, the value returned by the VLA-DUMP-OBJECT function is shown in bold.

Another way that the developer can determine the properties associated with an object is by using the VLAX-PROPERTY-AVAILABLE-P (vlax-property-available-p obj prop [check-modify]) function. This function, when supplied with an object and the name of a property, determines if the specified property belongs to the object and if that property can be modified. AutoLISP also provides a VLAX-METHOD-APPLICATION-P (vlax-method-applicable-p obj method) function that can be used to determine if an object can employ a method. Both functions return T if the property or method does exist for the specified object and nil if it does not.

WORKING WITH COLLECTIONS

It may often be necessary to perform an action on an entire collection of objects. When this occurs, the AutoLISP VLA-MAP-COLLECTION (vlax-map-collection obj function) function can be used. This function applies the specified function to all objects contained within the specified object. To generate a list of the properties associated with the two circles shown in Figure 9–11, the following expressions would be used.

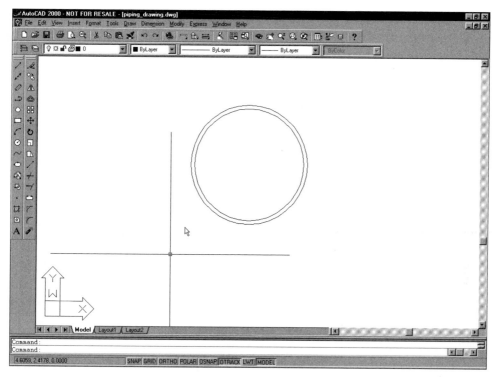

Figure 9–11　*Circle in an AutoCAD drawing*

```
(vl-load-com)
(SETQ application (VLAX-GET-ACAD-OBJECT)
                        ;assigns a pointer to the AutoCAD Application
       document    (VLA-GET-ACTIVEDOCUMENT application)
                        ;Connects to the current active document
      modelspace  (VLA-GET-MODELSPACE document)
)
(vlax-map-collection
   (vla-get-ModelSpace Document)
   'vlax-dump-object
)
```

The list of properties generated:

```
;  IAcadCircle: AutoCAD Circle Interface          ;Outer Circle
;  Property values:
;    Application (RO) = #<VLA-OBJECT IAcadApplication 00ac8928>
```

```
;   Area = 12.5664
;   Center = (7.29982 5.43683 0.0)
;   Circumference = 12.5664
;   Color = 256
;   Diameter = 4.0
;   Document (RO) = #<VLA-OBJECT IAcadDocument 00ebf6f4>
;   Handle (RO) = "2E"
;   HasExtensionDictionary (RO) = 0
;   Hyperlinks (RO) = #<VLA-OBJECT IAcadHyperlinks 019ab304>
;   Layer = "0"
;   Linetype = "ByLayer"
;   LinetypeScale = 1.0
;   Lineweight = -1
;   Normal = (0.0 0.0 1.0)
;   ObjectID (RO) = 22711664
;   ObjectName (RO) = "AcDbCircle"
;   OwnerID (RO) = 22711544
;   PlotStyleName = "ByLayer"
;   Radius = 2.0
;   Thickness = 0.0
;   Visible = -1
; IAcadCircle: AutoCAD Circle Interface                 ;Inner Circle
; Property values:
;   Application (RO) = #<VLA-OBJECT IAcadApplication 00ac8928>
;   Area = 11.0447
;   Center = (7.29982 5.43683 0.0)
;   Circumference = 11.781
;   Color = 256
;   Diameter = 3.75
;   Document (RO) = #<VLA-OBJECT IAcadDocument 00ebf6f4>
;   Handle (RO) = "2F"
;   HasExtensionDictionary (RO) = 0
;   Hyperlinks (RO) = #<VLA-OBJECT IAcadHyperlinks 019ab0f4>
;   Layer = "0"
;   Linetype = "ByLayer"
;   LinetypeScale = 1.0
;   Lineweight = -1
;   Normal = (0.0 0.0 1.0)
;   ObjectID (RO) = 22711672
;   ObjectName (RO) = "AcDbCircle"
;   OwnerID (RO) = 22711544
```

```
;    PlotStyleName = "ByLayer"
;    Radius = 1.875
;    Thickness = 0.0
;    Visible = -1
; 3 forms loaded from #<editor "<Untitled-16> loading...">
_$
```

When a series of expressions is to be evaluated by each object in a collection, then the VLAX-FOR (vlax-for symbol collection [expression1 [expression2 ...]]) function must be used. The first argument associated with this function is the symbol that is to be assigned to each object in the collection. This is followed by the variable representing the VLA-Object for the collection. Finally, the function is supplied with the expressions used to evaluate each object. For example:

```
(vl-load-com)
(SETQ application (VLAX-GET-ACAD-OBJECT)
                        ;assigns a pointer to the AutoCAD Application
      document     (VLA-GET-ACTIVEDOCUMENT application)
                        ;Connects to the current active document
      modelspace  (VLA-GET-MODELSPACE document)
)
(vlax-for for-item
              (vla-get-modelspace
                 (vla-get-activedocument (vlax-get-acad-object))
              )
   (vlax-dump-object for-item)    ; lists object properties
   (princ "\n")
   (princ "\n————       Next Entity        ———— ")
   (princ "\n")
)
```

The resulting evaluation:

```
_$
; IAcadCircle: AutoCAD Circle Interface
; Property values:
;    Application (RO) = #<VLA-OBJECT IAcadApplication 00ac8928>
;    Area = 12.5664
;    Center = (7.29982 5.43683 0.0)
;    Circumference = 12.5664
;    Color = 256
;    Diameter = 4.0
```

```
;   Document (RO) = #<VLA-OBJECT IAcadDocument 00ebf6f4>
;   Handle (RO) = "2E"
;   HasExtensionDictionary (RO) = 0
;   Hyperlinks (RO) = #<VLA-OBJECT IAcadHyperlinks 019b3744>
;   Layer = "0"
;   Linetype = "ByLayer"
;   LinetypeScale = 1.0
;   Lineweight = -1
;   Normal = (0.0 0.0 1.0)
;   ObjectID (RO) = 22711664
;   ObjectName (RO) = "AcDbCircle"
;   OwnerID (RO) = 22711544
;   PlotStyleName = "ByLayer"
;   Radius = 2.0
;   Thickness = 0.0
;   Visible = -1
──────────          Next Entity          ──────────
; IAcadCircle: AutoCAD Circle Interface
; Property values:
;   Application (RO) = #<VLA-OBJECT IAcadApplication 00ac8928>
;   Area = 11.0447
;   Center = (7.29982 5.43683 0.0)
;   Circumference = 11.781
;   Color = 256
;   Diameter = 3.75
;   Document (RO) = #<VLA-OBJECT IAcadDocument 00ebf6f4>
;   Handle (RO) = "2F"
;   HasExtensionDictionary (RO) = 0
;   Hyperlinks (RO) = #<VLA-OBJECT IAcadHyperlinks 019b3924>
;   Layer = "0"
;   Linetype = "ByLayer"
;   LinetypeScale = 1.0
;   Lineweight = -1
;   Normal = (0.0 0.0 1.0)
;   ObjectID (RO) = 22711672
;   ObjectName (RO) = "AcDbCircle"
;   OwnerID (RO) = 22711544
;   PlotStyleName = "ByLayer"
;   Radius = 1.875
;   Thickness = 0.0
;   Visible = -1
```

```
            ————————          Next Entity          ————————
; 3 forms loaded from #<editor "<Untitled-17> loading...">
_$
```

FREEING MEMORY AND RELEASING OBJECTS

Once an application has completed its assigned task, all references to objects within the drawing should be removed. In AutoLISP this is accomplished by releasing the object using the VLAX-RELEASE-OBJECT (vlax-release-object obj) function. This function, when supplied with an object pointer, removes that pointer, thus allowing AutoCAD to reclaim the once reserved memory when needed. Once a link to an object has been removed, the only way that object can accessed again is by reestablishment of the link.

ACCESSING APPLICATIONS OTHER THAN AUTOCAD WITH ACTIVEX

Although ActiveX offers the AutoLISP developer a means of creating and manipulating the AutoCAD drawings using functions that are must faster than traditional AutoLISP, the real advantage of using ActiveX is its ability to access other applications. For example, the developer could use Microsoft Excel to create a bill of materials and then import it into an AutoCAD drawing.

Before an AutoLISP expression can be written that will interact with other ActiveX applications, the developer will need to refer to the documentation provided for the other applications to learn the application object, method, and property names and how to access them. Normally, this information can be obtained from the application's online help file.

Establishing a Connection to an Application

Before information can be exchanged between an AutoLISP program and an external application, a connection to that application object must be established. Currently in AutoLISP there are three functions that enable an AutoLISP program to connect to an external application object: VLAX-GET-OBJECT (vlax-get-object prog-id), VLAX-CREATE-OBJECT (vlax-create-object prog-id), and VLAX-GET-OR-CREATE-OBJECT (vlax-get-or-create-object prog-id).

The VLAX-GET-OBJECT function establishes a connection to a currently running instance of an external application. If a developer is creating an AutoLISP application that writes out information to an Excel spreadsheet, then a connection to the Excel object can be made using the VLAX-GET-OBJECT function as long as Excel is currently running. If the application is not currently running or if the VLAX-GET-OBJECT function should return nil, then a connection to the specified application can be created through the AutoLISP function VLA-CREATE-OBJECT. This function creates a new instance of the target application object, after which it creates a connection to the instance.

An alternative to the get and create VLAX functions is the VLAX-GET-OR-CREATE-OBJECT function. This function attempts to make a connection to the specified application object. If the connection cannot be established, then the function creates an instance of the application and then establishes a connection.

Once an instance of an application object has been created, the application will not be visible until the developer instructs the application to make it visible. This is done by the VLA-PUT-VISIBLE function. To make an application visible, the function is set to :VLA-TRUE.

Accessing an Application Object's Methods, Properties, and Constants

Before the methods, properties, or constants associated with an application's object can be utilized by an AutoLISP application, their supporting code must be made available to AutoLISP. This can be achieved by either importing the application's type library or using AutoLISP to make direct contact with that object's methods, properties, or constants. When an object's type library is imported into Visual LISP, AutoCAD automatically generates a set of wrapper functions for the imported features. A wrapper function is a standard function that encapsulates the code used by the application object (the VLA functions in Visual LISP are merely wrapper functions created from the AutoCAD type library). In AutoLISP this is done by using the VLAX-IMPORT-TYPE-LIBRARY (vlax-import-type-library :tlb-filename filename [:methods-prefix mprefix :properties-prefix pprefix :constants-prefix cprefix]) function. The VLAX-IMPORT-TYPE-LIBRARY function has one required argument, the file name. The *filename* argument is a string that specifies the name of the type library to import. Legal file types that this function can import are Type Libraries (tlb), Object Libraries (olb), Executables (exe), Dynamic Linked Library (dll), compound documents holding a type library, or any file format understood by the LoadTypeLib API. The remaining arguments specify the prefixes to append to the object's methods, properties, and constants. For example, to import the object library associated with Microsoft Excel, the following syntax would be used:

```
(vlax-import-type-library
      :tlb-filename
        "C:\\Program Files\\Microsoft Office\\Office\\excel8.olb"
      :methods-prefix
    "MS-Excelm-"
      :properties-prefix
    "MS-Excelp-"
      :constants-prefix
    "MS-Excelc-"
    )
```

Once the application object's library has been imported into Visual LISP, then the wrapper function created can be accessed by the AutoLISP application and Visual LISP's symbol features (inspection windows, watch windows, Apropos, and symbol service). Importing an applications object's library can be beneficial to the developer; however, it can also have a dramatic impact on the application being developed by incorporating unnecessary methods, properties, and constants, thus adding to the amount of memory required by the application. To circumvent the adverse affects of loading an entire application object's library, AutoLISP has three functions that allow the developer to access an application's object without loading the application's object library. They are VLAX-INVOKE-METHOD (vlax-invoke-method obj method arg [arg...]), VLAX-GET-PROPERTY (vlax-get-property object property), and VLAX-PUT-PROPERTY (vlax-put-property obj property arg). The VLAX-INVOKE-METHOD function provides the developer with a means of calling ActiveX methods, while the VLAX-GET-PROPERTY and VLAX-PUT-PROPERTY functions allow the developer to retrieve and set an object's property.

APPLICATION - USING ACTIVEX TO WRITE THE RESULTS FROM AN AUTOLISP APPLICATION TO WORD

To better illustrate how ActiveX may be incorporated into an AutoLISP application, the following program automatically calculates the equivalent resistance of a series circuit and then prints the results to the screen (using the ALERT function) and writes them to a Microsoft Word file. The program uses the VLA-IMPORT-TYPE-LIBRARY function to create the wrapper functions from the Microsoft Word type library. Once the wrapper functions have been created, an instance of the object application is either generated or if the application is already running, a connection is established using VLAX-GET-OR-CREATE-OBJECT. After the connection has been established, the object is made visible using the expression (vla-put-visible MS-Word :vlax-true). Finally, the results of the calculations are written to the Word file using the expression (MS-Wordm-InsertAfter range string).

```
;;;******************************************************************
;;; Program Name: ActiveXSeries.lsp
;;; Program Purpose: Calculate the voltage drop for a group of
;;;                  resistors, equivalent resistance,
;;;                  Total Current
;;;
;;; Programmed By: James Kevin Standiford
;;; Date: 3/21/2000
;;;******************************************************************
```

```
(defun c:series    (/ volt l_length)
  (setvar "cmdecho" 0)
;;;***~~~~~~New~~~~~~***~~~~~~~NEW~~~~~~~~***~~~~~~~~NEW~~~~~~
  (vl-load-com)
  (if (equal nil MS-Wordc-wd100Words)
     (vlax-import-type-library
      :tlb-filename
       "C:\\Program Files\\Microsoft Office\\Office\\msword8.olb"
      :methods-prefix
      "MS-Wordm-"
       :properties-prefix
      "MS-Wordp-"
       :constants-prefix
      "MS-Wordc-"
     )
  )
  (setq MS-Word (vlax-get-or-create-object "Word.Application.8"))
  (vla-put-visible MS-Word :vlax-true)
  (setq    documents      (vla-get-documents MS-Word)
   new-document (MS-Wordm-add documents)
   paragraphs    (MS-Wordp-get-paragraphs new-document)
   page          (MS-Wordp-get-last paragraphs)
   range         (MS-Wordp-get-range page)
   res           (ssget "x" (list (cons 0 "insert") (cons 2 "resistor")))
  )
;;;***~~END New~~~~~~***~~~~END NEW~~~~~~~~***~~~~END NEW~~~~~~
  (if (/= res nil)
    (progn
      (setq rcnt      0
        volt          (getreal "\nEnter voltage : ")
        l_length      (sslength res)
        ent           (entget
                         (entnext (cdr (assoc -1 (entget (ssname res
  rcnt)))))
                      )
        r_list        (list (atof (cdr (assoc 1 ent))))
        ent_name_list (list
                        (cdr
                          (assoc -1
                             (entget (entnext (cdr (assoc -1 ent)))))
```

```
                                        )
                                      )
                                    )
              r_num
                              (list
                                (cdr
                                  (assoc
                                    1
                                    (entget
                                      (entnext
                                        (cdr
                                          (assoc
                                            -1
                                            (entget (entnext  (cdr  (assoc  -1 ent)))
                                            )
                                          )
                                        )
                                      )
                                    )
                                  )
                                )
              rcnt              (1+  rcnt)
         )
         (while  (<  rcnt  l_length)
       (setq
         ent              (entget
                              (entnext  (cdr  (assoc  -1  (entget  (ssname  res
rcnt)))))
                            )
         r_list
                            (append r_list
                                  (list  (atof  (cdr  (assoc  1  ent))))
                            )
         ent_name_list
                            (append
                              ent_name_list
                              (list (cdr
                                    (assoc
                                      -1
```

```
                                              (entget (entnext (cdr (assoc -1 ent)))
                                        )
                                   )
                                 )
                              )
                            )
          r_num
                        (append
                          r_num
                          (list
                            (cdr
                              (assoc
                                1
                                (entget
                                  (entnext
                                    (cdr (assoc
                                            -1
                                            (entget
                                              (entnext
                                                (cdr (assoc -1 ent))
                                              )
                                            )
                                          )
                                      )
                                  )
                                )
                              )
                            )
                          )
                        )
          rcnt          (1+ rcnt)
        )
      )
    (setq list_cnt (- (length r_list) 1)
      eq_res
                (nth 0 r_list)
      l_cnt     1
    )
    (while (<= l_cnt list_cnt)
(setq resistor (nth l_cnt r_list))
```

```
(if (/= resistor nil)
  (setq eq_res
        (+ eq_res resistor)
  )
)
(setq l_cnt (1+ l_cnt))
  )
  (setq name        (GETVAR "dwgname")
    length_string   (STRLEN name)
    name_truncated  (SUBSTR name 1 (- length_string 3))
    result_file             (open (STRCAT (getvar "dwgprefix")
                                   name_truncated
                                   ".RLT"
                            )
                            "a"
                   )
    amperage        (/ volt eq_res)
  )
  (princ "\n" result_file)
  (setq string
    (strcat
      (princ
        (strcat
          "\nThe equivalent resistance for the circuit on Drawing :
"
          (getvar "dwgname")
          "  is "
          (rtos eq_res)
          " ohms"
        )
        result_file
      )
      (princ
        (strcat "\nThe voltage for this circuit is " (rtos volt))
        result_file
      )
      (princ (strcat "\nThe total current for this circuit is "
                     (rtos amperage)
                     " amps"
            )
```

```
                          result_file
                  )
              )
          )
        (setq l_cnt 0)
        (while (<= l_cnt list_cnt)
     (setq
        string (strcat
                  string
                  (princ (strcat
                            "\nThe voltage drop for resistor "
                            (nth l_cnt r_num)
                            " is "
                            (setq g (rtos (* (nth l_cnt r_list)
                                             amperage
                                       )
                                  )
                            )
                            " volts"
                         )

                         result_file
                )
            )
      )
      (setq resistor_ent (nth l_cnt ent_name_list)
            entg         (entget resistor_ent)
            entg         (subst (cons 1 g) (assoc 1 (entget resistor_ent))
     entg)
       )
       (entupd (cdr (assoc -1 (entmod entg))))
       (setq l_cnt (1+ l_cnt))
          )
;;;***~~~~~~New~~~~~~***~~~~~~~NEW~~~~~~~~***~~~~~~~~NEW~~~~~~
        (MS-Wordm-InsertAfter range string)
;;;***~~END New~~~~~~***~~~END NEW~~~~~~~~***~~~~END NEW~~~~~~
        (alert string)
        (close result_file)
        (command "modemacro"
             (strcat "The voltage used for this circuit was "
```

```
                        (rtos volt)
                        " volts"
                )
            )
          (princ)
        )
      )
    (if (= res nil)
        (alert "No Resistors Found : ")
    )
    (princ)
)
(princ "\nEnter series to start program :")
(princ)
```

SUMMARY

Structured programming was developed to allow programmers to create applications that are more reliable and easier to maintain. Structured programming accomplishes this by allowing the developer to create applications in a modular fashion. The program is broken up into manageable portions; each portion can then be defined as a subroutine that can be called by the main program. Structured programming also provides the developer with an avenue for making decisions based upon information that has been tested against a specific condition. Finally, structured programming standardized the use of loops in computer programs.

In event-driven programming, a program remains idle until the operator performs some action that triggers an event, such as the moving of the mouse, or the pressing of a button. Event-driven programming not only extends the possibilities of programming, it also makes it possible for programmers to develop applications that are user friendly.

In object oriented programming (OOP), the data becomes the principal items in the program instead of the process. OOP technology is based upon the concepts of classes, objects, methods, and properties. In OOP, a class is a template used to create an object. The class defines the properties and methods that will be common to all objects belonging to a particular class. Objects consist of two primary parts, the properties (also called data) and methods. The methods are the procedures or functions used to manipulate the data associated with an object. The information regarding the characteristics of the object is called the properties.

All OOP languages use a procedure known as encapsulation to ensure the integrity of the object's properties from unauthorized or even inappropriate access. When a program treats different objects as if they were identical and each object reacts as it was originally intended, then this is called polymorphism. When an object is created from a particular class, then that object inherits the methods and properties of that class. In OOP an object can be created from other objects and therefore inherit the methods and properties of the original object's class, as well as any special methods and properties that may be associated with the original object (also called a parent). Inheritance is the factor in OOP that allows for program reusability. There are two types of inheritance associated with OOP: implementation inheritance and interface inheritance.

ActiveX is an object-oriented technology with its foundation based on the concept of compound object modeling. Before ActiveX controls can be added to an AutoLISP application, the supporting code must be loaded into memory. To load ActiveX controls into memory, the VL-LOAD-COM function must be called before any other function calls are made to ActiveX controls. Once the supporting source code has been loaded into memory, a connection to the AutoCAD object model can be established by the VLAX-GET-ACAD-OBJECT function. After a connection has been made to the AutoCAD application, a pointer must be established to the current drawing. In AutoLISP this is accomplished by the VLA-GET-ACTIVEDOCUMENT function. Before a graphic object can be modified, the space (paper or model) where the object resides must be obtained. This is done by the VLA-GET-MODELSPACE or VLA-GET-PAPERSPACE function.

Before information can be exchanged between an AutoLISP program and an external application, a connection to that application object must be established. Currently in AutoLISP there are three functions that enable an AutoLISP program to connect to an external application object: VLAX-GET-OBJECT, VLAX-CREATE-OBJECT, and VLAX-GET-OR-CREATE-OBJECT.

The VLAX-GET-OBJECT function establishes a connection to a currently running instance of an external application. If the application is not currently running or if the VLAX-GET-OBJECT function should return nil, then a connection to the specified application can be created by the AutoLISP function VLA-CREATE-OBJECT. An alternative to the get and create VLAX functions is the VLAX-GET-OR-CREATE-OBJECT function. This function attempts to make a connection to the specified application object. If the connection cannot be established, then the function creates an instance of the application and then establishes a connection.

Once an instance of an application object has been created, the application will not be visible until the developer instructs the application to make it visible by using the VLA-PUT-VISIBLE function. To make an application visible, the function is set to :VLA-TRUE.

Before the methods, properties, or constants associated with an application's object can be utilized by an AutoLISP application, their supporting code must be made available to AutoLISP. This can be done by either importing the application's type library or using AutoLISP to make direct contact with that object's methods, properties, or constants. When an object's type library is imported into Visual LISP, AutoCAD automatically generates a set of wrapper functions for the imported features. A wrapper function is a standard function that encapsulates the code used by the application object.

REVIEW QUESTIONS AND EXERCISES

1. Define the following terms:
   Object Oriented Programming
   ActiveX
   Event-Driven Programming
   Structured Programming
   Class
   Object
   Method
   Properties
   Data
   Spaghetti Programming
   Array
   SafeArray
   Encapsulation
   Polymorphism
   Inheritance
   Object Linking and Embedding
   Compound Object Modeling
   Parent Object
   Child Object
   Inspect Tool
   Visual Basic
   Visual Basic Application
   Variant
   Wrapper Functions
   Collection

2. What are the advantages and disadvantages of creating wrapper functions?

3. How are wrapper functions created?

4. What is the purpose of the VLAX-INVOKE-METHOD function?

5. What is the difference between spaghetti programming and structured programming?

6. What is the difference between event-driven programming and object oriented programming?

7. What is a method? How are methods called?

8. What is the difference between OLE and ActiveX?

9. How does the VLAX-GET-OBJECT function differ from the VLAX-CREATE-OBJECT function?

10. True or False: There is no advantage to using the VLAX-GET-OR-CREATE-OBJECT function over the VLAX-GET-OBJECT function. (If the answer is false, explain why.)

11. True or False: The AutoCAD application does not need to be running before an AutoLISP application can connect to it. (If the answer is false, explain why.)

12. True or False: The VLA-LOAD-COM function must be loaded before ActiveX controls can be used in an AutoLISP application. (If the answer is false, explain why.)

13. List two methods that can be used to access an external application object's methods. What are the advantages and disadvantages of these two methods?

14. Why is it necessary to release an object once a program is finished running?

15. True or False: When an AutoCAD object is created using ActiveX functions, Visual LISP stores that object as a VLA object type. (If the answer is false, explain why.)

16. True or False: When an object is supplied to an ActiveX function, that object must be supplied as a VLA-Object.

17. True or False: Entity name extraction using the ENTGET function can be used with ActiveX functions unless they are converted using the VLAXNAME->VLA-OBJECT function. (If the answer is false, explain why.)

18. True or False: Repeated calls to VLA-GET-ACAD-OBJECT, VLA-GET-ACTIVEDOCUMENT, VLA-GET-MODELSPACE, or VLA-GET-PAPERSPACE can have a negative impact on the performance of AutoCAD. (If the answer is false, explain why.)

19. Write an AutoLISP expression that will connect to the AutoCAD application object.

20. Write an AutoLISP application that will connect to an external application.

21. Rewrite the Gear.lsp program in Chapter 5 so that the gear train is constructed using ActiveX instead of ENTMAKE.

22. Rewrite the resistance program in the application section so that the values associated with each resistor are written to the Word file in addition to the overall summary.

CHAPTER 10

Reactors

OBJECTIVES

Upon completion of this chapter the student will be able to:

- Identify the five basic categories of reactor types available in AutoCAD 2000
- Distinguish between an AutoCAD reactor and an object reactor
- Use AutoLISP to define a callback function
- Use AutoLISP to create an AutoCAD reactor
- Use AutoLISP to create an object reactor
- Convert entity names to VLA-Objects using the VLAX-ENAME->VLA-OBJECT function
- Use AutoLISP to obtain and modify a reactor's definition data
- Determine if a reactor is persistent in a drawing

KEY WORDS AND AUTOLISP FUNCTIONS

:VLR-Toolbar-Reactor	:VLR-Lisp-Reactor
:VLR-Undo-Reactor	:VLR-Miscellaneous-Reactor
:VLR-Wblock-Reactor	:VLR-Mouse-Reactor
:VLR-Window-Reactor	:VLR-Object-Reactor
:VLR-XREF-Reactor	:VLR-SysVar-Reactor
:VLR-AcDb-Reactor	Callback Events
:VLR-Command-Reactor	Database Reactors
:VLR-DeepClone-Reactor	Document Reactors
:VLR-DocManager-Reactor	Editor Reactors
:VLR-DXF-Reactor	Linker Reactors

:VLR-Editor-Reactor   **Object Reactors**

:VLR-Insert-Reactor   **Reactor**

:VLR-Linker-Reactor   **VL-LOAD-COM**

INTRODUCTION TO REACTORS

Before the release of AutoCAD 14.01 and the introduction of Visual LISP, having an AutoCAD entity contact an AutoLISP application when an event occurred was nothing more than a pipe dream shared by almost all AutoLISP programmers at one time or another. With the introduction of Visual LISP this pipe dream has become a reality that is accomplished by using reactors. A reactor is an object that is attached to an AutoCAD drawing entity for the express purposes of having AutoCAD contact an application when a specified event occurs. Suppose that an application is created that generates a gear train based on the design parameters supplied by the user. If it is critical for the gear train to be modified as a whole, then by using a reactor AutoCAD notifies the gear train application and takes the appropriate action when the user modifies the part.

Recall from Chapter 7 that an AutoLISP expression or a group of expressions can be associated with either a tile or an entire dialog box. This type of expression (called an action expression) is dormant until the user performs an action; then information regarding this action is returned to the action expression in the form of a callback. Reactors use the same technology to communicate with an application. The application is executed through a callback function. Callback functions associated with reactors are like other functions used in creating AutoLISP programs. The function becomes a callback function once it has been attached to a reactor event.

CREATING A REACTOR

In AutoLISP the functions that are associated with reactors start with the VLR-prefix. Because reactors use the extended AutoLISP functions provided in Visual LISP, the supporting code that enables reactor calls must first be loaded. In Visual LISP the supporting code for reactors is loaded using the VL-LOAD-COM (vl-load-com) function. This function first checks to see if the extended AutoLISP functions have been loaded into memory. If the functions have not been loaded, then the function loads the extended AutoLISP functions. If the extended functions have already been loaded, then calling the VL-LOAD-COM function will result in no action being taken.

Note: All applications that use reactors (including callback functions) should begin by calling VL-LOAD-COM. If an application calls an extended AutoLISP function and the supporting code has not been loaded, then an error will occur, causing the application to fail.

Before a reactor can be employed in an AutoLISP program, the type of AutoCAD event to be monitored must be determined. In AutoCAD there are five basic categories of reactor types; each type responds to one or more specific events. With the exception of one category, all categories have one reactor type and AutoLISP identification associated with them. These categories are database, document, editor, linker, and object reactors.

Database Reactors

:VLR-AcDb-Reactor   (VLR-ACDB-REACTOR)

When an event occurs that performs a generic change to the AutoCAD drawing database, then a special type of reactor known as a *database reactor* can be used to notify an application that an event has taken place. Events that trigger a database reactor include:

- Adding a new object to the database

- Erasing an object from the database

- Restoring an object that has been erased from the database

- Changing an existing object inside the database

- Opening an object by a process for the purpose of either updating or creating (writing)

- Removing an object from the database using the UNDO command

- Reestablishing an object to the database (using the REDO command) that has been previously removed using the UNDO command)

Document Reactors

:VLR-DocManager-Reactor   (VLR-DOCMANAGER-REACTOR)

When the user changes the status of a document (drawing), then a *document reactor* can be used to notify the application that such an event has occurred. Events that fall under this category:

- Opening a drawing

- Closing a drawing

- Creating a new drawing

- Activating different document windows

- Changing the lock status of a drawing (document)

Editor Reactors

:VLR-Editor-Reactor (VLR-EDITOR-REACTOR)

When the user activates an AutoCAD command or activates an AutoLISP application, then an *editor reactor* can be used to notify an application that a command has been issued. Events that fall into this category are:

- Opening and closing a drawing

- Saving a drawing

- Importing and exporting a DXF file

- Changing a system or environment variable

In AutoCAD 14.01, editor reactors were grouped together under one reactor type, VLR-EDITOR-REACTOR. Starting with AutoCAD 2000, this category has been expanded and more specific reactor types have been added. Table 10–1 provides a list of the new identifiers that have been included in AutoCAD 2000.

 Note: The :VLR-Editor-Reactor type is still available in AutoCAD 2000 for backward compatibility. However, not all new Editor reactors introduced in AutoCAD 2000 can be accessed through the VLR-EDITOR-REACTOR identifier; they must be accessed through their own identifiers.

Table 10–1   Editor Reactor Identification

Reactor Type	Description
:VLR-Command-Reactor	Notifies of a command event.
:VLR-DeepClone-Reactor	Notifies of a deep clone event.
:VLR-DXF-Reactor	Notifies of an event related to reading or writing of a DXF file.
:VLR-Insert-Reactor	Notifies of an event related to block insertion.
:VLR-Lisp-Reactor	Notifies of a LISP event.
:VLR-Miscellaneous-Reactor	Does not fall under any of the other editor reactor types.
:VLR-Mouse-Reactor	Notifies of a mouse event (for example, a double click).

Reactor Type	Description
:VLR-SysVar-Reactor	Notifies of a change to a system variable.
:VLR-Toolbar-Reactor	Notifies of a change to the bitmaps in a toolbar.
:VLR-Undo-Reactor	Notifies of an undo event.
:VLR-Wblock-Reactor	Notifies of an event related to writing a block.
:VLR-Window-Reactor	Notifies of an event related to moving or sizing an AutoCAD window.
:VLR-XREF-Reactor	Notifies of an event related to attaching or modifying XREFs.

Courtesy of the AutoLISP Developers Guide

Linker Reactors
:VLR-Linker-Reactor

AutoCAD also provides a means of notifying an application whenever an ObjectARX application has been loaded and/or unloaded by using *linker reactor* identifier. ObjectARX is a programming environment that provides the developer a means of constructing object oriented programs for the customization and expansion of the AutoCAD working environment using the C++ programming application interface.

Object Reactors
:VLR-Object-Reactor   (VLR-OBJECT-REACTOR)

Object reactors notify an application whenever a specific object in the database has been changed, copied, deleted or modified in any way.

GENERATING A LIST OF REACTOR TYPES

A complete list of reactor types available in AutoCAD can be generated using the AutoLISP function VLR-TYPES (vlr-types). For example:

```
_$ (vlr-types) (ENTER)
```

The list returned:

```
(
        :VLR-Linker-Reactor
        :VLR-Editor-Reactor
        :VLR-AcDb-Reactor
        :VLR-DocManager-Reactor
        :VLR-Command-Reactor
        :VLR-Lisp-Reactor
        :VLR-DXF-Reactor
```

```
                        :VLR-DWG-Reactor
                        :VLR-Insert-Reactor
                        :VLR-Wblock-Reactor
                        :VLR-SysVar-Reactor
                        :VLR-DeepClone-Reactor
                        :VLR-XREF-Reactor
                        :VLR-Undo-Reactor
                        :VLR-Window-Reactor
                        :VLR-Toolbar-Reactor
                        :VLR-Mouse-Reactor
                        :VLR-Miscellaneous-Reactor
                        :VLR-Object-Reactor
             )
```

 Note: A description of each one of the different reactor types listed above can be obtained in the *AutoLISP Reference Guide*.

CALLBACK EVENTS

Each reactor type supports numerous events that cause that particular reactor to call an application. For example, when a command is issued at the command prompt, AutoCAD responds with a `:VLR-commandWillStart` event. Likewise, when the user cancels a command, AutoCAD responds by issuing a `:VLR-commandCancelled` event. The events associated with a reactor type are known as *callback events*. Their main purpose is to call the function associated with that particular event. The exact action that will be monitored must be determined first because it will determine which callback function will be selected.

Returning a List of Available Events

A list of available events can be generated for a particular reactor type through the AutoLISP function VLR-REACTION-NAMES (vlr-reaction-names reactor type). This function has only one required argument, the reactor type or reactor object for a list of events. For example, to generate a list of events associated with document reactors, the following expression would be used.

```
_$ (VLR-REACTION-NAMES :VLR-DocManager-Reactor) (ENTER)

    (
            :VLR-documentCreated
            :VLR-documentToBeDestroyed
```

```
                    :VLR-documentLockModeWillChange
                    :VLR-documentLockModeChangeVetoed
                    :VLR-documentLockModeChanged
                    :VLR-documentBecameCurrent
                    :VLR-documentToBeActivated
                    :VLR-documentToBeDeactivated
          )
```

Note: A list of all available reactor events is provided in the *AutoLISP Reference Guide*.

Caution: Make sure that the extended AutoLISP functions have been loaded into memory with the VL-LOAD-COM function before issuing VLR-REACTION-NAMES or any other reactor function. Otherwise an error will result with AutoLISP displaying the following message "no function definition".

CALLBACK FUNCTIONS

Before a reactor can be used with an AutoLISP application, a callback function must first be defined. The callback function specifies the action that is to be taken once a specific reactor event has occurred. Once the reactor callback function has been defined, the function is linked to a reactor event by the creation of a reactor object. Callback functions for reactors are a standard AutoLISP application defined by the DEFUN function. Nevertheless, unlike the AutoLISP applications that have been presented up to this point, reactor callback functions have a few restrictions and guidelines that must be observed or serious consequences could result. These restrictions and guidelines are listed below.

- A callback function cannot call an AutoCAD command using the COMMAND function.

- All access to drawing entities can be gained only by incorporating ActiveX controls (VLAX- functions). Therefore the use of ENTGET and ENTMOD functions are not permitted.

- Interactive functions should not be used in a callback function. This includes dialog boxes, GETPOINT, and GETKWORD as well as selection set functions. Any attempt to execute an interactive function can result in a serious problem occurring. Message and alert boxes can be employed in a reactor callback function because these actions are not considered interactive. Only compiled applications can issue an event while AutoCAD is displaying either a modal or message dialog box.

- A callback function should not attempt to update or make changes to the object where the event notification was triggered. However, a callback function can read information from the object that triggered the event.

- A callback function cannot perform an action that will trigger a duplication of the same event. If this should occur, then an infinite loop is created, causing AutoCAD to stop responding.

- Multiple callback functions for the same event should be avoided. This can be accomplished by first verifying that a reactor is not already set before attempting to set it.

- All callback functions, not including object reactors, must be defined to accept two arguments. The first argument is used to identify the reactor object that calls the function. The second argument supplies a list of arguments set by AutoCAD.

- Callback functions for object reactors must be defined to accept three arguments. The first argument identifies the object that fired the notification. The second argument identifies the reactor object that called the function, while the third argument supplies a list of parameters specific to the callback condition.

Courtesy of the AutoLISP Developers Guide

The types of parameters that are passed to a callback function are dependent upon the type of event associated with the reactor type. For example, the documentCreated event, which is associated with THE VLR-DOCMANAGER-REACTOR function, passes the name of the affected document to the callback function in the form of a VLA-Object. Likewise, the unknownCommand event, which is associated with the VLR-EDITOR-REACTOR, passes the name of the unknown command to the callback function in the form of a text string. A complete list of the parameters associated with a particular event can be found in the *AutoLISP Reference Guide*. An example of a callback function is provided below; this function is activated whenever the user enters an unknown command:

```
(defun unknown (reactor unknowncommand_info)
   (vl-load-com)
   (alert
     (strcat
       "The command "
       (car unknowncommand_info)
       " is not a valid AutoCAD command.  Please reenter command or press
F1 for help : "
```

```
      )
    )
    (princ)
  )
```

In this example, the reactor that called the function is supplied to the function as the argument *reactor*. The unknown command is supplied to the function as the *unknowncommand_info* argument. Once the name of the command is extracted, it is then combined with another message and displayed in an alert box.

 Note: There are two predefined callback functions available in AutoLISP for testing and debugging purposes: VLR-BEEP-REACTION and VLR-TRACE-REACTION. The VLR-BEEP-REACTION function causes the computer to beep. The VLR-TRACE-REACTION function prints a list of arguments to the Visual LISP trace window each time a function call is made.

Object Reactor Callback Functions

Unlike the previous reactors in AutoCAD, an object reactor is attached to a specific AutoCAD entity. Therefore, when an object reactor is defined, the object to which the reactor is to be attached must be identified. This also affects the callback function associated with an object reactor. As mentioned earlier, a callback function for an object reactor must be defined so that the function will accept three arguments. For example:

```
;;;*****************************************
;;;
;;; Program Name: PrintChangeLength.lsp
;;;
;;; Program Purpose: Example Object Reactor Callback Function
;;;
;;; Programmed By: James Kevin Standiford
;;;
;;; Program Date: 01/19/2000
;;;
;;;*****************************************
(defun printchangedLength (notifier reactor parameter)
  (vl-load-com)
  (cond
    (
      (vlax-property-available-p
       notifier
```

```
           "Length"
       )
       (alert (strcat "The new line length is   "
                  (rtos (vla-get-length notifier)
                  )
              )
       )
     )
   )
 )
)
```

Alternatively:

```
;;;*****************************************
;;;
;;; Program Name:  PrintChangeLength.lsp
;;;
;;; Program Purpose: Example Object Reactor Callback Function
;;;
;;; Programmed By: James Kevin Standiford
;;;
;;; Program Date: 01/19/2000
;;;
;;;*****************************************
(defun printchangedLength (notifier reactor parameter)
  (vl-load-com)
  (alert (strcat "The new line length is   "
            (rtos (vla-get-length notifier)
            )
       )
   )
)
```

CREATING AN AUTOCAD REACTOR

Once the callback function has been defined, then the function is linked to the specific event the developer wishes to monitor. As mentioned earlier, there is a specific AutoLISP function already provided to create each reactor type. With the exception of the object reactors, these reactor creation functions require that a list of dotted pairs be supplied where the name of the event and callback function is supplied. To link the callback function unknown to the event unknownCommand, the following expression would be executed.

```
_$ (VLR-EDITOR-REACTOR nil '((:vlr-unknownCommand . unknown))) (ENTER)
#<VLR-Editor-Reactor>
```

In this example the VLR-EDITOR-REACTOR (vlr-editor-reactor data callbacks) function is used to create the reactor that links the callback function unknown to the event unknownCommand (see Figure 10–1). The first argument associated with this function is the AutoLISP data to be associated with the reactor or nil, if no data is to be associated. The second argument is the dotted pair containing the name of the event to associate with the reactor and its callback function.

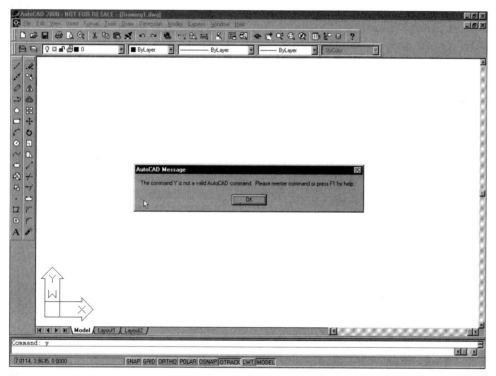

Figure 10–1 *Unknown AutoCAD command message*

CREATING AN OBJECT REACTOR

The type of reactor created in the previous section is considered an AutoCAD reactor and is not associated with a particular entity. However, when a reactor is to be confined to an entity, an object reactor must be constructed. Object reactors are unlike AutoCAD reactors in that the reactor itself is attached to a specific object. In addition to the event and the callback functions normally associated with

AutoCAD reactors, object reactors require that a VLA-Object be supplied to the Visual LISP function used to create the object reactor. To create an object reactor, the VLR-OBJECT-REACTOR (vlr-object-reactor owners data callbacks) function is used. The first argument associated with this function is the reactor's owner list. This is a list of VLA-Objects to be associated with that particular reactor. The second argument supplies any data to be associated with this particular reactor. The third argument is a dotted pair where the event and its corresponding callback function are supplied. To create an object reactor and attach that reactor to the object shown in Figure 10–2, use the following expression.

```
(setq vlr-object
        (vlr-object-reactor
      (list (vlax-ename->vla-object
            (cdr (assoc -1 (entget (car (entsel)))))))
          )
      )
      "Example Line Reactor"
      '((:vlr-modified . printchangedLength))
        )
)
```

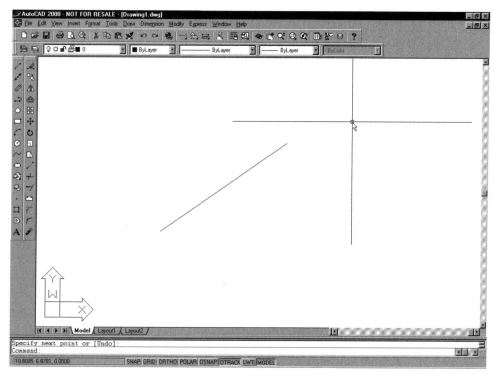

Figure 10–2 *Line object*

Figure 10–3 shows the object reactor message and Figure 10–4 shows the new line length.

412

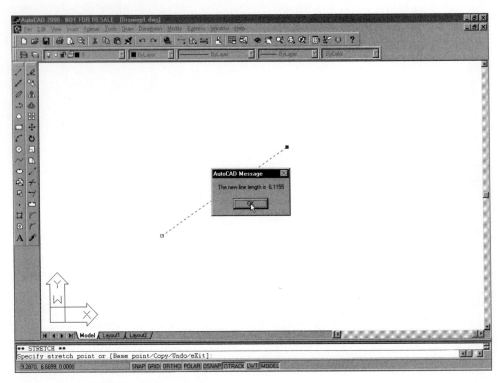

Figure 10–3 *Object reactor message*

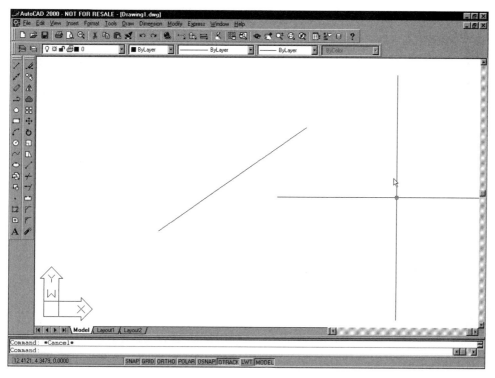

Figure 10–4   *New line length*

 Caution: Once an object has been included in the object reactor owner list, that object cannot be modified by that reactor's callback function. Doing so will cause AutoCAD to generate an error and could even cause AutoCAD to perform a fatal error. Only VLA-Objects can be supplied in the reactor's owner list. If an entity is selected using any of the ENTXXX functions, then the entity name must first be converted using the VLAX-ENAME->VLA-OBJECT (vlax-ename->vla-object entname)function before that entity can be added to the reactor's owner list. Callback functions can modify only an AutoCAD entity using ActiveX controls, and therefore ActiveX controls work only on VLA-Objects.

Because an object reactor is supplied with an owner's list, more than one VLA-Object can be supplied to the VLR-OBJECT-REACTOR function at a time. With a few modifications to the previous expression, in one step multiple entities can be attached to the previous reactor (see Figure 10–5).

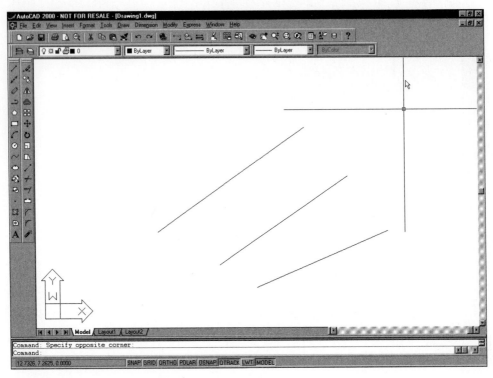

Figure 10–5 *Multiple line objects*

```
;;;*******************************************
;;;
;;; Program Name: define_reactor.lsp
;;;
;;; Program Purpose: Defines object reactors
;;;
;;; Programmed By: James Kevin Standiford
;;;
;;; Program Date: 01/19/2000
;;;
;;;*******************************************
(defun c:define_reactor   ()
  (vl-load-com)
  (setq ent (entsel))
  (while (/= ent nil)
    (if    (/= ent nil)
      (progn
```

```
(setq
  e_list
   (append
     e_list
     (list
        (vlax-ename->vla-object
        (cdr (assoc -1 (entget (car ent))))
       )
     )
   )
 )
   )
 )
  (setq ent (entsel))
)
(setq    vlr-object
   (vlr-object-reactor
      (list (vlax-ename->vla-object
             (cdr (assoc -1 (entget (car (entsel))))))
          )
     )
      "Example Line Reactor"
      '((:vlr-modified . printchangedLength))
   )
 )
)
```

Figures 10–6, 10–7, and 10–8 show the object reactor messages for each line.

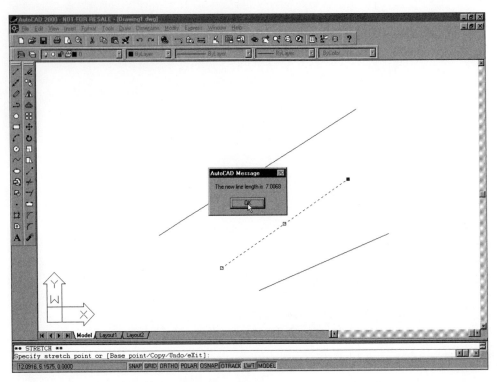

Figure 10–6 *Object reactor message for first line*

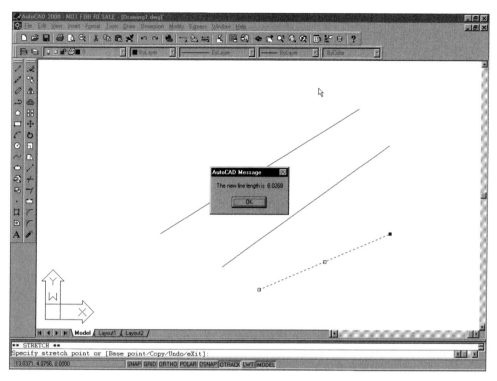

Figure 10–7 *Object reactor message for second line*

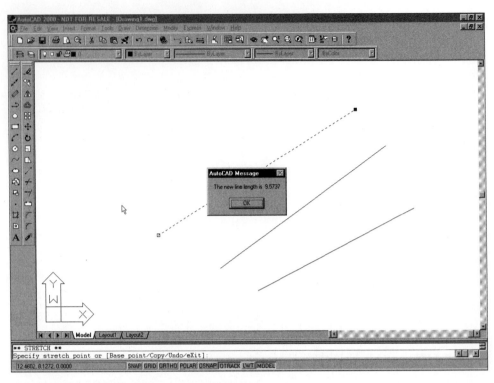

Figure 10–8 *Object reactor message for third line*

 Note: Because AutoCAD entities can have multiple reactors associated with them, the data argument can be used to associate a description text string or identification string to a particular reactor. This was illustrated in the previous example where the text string "Example Line Reactor" was supplied as the data argument.

USING REACTORS WITH MULTIPLE DRAWINGS

Prior to AutoCAD 2000, the ability to work in more than one drawing in a single AutoCAD session was not possible. However, AutoCAD 2000 has added this feature and it must now be taken into consideration when an application is designed. AutoLISP is currently designed to work in one drawing at a time, unlike other applications that support simultaneous execution in multiple documents (ObjectARX and VBA). This ability allows an application to modify open drawings that are not currently active. Although AutoLISP does not fully support this activity right now, it does supply limited support for reactor callback functions to operate in one or more drawings that are currently open but not active. By default, a reactor callback function is executed only if the notification event associated with that particular reactor is fired in an active drawing where the reactor was defined. To modify a reactor's behavior so that its callback functions can be executed even when the parent document containing the reactor is not active, the VLR-SET-NOTIFICATON (vlr-set-notification reactor 'range) function should be used. The *'range* argument is a symbol that is used to specify whether a reactor's callback function can be executed in the active parent drawing ('active-document-only) or a non-parent ('all-documents) active drawing. To activate the callback function associated with an AutoCAD reactor that notifies the user that an unknown command has been entered, the following expressions would be used:

```
_$ (SETQ unknown_reactor (VLR-EDITOR-REACTOR nil '((:vlr-unknownCommand .
   unknown)))) (ENTER)
#<VLR-Editor-Reactor>
_$ (VLR-SET-NOTIFICATION unknown_reactor 'all-documents) (ENTER)
#<VLR-Editor-Reactor>
_$
```

Figure 10–9 shows the parent document active while Figure 10–10 shows the non-parent document active.

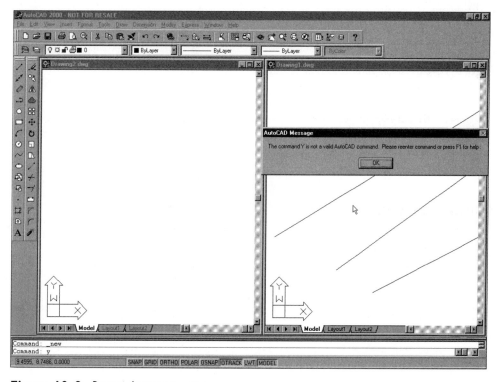

Figure 10–9 *Parent document active*

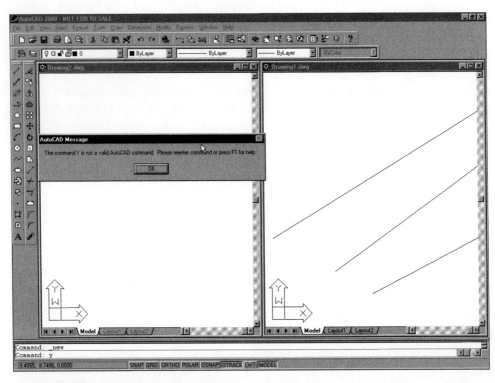

Figure 10–10 *Non-parent document active*

 Caution: If a reactor's notification setting is set to 'all-documents, then the reactor's callback function should not do anything that would cause the function to modify an AutoCAD entity. This action could cause the system to become unstable.

DETERMINING A REACTOR'S NOTIFICATION SETTING

A reactor's notification setting can be reset; for this reason it often becomes necessary to determine the current setting of a particular reactor's notification. In AutoLISP this is accomplished by the VLR-NOTIFICATON (vlr-notification reactor) function. This function, when supplied with the VLA-Object representing a reactor, returns the current state of that reactor's notification in the form of a symbol. To check the current notification setting of the previous reactor, the following expression would be used:

```
_$ (VLR-NOTIFICATION unknown_reactor) (ENTER)
ALL-DOCUMENTS
```

OBTAINING AND MODIFYING INFORMATION ABOUT REACTORS

Inside Visual LISP are an abundance of inspection tools and AutoLISP functions used to obtain various information regarding a reactor. This information includes a list of all reactors defined, as well as the type of reactors defined in a drawing, the objects that own reactors, events and their associated callback functions, a reactor's active status, and user-defined data attached to a reactor.

Using AutoLISP to List the Reactors Defined in a Drawing

When working with reactors it is occasionally necessary for the developer to generate a list of the reactors that are already present in an AutoCAD drawing. In AutoLISP this is accomplished by the VLR-REACTOR (vlr-reactors [reactor-type...]) function. This function, when executed without the optional argument, returns a list of all reactors defined in the current drawing. The list is comprised of one of more sub-lists depending upon the number of different reactor types used. The sub-list begins with a symbol identifying the reactor type, followed by pointers to each reactor type. To obtain a listing of all the reactors defined in Figure10–5, the following expressions would be used:

```
_$ (vlr-reactors) (ENTER)
((:VLR-Object-Reactor #<VLR-Object-Reactor> #<VLR-Object-Reactor>#<VLR-
    Object-Reactor>) (:VLR-Editor-Reactor #<VLR-Editor-Reactor>))
_$
```

If the optional *reactor-type* argument is supplied to the function, then all reactors defined in the drawing for that particular reactor type would be listed. To obtain a listing of all object reactors defined in Figure 10–5, the following expression would be used:

```
_$ (vlr-reactors :vlr-object-reactor) (ENTER)
((:VLR-Object-Reactor #<VLR-Object-Reactor> #<VLR-Object-Reactor>#<VLR-
    Object-Reactor> #))
_$
```

Using AutoLISP to Identify a Specified Reactor Type

The VLR-TYPE (vlr-type reactor) function can be used to determine a specific reactor's type when an application is developed. This function, when supplied with a VLR-Object, returns a symbol identifying the reactor type. To verify the type of reactor created in the section "Creating an Object Reactor," the following expression would be used:

```
_$ (vlr-type vlr-object) (ENTER)
:VLR-Object-Reactor
_$
```

422

Using AutoLISP to Extract and Modify Application-Specific Data Associated with a Specific Reactor

Again, when a developer is creating a reactor, application-specific data can be supplied along with the remaining reactor's definition. Application-specific data is not critical to the performance of the reactor in any way. Its purpose is to allow the developer to store information regarding a reactor with a reactor. Since more than one reactor can be attached to an object, this option is often used to supply a reactor with a means of differentiating it from other reactors that might be attached to the same object. Once a reactor has been defined, then the application-specific data can be accessed by the VLR-DATA (vlr-data obj) function. This function, when supplied with a VLR-Object that represents the reactor for which the application-specific data is needed, returns that data in the form of a string. To extract the application-specific data associated with the reactor created in the previous section, the following expression would be used:

```
_$ (vlr-data example) (ENTER)
"Example Line Reactor"
_$
```

Application-specific data can also be changed through the AutoLISP function VLR-DATA-SET (vlr-data-set obj data). This function, when supplied with the VLR-Object representing the reactor, replaces the existing application-specific data with application-specific data supplied to the function. To change the application-specific data in the previous example, the following expression would be used.

```
_$ (vlr-data-set example "Example of how to change the application-
    specific data associated with a reactor") (ENTER)
"Example of how to change the application-specific data associated with a
    reactor"
```

Using the VLR-DATA function, the change can be verified.

```
_$ (vlr-data example) (ENTER)
"Example of how to change the application-specific data associated with a
    reactor"
_$
```

Using AutoLISP to Generate and Modify a List of Owners

When an object is associated with a reactor, then that object is considered to be the reactor's owner. Because objects can have multiple reactors and reactors can have multiple owners, it often becomes necessary to obtain a listing of that reactor's owners.

This is accomplished by the VLR-OWNER (vlr-owners reactor) function. This function, when supplied with a VLR-Object representing a reactor, generates a list of that reactor's owners. To view a list of owners for the reactor in the previous example, the following expression would be used.

```
_$ (vlr-owners example) (ENTER)
(#<VLA-OBJECT IAcadLine 01a0c604> #<VLA-OBJECT IAcadLine 01a2de64>
#<VLA-OBJECT IAcadLine 01a0fe54> #<VLA-OBJECT IAcadLine 01a0fd94>
#<VLA-OBJECT IAcadLine 01a0ce94> #<VLA-OBJECT IAcadLine 01a0ff14>
#<VLA-OBJECT IAcadLine 01a0cdd4> #<VLA-OBJECT IAcadLine 01a0ffb4>)
_$
```

Objects can also be added to and removed from a reactor's owner list. This is accomplished by the VLR-OWNER-ADD (vlr-owner-add reactor owner) and VLR-OWNER-REMOVE (vlr-owner-remove reactor owner) functions. The VLR-OWNER-ADD function, when supplied with a VLR-Object representing the reactor and a VLR-Object representing the object to add, adds the object to the reactor's owner list. To add the line to the previous reactor's owner list, the following expressions would be used:

```
(setq ent         (entget (car (entsel)))
      object-name (vlax-ename->vla-object (cdr (assoc -1 ent)))
)
(vlr-owner-add example object-name)
```

To remove an object from a reactor's owner list, the following expression would be used:

```
(setq ent         (entget (car (entsel)))
       object-name (vlax-ename->vla-object (cdr (assoc -1 ent)))
)
(vlr-owner-remove example object-name)
```

Using AutoLISP to Identify and Modify a Specific Reactor Callback Function

AutoLISP also provides the developer with a means of extracting a reactor's callback function and associated event. The VLR-REACTION (vlr-reactions reactor) function, when supplied with a VLR-Object representing a reactor, returns the event and associated callback function as a dotted pair. For example:

```
_$ (vlr-reactions example) (ENTER)
((:VLR-modified . PRINTCHANGEDLENGTH))
_$
```

To change the callback function associated with a reactor, the VLR-REACTION-SET (vlr-reaction-set reactor event function) function would be used. The first argument supplied to this function is the VLR-Object representing the reactor to change, while the second argument represents that reactor's notification event. The last argument required by this function is the new callback function to be associated with the reactor. To change the callback function associated with the reactor in the previous example, the following expressions would be used:

```
_$ (vlr-reaction-set example :vlr-modified 'new_callback_function) (ENTER)
"NEW_CALLBACK_FUNCTION"
_$
```

Using Visual LISP to Examine a Reactor

Visual LISP provides a very useful tool for examining a reactor, the inspection tool. To examine a reactor using the inspection tool, the reactor must first be set to a variable. The variable is highlighted and the inspection tool selected from the toolbar (see Figures 10–11, 10–12, and 10–13).

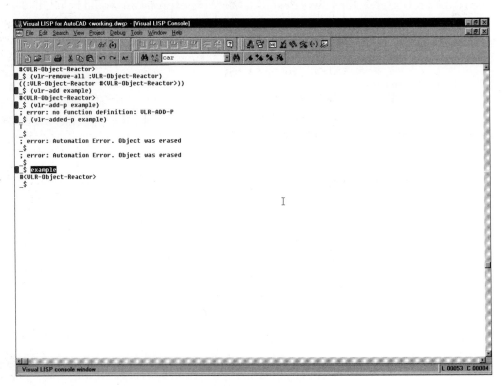

Figure 10–11   *Variable name highlighted*

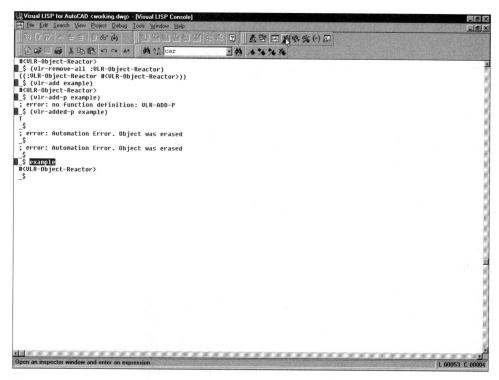

Figure 10–12 *Inspection tool toolbar*

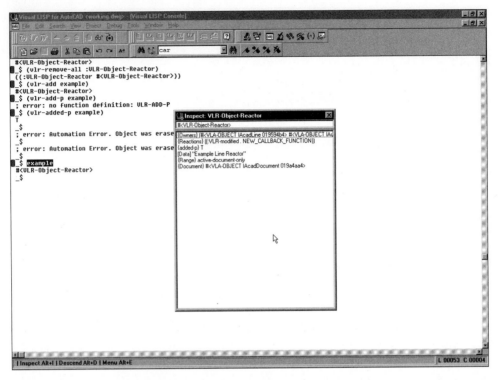

Figure 10–13 *Inspection tool*

The inspection tool displays the following information concerning a reactor:

- Objects owning the reactor

- Event and associated callback function

- Whether or not the reactor is active

- User data attached to the reactor

- Document range where the reactor will fire

- The AutoCAD document attached to the object reactor

Double-clicking on an item will expand the information contained within the inspection tool. Double-click on the {Owner} item listed in the inspection window to view a complete list of the objects owning the reactor *example* (see Figure 10–14).

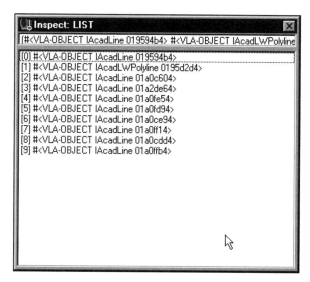

Figure 10–14   *Owner list expanded*

Items can be expanded in the inspection window by a right mouse click. This activates a floating menu where Inspect is one of the options (see Figures 10-15 and 10-16).

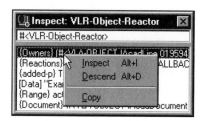

Figure 10–15   *Floating menu*

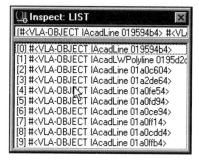

Figure 10–16   *Selected item expanded*

The inspection tool can also be launched by selecting the variable name representing the reactor to inspect followed by a right mouse click (see Figures 10–17 and 10–18).

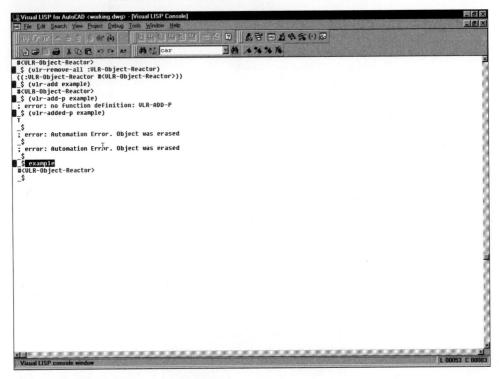

Figure 10–17 *Variable name highlighted*

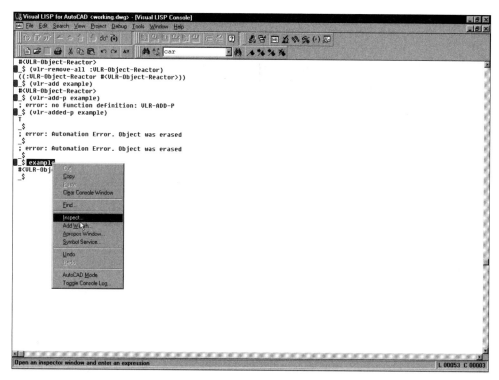

Figure 10-18 *Shortcut menu*

DISABLING AND ENABLING REACTORS

Once a reactor has been created, that reactor is active by default. AutoLISP offers the developer a means of disabling a reactor by using the VLR-REMOVE (vlr-remove reactor) function. This function, when supplied with a VLR-Object representing a reactor, disables that reactor. The reactor is not removed from the drawing and can be reactivated at any time with the VLR-ADD (vlr-add obj) function. To disable the reactor in the previous example, the following expression would be used:

```
_$ (vlr-remove example) (ENTER)
#<VLR-Object-Reactor>
_$
```

To disable all reactors in a drawing, the VLR-REMOVE-ALL (vlr-remove-all [reactor-type]) function would be employed. This function has one optional argument associated with it, the reactor type. If this argument is furnished, then only the reactor belonging to the specified type will be disabled; otherwise the function disables all reactors in a drawing. To disable only the object reactors in a drawing, the following expression would be used:

```
_$ (vlr-remove-all :VLR-Object-Reactor) (ENTER)
((:VLR-Object-Reactor #<VLR-Object-Reactor>))
_$
```

To reactivate a reactor in a drawing, the VLR-ADD (vlr-add obj) function is used. This function, when supplied with a VLR-Object representing a reactor, reactivates that reactor. For example:

```
_$ (vlr-add example) (ENTER)
#<VLR-Object-Reactor>
_$
```

The previous function makes it necessary to determine whether the reactor in enabled or disabled. To accomplish this, the VLR-ADDED-P (vlr-added-p obj) function can be used. This function, when supplied with a VLR-Object representing the reactor to test, determines if the specified reactor is enabled or disabled. If a reactor is enabled, the function returns the symbol T; otherwise the function returns nil. To determine if the reactor in the previous example is already enabled, the following expression would be used:

```
_$ (vlr-added-p example) (ENTER)
T
_$
```

MAKING A REACTOR TRANSIENT OR PERSISTENT

When a reactor is first defined, it is considered to be transient by default. A transient reactor is one where the definition is lost once the drawing in which the reactor is defined is terminated. The reactor definition is not saved with the drawing; when a reactor is transient, it must be redefined every time the drawing session is begun. On the other hand, when a reactor is persistent, then the reactor's definition is saved with the drawing and will be functional every time the drawing is reopened. To make a reactor persistent in a drawing, the VLR-PERS (vlr-pers reactor) function should be used. This function, when supplied with a VLR-Object representing a reactor, makes the specified reactor persistent. For example:

```
_$ (vlr-pers example) (ENTER)
#<VLR-Object-Reactor>
_$
```

When a reactor is made persistent, that reactor is merely a link connecting the event and the callback function. With persistent reactors, this link is saved as a part of the drawing; however, the callback function is not and must be reloaded each time the drawing is reopened. If a reactor is made persistent and the associated callback function is not automatically loaded each time the drawing is reopened, then AutoCAD displays an error message on startup (see Figure 10–19).

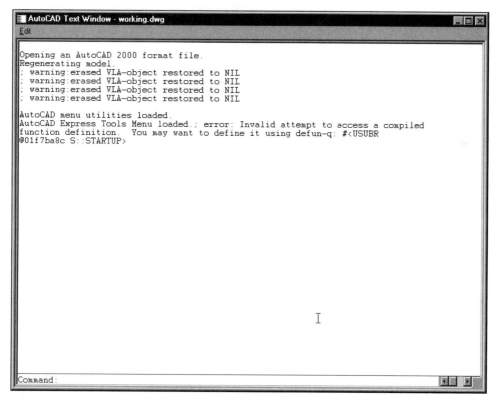

Figure 10–19 *AutoCAD error message*

The callback function associated with a reactor can be made to automatically load by either appending the callback definition to the acad.lsp file or defining the reactor and callback function in a separate NameSpace (VLX). (Additional information regarding NameSpaces can be found in Chapter 11.)

AutoLISP offers two functions for determining whether or not a reactor is persistent in a drawing. They are the VLR-PERS-LIST (vlr-pers-list [reactor]) and VLR-PERS-P (vlr-pers-p reactor). The VLR-PERS-LIST function returns a list of persistent reactors contained within a drawing. The function has one option argument, *reactor*. This argument, if supplied, allows the developer to list only persistent reactors of a specified type. To generate a list of all persistent reactors in Figure 10–20, the following expression would be used:

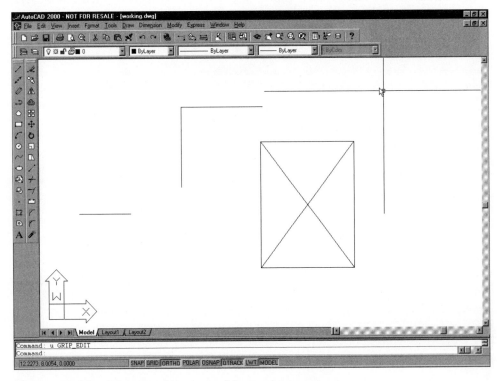

Figure 10–20 *Drawing to be examined for persistent reactors*

```
_$ (vlr-pers-list) (ENTER)
(#<VLR-Object-Reactor> #<VLR-Object-Reactor>)
_$
```

To determine if a specific reactor is persistent or not, the VLR-PERS-P (vlr-pers-p reactor) function can be used. This function, when supplied with a VLR-Object representing the reactor to test, returns T if the reactor is persistent and nil if the reactor is not. For example:

```
_$ (vlr-pers-p example)
nil
_$
```

Once a reactor has been made persistent, that reactor can be made transient again by the AutoLISP function VLR-PERS-RELEASE (vlr-pers-release reactor). For example:

```
_$ (vlr-pers example) (ENTER)              ;Reactor  is  made  persistent
#<VLR-Object-Reactor>
_$ (vlr-pers-release example) (ENTER)      ;Reactor  is  made  transient
#<VLR-Object-Reactor>
_$
```

APPLICATION – INCORPORATING A REACTOR

To illustrate how a reactor can be incorporated into an engineering design application, the following program uses a :VLR-ERASED event to notify the AutoLISP application gear.lsp (from Chapter 5) when the user has removed the gear train from the drawing. Once the application is notified, the user is prompted to reenter the design parameters and a new gear train is created. The new gear train is then appended to the reactor's owner list so that the process may be repeated if the user should delete the gear train again.

Other than the addition of the reactor creation function at the end of the gear.lsp application, very few modifications were required. A new main function was added to allow the user to activate the application from the AutoCAD command prompt as if the application were an AutoCAD command. A new function call was added at the end of the existing gear.lsp application to allow the program to call the reactor creation function. The reactor creation function, when activated, extracts the last entity created (the gear) and appends that entity to the reactor's owner list.

```
;;;************************************
;;;
;;;    Program Name  :  Gear.lsp
;;;
;;;    Program Purpose : Create a gear train given
;;;             RPM's, gear ratio, Diametral Pitch,
;;;             Shaft Diameter
;;;
;;;    Program Date  :  03-19-2000
;;;
;;;    Programmed By : James Kevin Standiford
;;;
;;;***********************************
;;;****NEW**********NEW*********NEW******
;;;
;;; Main
```

```
;;;
;;;****NEW**********NEW*********NEW******
(DEFUN c:gear ()
  (gear_reactor 1 1 1)
)
(DEFUN gear_reactor (notifier reactor parameter)
  (SETQ    rpm            (GETREAL "\nEnter RPM of Pinion Gear : ")
   number_teeth (GETREAL "\nEnter Number of Teeth : ")
   gear_ratio   (/ 1 (GETREAL "\nEnter Gear Ratio : "))
   dia_pitch    (GETREAL "\nEnter Diametral Pitch : ")
   shaft_dia    (GETREAL "\nEnter Shaft Diameter : ")
   PT           (GETPOINT "\nSelect Point : ")
   pitch_dia    (/ number_teeth dia_pitch)
   gear_teeth   (* gear_ratio number_teeth)
   gear_out     (+ pitch_dia (* 2 (/ 1 (/ gear_teeth pitch_dia))))
   pinion_out   (+ pitch_dia (* 2 (/ 1 (/ number_teeth pitch_dia))))
   gear_rpm     (* gear_ratio rpm)
   dist2Dist    (+ (* 0.5 pinion_out) (* 0.5 gear_out))
   pt_x         (+ dist2Dist (CAR pt))
   pt_y         (CADR pt)
   pinion       (LIST (CONS 0 "CIRCLE")
                      (CONS 10 pt)
                      (CONS 40 (/ pinion_out 2))
                      (CONS 8 "pinion")
                )
   gear         (LIST (CONS 0 "CIRCLE")
                      (CONS 10 (LIST pt_x pt_y))
                      (CONS 40 (/ gear_out 2.0))
                      (CONS 8 "gear")
                )
   shaft_pinion (LIST (CONS 0 "CIRCLE")
                      (CONS 10 pt)
                      (CONS 40 (/ shaft_dia 2.0))
                      (CONS 8 "pinion")
                )
   shaft_gear   (LIST (CONS 0 "CIRCLE")
                      (CONS 10 (LIST pt_x pt_y))
                      (CONS 40 (/ shaft_dia 2.0))
                      (CONS 8 "pinion")
                )
  )
```

```
(ENTMAKE (list (cons 0 "block")
          (CONS 2 "pinion")
          (cons 10 (LIST pt_x pt_y))
          (cons 70 64)
    )
)
(ENTMAKE pinion)
(ENTMAKE shaft_pinion)
(ENTMAKE (list (cons 0 "endblk")))
(ENTMAKE (list (CONS 0 "INSERT")
          (cons 2 "pinion")
          (cons 10 (LIST pt_x pt_y))
    )
)
(setq lastent (entget (entlast)))
(regapp "pinion")
(setq   exdata1
    (LIST
      (LIST "pinion"
          (CONS 1000 (STRCAT "Pinion's RPM " (RTOS rpm)))
          (CONS 1041 gear_ratio)
          (CONS 1042 number_teeth)
      )
    )
)
(setq   newent1
    (append lastent (list (append '(-3) exdata1)))
)
(entmod newent1)
(ENTMAKE (list (cons 0 "block")
          (CONS 2 "gear")
          (cons 10 (LIST pt_x pt_y))
          (cons 70 64)
    )
)
(ENTMAKE gear)
(ENTMAKE shaft_gear)
(ENTMAKE (list (cons 0 "endblk")))
(ENTMAKE (list (CONS 0 "INSERT")
          (cons 2 "gear")
          (cons 10 (LIST pt_x pt_y))
```

```
        )
    )
    (setq lastent (entget (entlast)))
    (regapp "gear")
    (setq    exdata
      (LIST
        (LIST "gear"
               (CONS 1000 (STRCAT "Mating Gear's RPM " (RTOS rpm)))
               (CONS 1041 gear_ratio)
               (CONS 1042 gear_teeth)
        )
      )
    )
    (setq    newent
      (append lastent (list (append '(-3) exdata)))
    )
    (entmod newent)
    (princ)
;;;******NEW*******NEW*********

    (define_reactor)

;;;*****NEW********NEW*********
)
;;;****NEW********NEW*******NEW*************
;;;
;;; defines an object reactor
;;;
;;;*****NEW********NEW*******NEW************
(defun define_reactor ()
    (vl-load-com)
    (setq    vlr-object
      (vlr-object-reactor
        (list (vlax-ename->vla-object
                (cdr (assoc -1 (entget (entlast)))))
            )
        )
        "Example Line Reactor"
        '((:VLR-erased . gear_reactor))
      )
    )
)
```

SUMMARY

Reactors provide the developer with a means of monitoring and responding to events triggered by the actions of the AutoCAD user. These events can be anything, from the opening of an existing drawing to the deleting of an AutoCAD entity. In AutoLISP, the functions that are associated with reactors start with the prefix VLR-. Before these functions can be used in an AutoLISP application, the supporting code must be loaded using the VL-LOAD-COM function. This function first checks to determine if the extended AutoLISP functions have been loaded. If the functions have been loaded, then the function does nothing. If the functions have not been loaded, then the function loads the extended functions.

Before a reactor can be employed in an AutoLISP application, the type of AutoCAD event to be monitored must be determined. In AutoCAD, there are five basic categories of reactor types, each responding to one or more specific events. These categories are database, document, editor, linker, and object reactor. A database reactor responds to generic changes in the AutoCAD database. The document reactor responds to changes made by the user to the status of the AutoCAD drawing. The editor reactor responds to AutoCAD commands activated by the user. The linker reactor provides a means of notifying an application whenever an ObjectARX application has been loaded. An object reactor notifies an application whenever a specific object in the database has been changed, copied, deleted, or modified.

Each reactor type supports numerous events that cause that particular reactor to call an application. The events associated with a reactor type are known as callback events. Their main purpose is to call the function associated with that particular event. Before a reactor can be used, a callback function must first be defined. The callback function specifies the action that is to be taken once a specific reactor event has occurred. Callback functions for reactors are standard AutoLISP applications that have been defined using the DEFUN function. Once the callback function has been defined, then the function is linked to the specific event that the developer wishes to monitor.

REVIEW QUESTIONS AND EXERCISES

1. Define the following AutoLISP functions and terms:
   :VLR-Toolbar-Reactor
   :VLR-Undo-Reactor
   :VLR-Wblock-Reactor
   :VLR-Window-Reactor
   :VLR-XREF-Reactor
   :VLR-AcDb-Reactor
   :VLR-Command-Reactor
   :VLR-DeepClone-Reactor
   :VLR-DocManager-Reactor
   :VLR-DXF-Reactor
   :VLR-Editor-Reactor
   :VLR-Insert-Reactor
   :VLR-Linker-Reactor
   :VLR-Linker-Reactor
   :VLR-Lisp-Reactor
   :VLR-Miscellaneous-Reactor
   :VLR-Mouse-Reactor
   :VLR-Object-Reactor
   :VLR-SysVar-Reactor
   Callback Events
   Database Reactors
   Document Reactors
   Editor Reactors
   General Editor Reactor
   Linker Reactors
   Linker Reactors
   Object Reactors
   Reactor
   VL-LOAD-COM

2. What are the five basic categories of reactor types available in AutoCAD 2000?

3. What is the difference between an AutoCAD reactor and an object reactor?

4. How are reactor callback functions created in AutoLISP?

5. What AutoLISP function is used to convert entity names to VLA-Objects?

6. True or False: AutoLISP cannot be used to obtain and modify a reactor's definition data. (If the answer is false, explain why.)

7. How do you determine if a reactor is persistent in a drawing?

8. What AutoLISP function is used to determine current notification settings?

9. True or False: Reactors are not part of the extended AutoLISP functions, and therefore it is not necessary to use the VL-LOAD-COM function before making calls to reactor functions. (If the answer is false, explain why.)

10. True or False: Once a reactor is defined it cannot be removed. (If the answer is false, explain why.)

11. What is an owner's list?

12. Using the drawing reactor_example located on the student CD-ROM and the AutoLISP application automatic_series.lsp, create a reactor that will automatically call the application when a resistor is erased.

13. Modify the reactor in the exercise above to call the automatic_series application when a new resistor is added.

14. Using the gear.lsp application located on the student CD-ROM, create a reactor that will prompt the user to rerun the application if the gear train has been deleted.

15. Create a reactor that writes out to a file all AutoCAD commands entered by the user.

CHAPTER 11

Multiple Document Interface

OBJECTIVES

Upon completion of this chapter the reader will be able to:

- Describe the difference between Single Document Interface and Multiple Document Interface

- Define the term NameSpace

- Load an AutoLISP application into multiple NameSpaces

- Make a VLX application that defines its own NameSpace available to the current document

- Retrieve and set the value of a variable defined in a different NameSpace

- Use AutoLISP to determine which VLX applications are currently loaded into a document's NameSpace

- Describe the importance of providing error trapping for an AutoLISP application

- Pass control from a VLX error-trapping function to a document error-trapping function

- Use the Visual LISP blackboard to transfer data from one NameSpace to another

KEY WORDS AND AUTOLISP FUNCTIONS

Blackboard	VL-DOC-SET
Error Trapping	VL-EXIT-WITH-ERROR
MDI	VL-EXIT-WITH-VALUE
NameSpace	VL-LIST-EXPORTED-FUNCTIONS
SDI	VL-LIST-LOADED-VLX
VL-DOC-EXPORT	VL-UNLOAD-VLX
VL-DOC-IMPORT	VL-VLX-LOADED-P
VL-DOC-REF	

INTRODUCTION

The ability to open multiple documents (drawings) in a single AutoCAD session, known as *Multiple Document Interface* (MDI), has been on the wish list of most AutoCAD users for years. However, until the release of AutoCAD 2000, Autodesk continued to support a *Single Document Interface* (SDI) environment. Running AutoCAD in a single document interface environment simplified the AutoLISP developer's task of developing applications by eliminating many of the hassles associated with multiple document interfaces, but at the same time it limited the potential of the applications being developed. However, with AutoCAD 2000, Autodesk stopped supporting single document interface and started supporting multiple document interface. This change in the AutoCAD environment not only affects the AutoCAD user, but it also has a major impact on the applications that are designed to run using the AutoCAD engine. It is essential that the developer have a good understanding of how the multiple document interface works, as well as its benefits and the problems associated with it.

MULTIPLE DOCUMENT INTERFACE

In a single document interface environment, applications and variables were confined to the current document. Once a document was closed and a new document was opened in its place, any applications that were previously loaded (not using ACAD.LSP) must be reloaded before being executed. This worked exceptionally well in preventing variables declared and set by one application from affecting the outcome of a program running in another drawing.

NAMESPACE

To prevent applications from affecting one another in an MDI environment, Autodesk introduced a new concept called *NameSpace*. In AutoCAD, a NameSpace is a LISP environment that contains an isolated set of symbols (variables and functions). Each open drawing has it own unique NameSpace. Variables and functions defined in one NameSpace are isolated from variables and functions defined in another (see Figure 11–1).

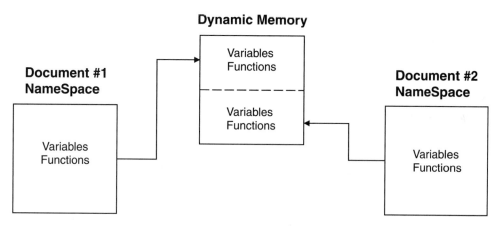

Figure 11–1 *NameSpace and dynamic memory*

This concept can be illustrated with the following example, where two different drawings are opened and displayed side by side (see Figure 11–2). In drawing #1 (the drawing on the left), the variable *namespace_variable* is set equal to the string "NameSpace example - Drawing #1" (see Figure 11–3) using the following expression:

Command: **(setq namespace_variable "NameSpace example - Drawing #1")** (ENTER)
"NameSpace example - Drawing #1"

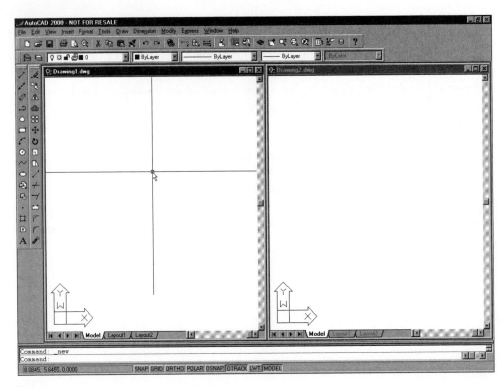

Figure 11–2 *Two drawings side by side*

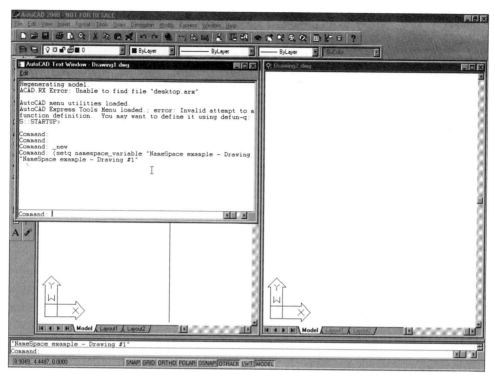

Figure 11–3 *Variable set in drawing #1*

Once the variable has been set in drawing #1, its value can be confirmed using the following expression, (see Figure 11–4).

> Command: **!NameSpace_variable** (ENTER)
> "NameSpace example - Drawing #1"

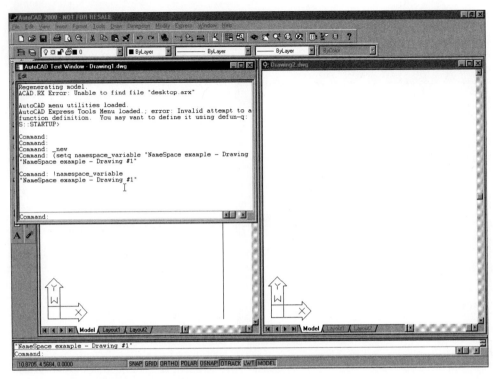

Figure 11–4 *Variable value confirmed*

After the value of the variable namespace_variable has been confirmed, the drawing #2 (the drawing on the right) is made active and the value of the variable is again checked (see Figure 11–5) using the following expression:

Command: **!NameSpace_variable** (ENTER)
nil

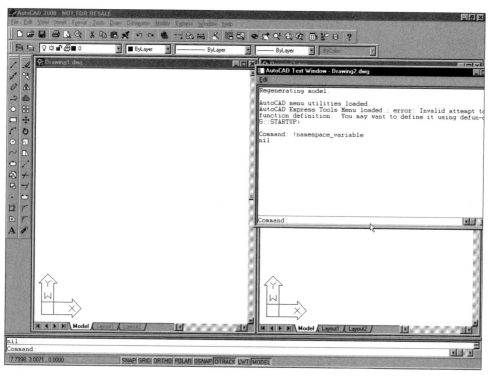

Figure 11–5 *Variable value checked*

The value of the variable returned in the second drawing is nil because drawing #2 has a different NameSpace from that of drawing #1. Any variables or functions that are defined in drawing #1 will be defined only in that drawing. The same exercise can be performed with user-defined functions and the results will remain the same.

Loading AutoLISP Applications into Multiple NameSpaces

As stated, when an AutoLISP application is loaded, the function is accessible only to the drawing where it was loaded. However, if the application is needed by multiple documents, then its contents can be automatically loaded by either appending the Acaddoc.lsp file or by using the VL-VLOAD-ALL (vl-load-all filename) function. This function, when supplied with the name of an AutoLISP application, loads the application in all open AutoCAD documents as well as any documents opened subsequently during the current AutoCAD session. Any variables that might be contained within the loaded application will be unique to the NameSpace where they are contained. By loading the application into multiple drawings, the variables set by the application in one drawing will be unique to that particular drawing and will have no effect on any of the other opened drawings. Using the VL-LOAD-ALL function, the AutoLISP application shown below can be loaded into all opened AutoCAD documents:

```
;;;**********************************************
;;;
;;;    Program Name: NameSpace.lsp
;;;
;;;    Program Purpose: Illustrate how MDI works
;;;
;;;    Program Date: 01-30-2000
;;;
;;;    Written By: Kevin Standiford
;;;
;;;**********************************************

(defun c:namespace ()
  (setq    variable (getstring "\nThis is an example of namespace")
  )
  (princ)
)
```

To load the application:

Command: **(VL-LOAD-ALL "namespace.lsp")** (ENTER)

NameSpace and VLX Applications

When an AutoLISP application is compiled into a Visual LISP executable (VLX) file using the Make Application procedure, the developer is given the choice of creating a separate NameSpace for the application (see Figure 11–6). This option allows the compiled application to create its own NameSpace upon being loaded for execution. The application maintains its own NameSpace separate from that of the document where it was loaded. By creating a separate NameSpace for compiled Visual LISP applications (VLX), the developer is guaranteed that the application's variables will not be accidentally or in some cases intentionally overwritten by other applications. Functions that are defined by an application can be exported to the document's NameSpace so that the application is able to access the contents of that document (see Figure 11–7).

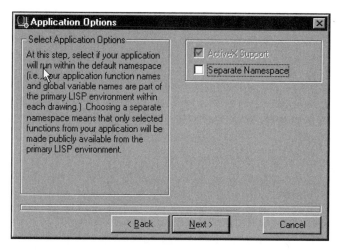

Figure 11–6 *Application NameSpace*

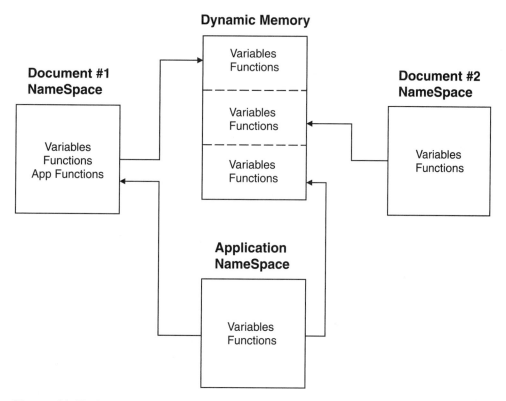

Figure 11–7 *Function exported to a document's NameSpace*

Each time a compiled Visual LISP application (VLX) in loaded into a different document, the application creates a new NameSpace to maintain its variables and functions. Once an application has been loaded into a document, any further attempts to load that application into the same document will result in AutoCAD's issuing an error message. If an application is to be reloaded, that application must first be unloaded through the VL-UNLOAD-VLX (vl-unload-vlx "appname") function. The *appname* argument associated with this function is the name of the application to be unloaded, without the application's extension and path. To unload the application in the previous example, the following expression would be used.

Command: **(VL-UNLOAD-VLX "namespace")** (ENTER)

 Note: When an application is compiled and the Separate Namespace option is not selected, then all variables and functions defined by the application are loaded in the document's NameSpace.

Making Functions Accessible to Documents and Applications

When an application is compiled using the Separate Namespace option, by default that application's functions are not exposed to the document where the application is loaded. The functions must be exported to the document through the VL-DOC-EXPORT (vl-doc-export 'function) function. When this function is supplied with a symbol representing a function name and is used within a VLX application where a separate NameSpace is defined, it exposes the specified function to any document's NameSpace that loads the application. For example, when the following program is compiled into a VLX application (with the Separate Namespace option), loaded into AutoCAD (with the APPLOAD command), and executed, AutoCAD responds with the error message "; error: no function definition: NAMESPACE_EXAMPLE".

```
;;;****************************************************************
;;;
;;;     Program Name : namespace_example.lsp
;;;
;;;     Program Purpose : To demonstrate the VL-DOC-EXPORT function
;;;
;;;     Programmed By : Kevin Standiford
;;;
;;;     Date : 02-01-2000
;;;
;;;****************************************************************
```

```
(DEFUN NameSpace_Example ()
  (SETVAR "CMDECHO" 0)
  (SETQ   number_one (GETREAL "\nEnter first number : ")
          number_two (GETREAL "\nEnter second number : ")
  )
  (WRITE-LINE "")
  (PRINC
    (STRCAT "The answer is " (RTOS (+ number_one number_two)))
  )
  (PRINC)
)
```

To make this application accessible to the document and any other documents in which the application is loaded, the expression (VL-DOC-EXPORT "namespace_example") must be added to the application. For example:

```
;;;****************************************************************
;;;
;;;     Program Name : namespace_example.lsp
;;;
;;;     Program Purpose : To demonstrate the VL-DOC-EXPORT function
;;;
;;;     Programmed By : Kevin Standiford
;;;
;;;     Date : 02-01-2000
;;;
;;;****************************************************************

(VL-DOC-EXPORT "namespace_example")        ; Exposes the
                                           ; namespace_example function
                                           ; to the current document's
                                           ; NameSpace.

(DEFUN NameSpace_Example ()
  (SETVAR "CMDECHO" 0)
  (SETQ   number_one (GETREAL "\nEnter first number : ")
      number_two (GETREAL "\nEnter second number : ")
  )
  (WRITE-LINE "")
  (PRINC
```

```
        (STRCAT "The answer is " (RTOS (+ number_one number_two)))
    )
  (PRINC)
)
```

 Note: Any attempt to call the VL-DOC-EXPORT function from outside the application where the separate NameSpace is defined will not expose that application's function to the document. In other words this action has no effect.

Visual LISP provides two functions for determining which applications define separate NameSpaces and which of their functions have been exposed to the current document's NameSpace: the VL-LIST-LOADED-VLX and VL-LIST-EXPORTED-FUNCTIONS functions. The VL-LIST-LOADED-VLX function generates a list of all applications that define a separate NameSpace and that are associated with the current document, while the VL-LIST-EXPORTED-FUNCTIONS function returns a list of the exposed functions associated while a particular application. For example:

Command: **(VL-LIST-LOADED-VLX)** (ENTER)
(NAMESPACE_EXAMPLE NAMESPACE_TEST)
Command:

Command: **(vl-list-exported-functions "namespace_example")** (ENTER)
("namespace_example")
Command:

When a function is defined having a separate NameSpace, that function by default is not exposed to any other applications that define a separate NameSpace. However, once an application has been exposed through the VL-DOC-EXPORT function, its functions can be made available to other applications using the VL-DOC-IMPORT (vl-doc-import application ['function...]) function. Like the VL-DOC-EXPORT function, this function should be placed at the beginning of an application's source code file and should not be executed within any functions. For example:

```
;;;****************************************************************
;;;
;;;    Program Name : namespace_example.lsp
;;;
;;;    Program Purpose : To demonstrate the VL-DOC-EXPORT function
;;;
;;;    Programmed By : Kevin Standiford
;;;
```

```
;;;    Date : 02-01-2000
;;;
;;;*****************************************************************
(vl-doc-import "namespace_test" 'namespace_test)
(VL-DOC-EXPORT "namespace_example")

(DEFUN NameSpace_Example ()
   (SETVAR "CMDECHO" 0)
   (SETQ    number_one (GETREAL "\nEnter first number : ")
            number_two (GETREAL "\nEnter second number : ")
   )
   (WRITE-LINE "")
   (PRINC
      (STRCAT "The answer is " (RTOS (+ number_one number_two)))
   )
   (PRINC)
)
```

Making Variables Accessible to Documents and Applications

Just like functions that are defined in applications that create their own NameSpace, variables contained within these applications are not known by the document's NameSpace. AutoLISP does provide three functions that allow variables that are defined within a document's NameSpace to be accessed by a VLX application: VL-DOC-REF (vl-doc-ref 'symbol), VL-DOC-SET (vl-doc-set 'symbol value) and VL-PROPAGATE (vl-propagate 'symbol).

To copy the value of a variable from a document's NameSpace into a VLX's NameSpace, the AutoLISP function VL-DOC-REF should be used. This function returns the value of a variable when supplied with a symbol representing a specific variable. The following VLX application extracts the value of the variable *example* from the current document's NameSpace and displays it in a dialog box. If the value of the variable is nil, then the application sets the variable to the error message "The value of variable example is nil :" (see Figures 11–8 and 11–9).

```
;;;*****************************************************************
;;;
;;;    Program Name : namespace_example.lsp
;;;
;;;    Program Purpose : To demonstrate the VL-DOC-EXPORT function
;;;
```

```
;;;     Programmed By : Kevin Standiford
;;;
;;;     Date : 02-01-2000
;;;
;;;****************************************************************
(vl-doc-import "namespace_test" 'namespace_test)
(VL-DOC-EXPORT "namespace_example")

(DEFUN NameSpace_Example ()
  (SETVAR "CMDECHO" 0)
  (setq example (vl-doc-ref 'example))
  (IF (= example nil)
    (setq example "The value of variable example is nil : ")
  )
  (ALERT example)
  (SETQ    number_one (GETREAL "\nEnter first number : ")
           number_two (GETREAL "\nEnter second number : ")
  )
  (WRITE-LINE "")
  (PRINC
    (STRCAT "The answer is " (RTOS (+ number_one number_two)))
  )
  (PRINC)
)
```

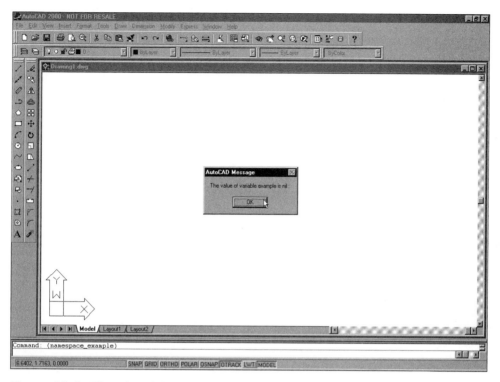

Figure 11–8 *The value of the variable example is equal to nil*

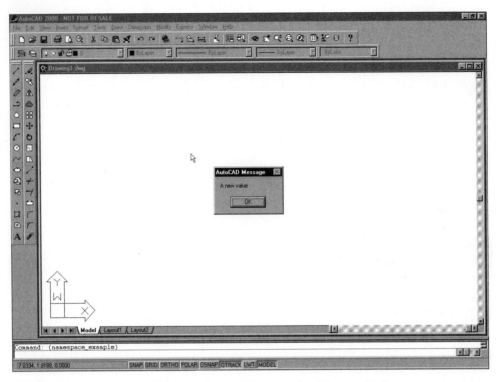

Figure 11–9 *The value of the variable example is equal to "A new value"*

 Note: When the function VL-DOC-REF is executed from within a document, the result is equivalent to executing the EVAL function. The expression is evaluated without the value being exported to the application's NameSpace.

To set the value of a variable contained within a document's NameSpace, the function VL-DOC-SET must be used. Like the VL-DOC-REF function, this function requires that the variable be furnished in the form of a symbol. Unlike the VL-DOC-REF function, this function also requires that the value of the document's variable be supplied. The following program sets the value of the variable *example_1* equal to the variable *example*:

```
;;;**************************************************************
;;;
;;;     Program Name : namespace_example.lsp
;;;
;;;     Program Purpose : To demonstrate the VL-DOC-EXPORT function
```

```
;;;
;;;    Programmed  By  :  Kevin  Standiford
;;;
;;;    Date  :  02-01-2000
;;;
;;;********************************************************************
(vl-doc-import  "namespace_test"  'namespace_test)
(VL-DOC-EXPORT  "namespace_example")

(DEFUN  NameSpace_Example  ()
   (SETVAR  "CMDECHO"  0)
   (setq  example  (vl-doc-ref  'example))
   (VL-DOC-SET  'example_1  example)
   (IF  (=  example  nil)
      (setq  example  "The  value  of  variable  example  is  nil  :  ")
   )
   (ALERT  example)
   (if  (/=  example_1  nil)
      (ALERT  example_1)
   )
   (SETQ     number_one  (GETREAL  "\nEnter  first  number  :  ")
             number_two  (GETREAL  "\nEnter  second  number  :  ")
   )
   (WRITE-LINE  "")
   (PRINC
      (STRCAT  "The  answer  is  "  (RTOS  (+  number_one  number_two)))
   )
   (PRINC)
)
```

To execute the program:

Command: **!example_1**
"A new value"
Command:

Note: When the function VL-DOC-SET is executed from within a document's NameSpace, the result is equivalent to executing the SETQ function.

Finally, to set the value of a variable in all open documents as well as any new documents created during the current AutoCAD session, the VL-PROPAGATE function should be used. The following expression copies the value of the variable *example_1* from document one's NameSpace into all open documents equal to the value of the variable *example* contained within the current document (see Figures 11–10 through 11–13).

```
Command:
Command: (VL-PROPAGATE 'example_1)
"Test"
Command:
```

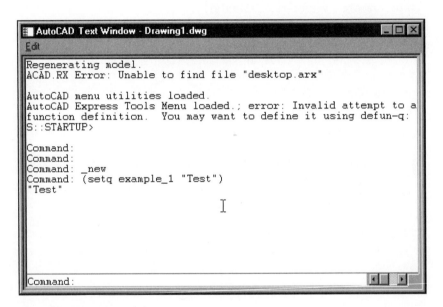

Figure 11–10   *Document #1*

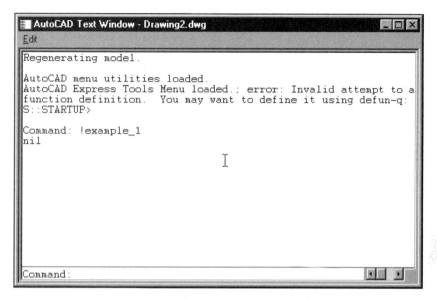

Figure 11–11 *Document #2 before the expression is executed*

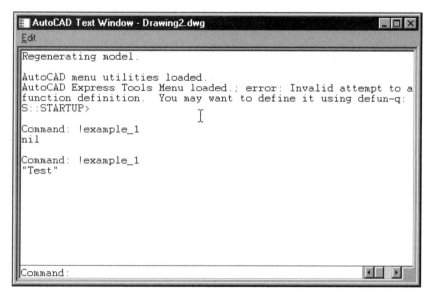

Figure 11–12 *Document #2 after the expression has been executed*

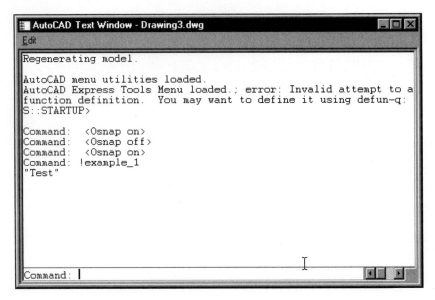

Figure 11–13 *New document created after the expression has been evaluated*

BLACKBOARDS

The value of a variable can also be transferred from one document's NameSpace to another using Visual LISP's *blackboard*. A blackboard is a NameSpace that is separate from documents and VLX applications. Its purpose is to provide documents and applications with a means of sharing data between NameSpaces.

Setting and Extracting Data from Visual LISP's Blackboard

To set a variable in the blackboard from any document or application, the VL-BB-SET (vl-doc-set 'symbol value) function should be used. Likewise, to retrieve the value of a variable from the blackboard, the VL-BB-REF (vl-bb-ref 'variable) function is used. To set the value of the variable *blackboard_example* from document #1 (Figure 11–14), the following expression would be used:

Command: **(VL-BB-SET 'blackBoardExample "This is an example of the Visual LISP**

Blackboard") (ENTER)
"This is an example of the Visual LISP Blackboard"
Command:

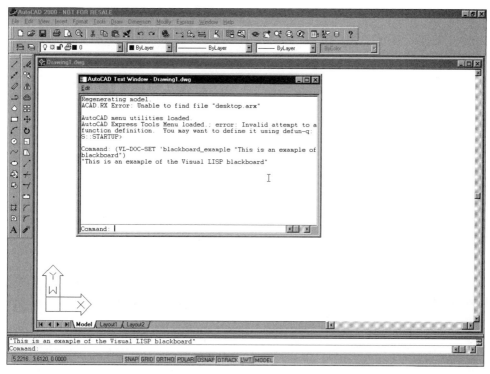

Figure 11–14   *Document #1*

To retrieve the value of the variable *blackboard_example* from document #2 (see Figure 11–15), the following expression would be used.

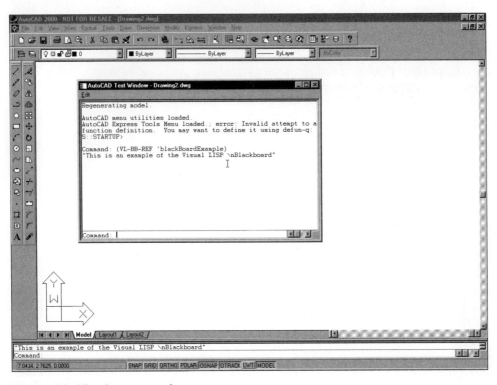

Figure 11–15 *Document #2*

Command: **(VL-BB-REF 'blackBoardExample)** (ENTER)
"This is an example of the Visual LISP \nBlackboard"
Command:

Note: Both the VL-BB-SET and VL-BB-REF functions require that a symbol representing the variable be passed to them. Values that are assigned to variables using the Visual LISP VL-BB-SET function will not affect the variables using the same NameSpace in the document or application where the operation was performed.

ERROR MESSAGES

At some point in time all programs are subjected to conditions that the developer had not originally anticipated. These conditions can cause problems ranging from the program returning incorrect results to the abrupt termination of the application altogether (fatal error). In previous versions of AutoLISP, when a fatal error occurred, AutoLISP would automatically print an error message followed by the remaining portion of the program. To prevent AutoLISP from scrolling the remaining portion of an application when an error does occur, an *error trap* must be defined. Error traps are user-defined functions that are created using the DEFUN function. Their purpose is to convey to the user information about the condition that caused the error to occur, restore the AutoCAD environment to a known condition, and continue program execution. A typical error trap can be defined as follows:

```
(DEFUN *error* (msg)
       (ALERT (STRCAT "\nError:" msg))
)
```

When an application encounters an error, AutoCAD responds by passing a text string describing the error condition to the currently defined *error* function. In the previous example, the error message provided by AutoCAD is displayed to the user in an AutoCAD alert dialog box.

 Note: A good error trap is designed to execute quickly and inform the user about the nature of the error.

Error Conditions and MDI

In the current release of AutoCAD, each document is provided with a predefined *error* function:

```
(defun *error* (msg)
                (princ "error: ")
                (princ msg)
                (princ)
)
```

When a VLX application is run within a document's NameSpace, that application shares the document's default error-trapping function. However, a VLX application designed to run in its own NameSpace can use either the default error function or one specifically designed for that application. To pass control from a VLX's error-trapping function to the document's error-trapping function, the AutoLISP VL-EXIT-WITH-ERROR (vl-exit-with-error msg) function can be used. To pass an error message to the documents NameSpace, the following expressions would be added to the VLX application's source code:

```
(defun *error* (msg)
  (vl-exit-with-error (strcat "The application has encountered the follow-
    ing error! " msg))
)
```

Using a VLX Error Trap to Return a Value

In addition to returning an error message from a VLX application's error trap to the document's default error-trapping function, AutoLISP also provides a means of returning a value to that document's NameSpace. To accomplish this, the FUNCTION VL-EXIT-WITH-VALUE (vl-exit-with-value value) is used. To return the integer value 8 to the document's NameSpace when an error occurs, the following expressions would be added to the VLX's source code:

```
(defun *error* (msg)
      (vl-exit-with-value    8)
)
```

Caution: The *VL-EXIT-WITH-ERROR* and *VL-EXIT-WITH-VALUE* functions can be executed only within an error trap of a VLX application. Any attempt to execute these functions in any other location will result in an error.

LIMITATIONS OF WORKING WITHIN MULTIPLE DOCUMENT INTERFACE USING AUTOLISP

When working within a multiple document interface environment, AutoLISP is limited to working with one document at a time. An AutoLISP application cannot issue any entity creation or modification functions in a NameSpace different from the one where the application is currently running. Although in theory, ActiveX controls can be employed in an AutoLISP application to access multiple document NameSpaces, this feature is currently not supported by AutoLISP and any attempt to use ActiveX in this capacity will result in creating an unstable AutoCAD environment and most likely will cause AutoCAD to crash.

SUMMARY

When an application has the ability to open more than one document in a single session, this is known as multiple document interface. When an application is developed for a MDI environment, special consideration must be taken, or the application may not function properly or may interfere with other applications. To prevent applications from affecting one another, Autodesk has developed the concept of NameSpace. A NameSpace is a LISP environment that contains an isolated set of symbols (variables and functions). Each open drawing has it own unique NameSpace. Variables and functions defined in one NameSpace are isolated from variables and functions defined in another. When an AutoLISP application is loaded, the function is accessible only to the drawing where it was loaded. If the application is needed for multiple documents, then its contents can be automatically loaded by either appending the application to the Acaddoc.LSP or by using the VL-VLOAD-ALL function.

When an AutoLISP application is compiled into a Visual LISP executable, the developer is given the choice of creating a separate NameSpace for the application. When the developer chooses this option, upon loading, the application creates its own NameSpace separate from the NameSpace of the document where the application was loaded. This guarantees that the variables used in the application will not accidentally be overwritten. Each time a compiled Visual LISP application is loaded into a different document, a new NameSpace is created to maintain the variables and functions associated with that application.

When an application is compiled with the Separate Namespace option, by default that application's functions are not exposed to the document where the application is launched. The functions must be exported to the document using the VL-DOC-EXPORT function. To make an application accessible to the document and any other documents where the application is loaded, the VL-DOC-EXPORT function must be used. To determine which applications have defined separate NameSpaces and which of their functions have been exposed to the current document's NameSpace, the VL-LIST-LOADED-VLS and VL-LIST-EXPOSED-FUNCTIONS functions can be used. Once an application has been exposed through the VL-DOC-EXPORT function, its functions can be made available to other applications through the VL-DOC-IMPORT function.

Variables contained within applications that define their own NameSpace are not known to the document's NameSpace where the application was loaded. Variables contained within a document's NameSpace may be accessed by a VLX application through the VL-DOC-REF, VL-DOC-SET, and VL-PROPAGATE functions. The value of a variable can be transferred from one document's NameSpace to another document's NameSpace using the Visual LISP blackboard. A blackboard is a NameSpace that is separate from documents and VLX applications. Its purpose is to provide documents with a means of sharing data between NameSpaces.

Currently AutoLISP is limited to working with one document at a time. An AutoLISP application cannot issue any entity creation or modification functions in a NameSpace different from the one where the application is currently running.

REVIEW QUESTIONS

1. Define the following terms:
   Blackboard
   Error Trapping
   MDI
   NameSpace
   SDI
   VL-ARX-IMPORT
   VL-DOC-EXPORT
   VL-DOC-IMPORT
   VL-DOC-REF
   VL-DOC-SET
   VL-EXIT-WITH-ERROR
   VL-EXIT-WITH-VALUE
   VL-LIST-EXPORTED-FUNCTIONS
   VL-LIST-LOADED-VLX
   VL-UNLOAD-VLX
   VL-VLX-LOADED-P

2. Describe the difference between Single Document Interface and Multiple Document Interface.

3. How is an AutoLISP application loaded into multiple NameSpaces?

4. How is a VLX application that defines its own NameSpace made available to the current document?

5. How are variables retrieved and set in different NameSpaces?

6. How can you determine which VLX applications are currently loaded into a document's NameSpace?

7. Describe the importance of providing error trapping for an AutoLISP application.

8. How is control passed from a VLX error-trapping function to a document error trapping-function?

9. What is the purpose of the Visual LISP blackboard?

10. What AutoLISP functions are used to transfer data from one document's NameSpace to another using the Visual LISP blackboard?

11. True or False: Currently there are no limitations on AutoLISP for working with MDI. (If the answer is false, explain why.)

CHAPTER 12

Windows 95, 98, and NT Registry, Loading and Executing VBA applications

OBJECTIVES

Upon completion of this chapter the reader will be able to:

- Define the terms Windows registry, initialization file, subtree, and subkey

- Describe the limitations associated with initialization files

- Describe the advantages of using the registry as opposed to an initialization file to store important program information

- List the six main subtrees associated with the Windows registry

- Use AutoLISP to read and write information to and from the registry

- Use AutoLISP to load and execute VBA applications

KEY WORDS AND AUTOLISP FUNCTIONS

Initialization Files	**vl-registry-delete**
Registry	**vl-registry-descendents**
Startup File	**vl-registry-read**
Subkey	**vl-registry-write**
Subtree	**vl-vbaload**
vlax-product-key	**vl-vbarun**

INTRODUCTION TO WINDOWS REGISTRY

Often computer applications rely on essential information (such as startup parameters) that is stored on the computer separately from the compiled application. Traditionally DOS and early versions of Windows (3.X) applications stored this information in initialization (.INI) files. Initialization files are ASCII based text files that are limited in length to 64K. This created a problem: because these files were ASCII text files, they were considered low security and therefore could be edited using any text editor. Anyone using a text editor could easily alter the contents of these files; if the person editing the file made a mistake, then the program would either not function properly or in most cases not start at all. Also, the lack of standardization in the early days of the software industry often led to initialization files that were extremely large, causing the application (including the operating system) to run slowly. The lack of standards caused initialization files from one application to be overwritten by another application, especially if the files shared the same name and location. Finally, because these files were not stored in a central location, they were often difficult to locate.

It wasn't until the introduction of Windows NT that Microsoft started addressing these problems. Their solution was to develop a system of database files (collectively known as the registry) stored in a centralized location that would house important startup information used by Windows and Windows-based applications. The concept has worked extremely well in general, and it has been used in all subsequent Windows operating systems.

THE WINDOWS REGISTRY

The registry is a system of proprietary binary database files whose primary purpose is to ensure the integrity of the computer's operating system and applications by storing important settings and information. In Windows 95, the database files that make up the registry are System.dat and User.dat. In Windows 98, the registry's database files are System.dat, User.dat and Policy.pol, while in Windows NT, the registry is comprised of the Sam and Sam.log, Security and Security.log, System and System.log, Ntuser.dat and Ntuser.dat.log, Default and Default.log. When these files are viewed collectively using the Microsoft REGEDIT utility, a hierarchical structure appears (see Figure 12–1).

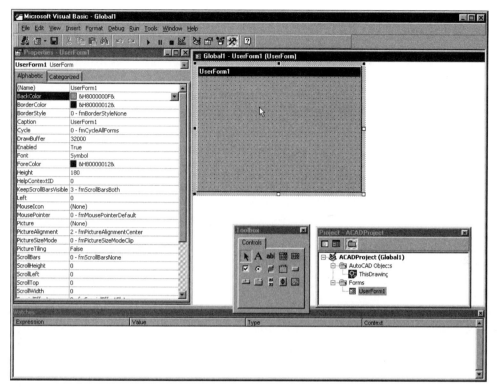

Figure 12–1 *Windows 98 registry as it appears through the REGEDIT utility*

 Note: Additional information regarding the registry can be found at www.microsoft.com.

The structure is comprised of six main components called subtrees: HKEY_LOCAL_MACHINE, HKEY_CLASS_ROOT, HKEY_CURRENT_ USER, HKEY_USERS, HKEY_CURRENT_CONFIG, and HKEY_ DYN_DATA. Each subtree is made up of subkeys. The subkeys contained in each subtree may also contain subkeys or they may contain active keys. It is the active keys that store the actual values associated with a particular application. Table 12–1 provides a brief description of each subtree.

Table 12–1   Registry Subtree

Subtree Name	Description
HKEY_LOCAL_MACHINE	Contains information regarding the local computer's hardware and software (bus type, system memory, device drivers, etc.).
HKEY_CLASSES_ROOT	Contains data for OLE and file class association.
HKEY_CURRENT_USER	Contains profile information for the current user (environment variables, desktop settings, network connections, etc.).
HKEY_USERS	Contains all actively loaded user profile information.
HKEY_CURRENT_CONFIG	Contains hardware information used by the local computer during startup.
HKEY_DYN_DATA	Contains points to dynamic information that constantly changes from session to session.

Out of the six subtrees located in the Windows registry, just two contain 90 percent of the data used by the registry, HKEY_LOCAL_MACHINE and HKEY_USER. The remaining subtrees (with the exception of the HKEY_DYN_DATA) are actually aliases that contain a copy of the contents of the other two keys. All information stored in the registry by an AutoLISP application should be placed in either the HKEY_LOCAL_MACHINE or HKEY_USER subtree.

AUTOLISP AND THE WINDOWS REGISTRY

Visual LISP has provided several extended functions that will allow the developer to interface with the Windows registry. Just as with the other extended functions presented in earlier chapters, before the extended registry functions can be used, their supporting code must be loaded through the VL-LOAD-COM function.

 Caution: Before any modifications are made to the registry, a backup copy should be created through the Microsoft utility REGEDIT. One mistake in the registry can render a computer useless.

Using AutoLISP to Read the Windows Registry

Currently there are three functions that allow an AutoLISP application to read the Windows registry: VL-REGISTRY-DESCENDENTS (vl-registry-descendents reg-key [val-names]), VL-REGISTRY-READ (vl-registry-read reg-key [val-name]), and VLAX-PRODUCT-KEY (vlax-product-key). The VL-REGISTRY-DESCENDENTS function returns a list of subkeys or values for a specified Windows registry key. The function has two arguments associated with it *reg-key* and *val-names*. The *reg-key* argument is a string used to specify the Windows registry key and/or subtree to search. The *val-names* argument represents the value associated with the specified key. If this argument is supplied and its value is not nil, then the function returns a list of all values associated with the specified key. If the argument is not supplied or its value is equal to nil, then the function returns a list of all subkeys associated with the specified key. If no subkeys are associated with the specified key, then the function returns nil. This is illustrated in the following example (see Figure 12–2).

```
_$ (vl-registry-descendents
   "HKEY_LOCAL_MACHINE\\Software\\Autodesk\\AutoCAD\\R15.0" "1" )
("CurVer")
_$ (vl-registry-descendents
   "HKEY_LOCAL_MACHINE\\Software\\Autodesk\\AutoCAD\\R15.0" nil )
("ACAD-1:409")
_$ (vl-registry-descendents
   "HKEY_LOCAL_MACHINE\\Software\\Autodesk\\AutoCAD\\R15.0\\ACAD-
   1:409\\AdLM" nil )
nil
_$
```

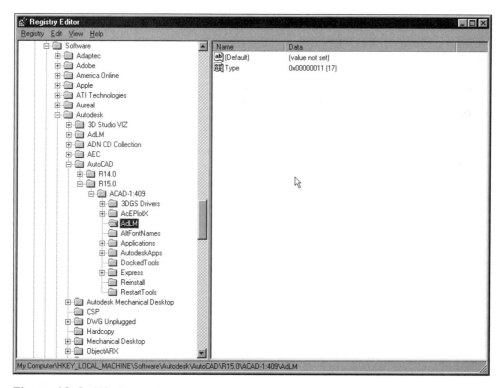

Figure 12–2 *Windows registry*

An example of how this function may be used in an application is illustrated in the following program. The program, when executed, displays a dialog box where the user is prompted to enter the registry key that will return a listing of all subkeys. Once the user has entered a key, the program searches the registry for the specified key. If a match is located, then the program returns a listing of all subkeys associated with the specified key (see Figures 12–3 and 12–4).

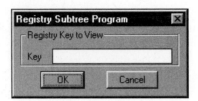

Figure 12–3 *Registry program dialog box*

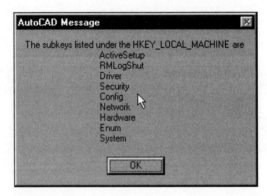

Figure 12–4 *Results of program*

AutoLISP Application:

```
(defun c:regview ()
  (vl-load-com)
  (setq    dclId (load_dialog
            "C:/WINDOWS/Desktop/Visual_Lisp/Ch12/registry.dcl"
              )
  )
  (if (not (new_dialog "registry" dclId))
    (exit)
  )
```

```
(action_tile "accept" "(get_info)(done_dialog 1)")
(action_tile "cancel" "(setq result 0) (done_dialog 0)")
(start_dialog)
(if (/= result 0)
  (progn
    (setq information
            (vl-registry-descendents hkey)
        cnt      0
        listLength
          (length information)
        string
          (strcat "The subkeys listed under the " hkey " are ")
    )
    (while (< cnt (- listLength 1))
    (setq line (nth cnt information))
    (if (/= line nil)
      (setq string (strcat string "\n" "\t" "\t" line))
    )
    (setq cnt (1+ cnt))
    )
    (alert string)
  )
 )
)
(defun get_info   ()
  (setq hkey (strcase (get_tile "key"))))
)
```

DCL Application:

```
registry : dialog {
  label = "Registry Subtree Program";
      fixed_width = true;
        : boxed_column {
        label = "Registry Key to View";
        : edit_box {
        label = "Key";
        key = "key";
        }
```

```
            }
        is_default = true;
        ok_cancel;
    }
```

The VL-REGISTRY-READ function returns the data stored for a specified Windows
registry key. Like VL-REGISTRY-DESCENDENTS, this function requires that the name of
a key be specified. However, unlike VL-REGISTRY-DESCENDENTS, when this function's
optional argument is supplied and its value is not nil, the data associated with the
specified value is returned. If the argument is omitted or its value is nil, then the
function returns the specified key and all values associated with it. This is illustrated
in the following example (see Figure 12–5):

```
_$ (vl-registry-read
   "HKEY_LOCAL_MACHINE\\Software\\Autodesk\\AutoCAD\\R15.0\\ACAD-
   1:409\\Reinstall" "folder")
"AutoCAD 2000"
_$ (vl-registry-read
   "HKEY_LOCAL_MACHINE\\Software\\Autodesk\\AutoCAD\\R15.0\\ACAD-
   1:409\\Reinstall" nil)
""

_$
```

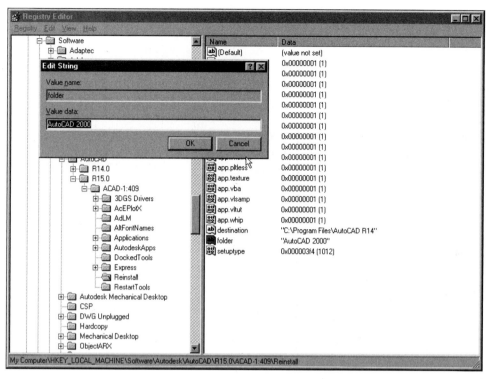

Figure 12–5 *Windows registry*

The last function associated with the Windows registry is the VLAX-PRODUCT-KEY function. Unlike the previous two functions, this function always returns the same value, the Windows registry path for AutoCAD where an application can be registered for demand loading. Demand loading allows AutoCAD to automatically load third-party applications that are not currently resident in AutoCAD. Demand loading can be accomplished when AutoCAD is first launched, when a third-party application is called, or if the drawing contains custom objects created by a third-party application.

Using AutoLISP to Modify the Windows Registry

Currently, there are two extended AutoLISP functions provided that allow the developer to alter the Windows Registry: VL-REGISTRY-DELETE (vl-registry-delete reg-key [val-name]) and VL-REGISTRY-WRITE (vl-registry-write reg-key [val-name val-data]). The VL-REGISTRY-DELETE function allows the developer to delete a specified registry key or a registry key value. If the developer specifies a value along with a registry key, then the function will purge the specified value from the key. If the developer does not specify a value, then the function will remove the specified key and all values associated with it.

The VL-REGISTRY-WRITE function allows the developer to create entries in the Windows registry. This function has one required argument, *reg-key*, the key name. The key name is a string value representing the Windows registry key. The remaining arguments associated with this function are optional. If the *val-name* argument is not supplied, then a default value (default) for the key is created, and an empty string ("") is created (see Figure 12–6).

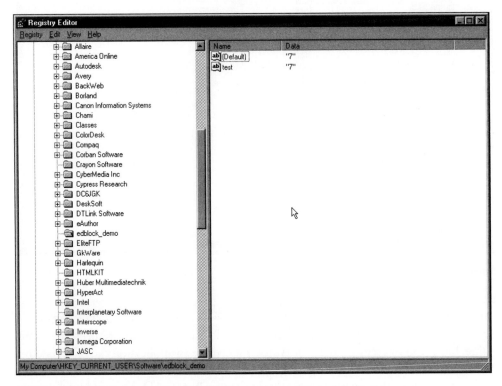

Figure 12–6 *Windows registry after the* VL-REGISTRY-WRITE *function is run*

```
_$ (VL-registry-write "HKEY_CURRENT_USER\\Software\\edblock_demo" nil "9")
"9"
_$ (VL-registry-write "HKEY_CURRENT_USER\\Software\\edblock_demo" "test"
  "10")
"10"
_$
```

If the value supplied to the function already exists, then any calls made to the VL-REGISTRY-WRITE function will overwrite that value's current data (see Figure 12–7). For example:

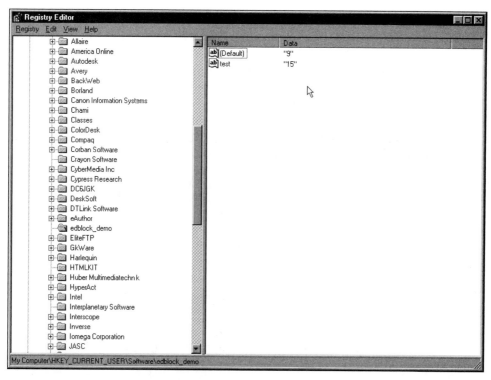

Figure 12–7 *Windows registry after the* VL-REGISRTY-WRITE *function is run a second time*

```
_$ (VL-registry-write "HKEY_CURRENT_USER\\Software\\edblock_demo" "test"
   "15")
"15"
_$
```

 Note: The VL-REGISTRY-WRITE function cannot be used to create entries in either the HKEY_USER or HKEY_LOCAL_MACHINE subtree.

APPLICATION - SPOT ELEVATION

With a few precautions, the registry can be a valuable asset to the AutoLISP developer for saving application settings. This is illustrated in the following example, where the spot elevation program featured in Chapter 7 has been modified to automatically advance the point number assigned to spot elevations with each execution. The point number for the next point to be labeled is stored in the Windows registry in the HKEY_CURRENT_USER\Software\spot_elevation subkey. Each time the program is executed, the application reads the registry for the point number. After the user has entered the required information and selected OK, the application calculates the next point number and stores that information in the registry for the application's next execution. The application also stores the text's rotation angle in the registry. This allows the application to recall how the user set the rotation angle for the last spot elevation labeled. If the application is executed for the first time or the information cannot be found in the registry, then the application uses default settings for the point number and the text rotation angle (see Figures 12–8 and 12–9).

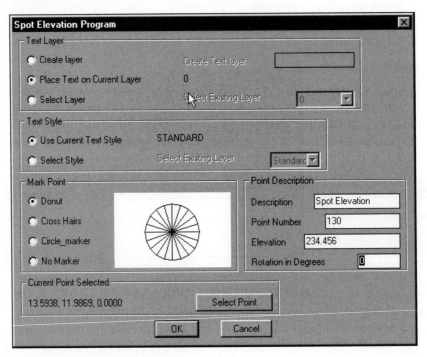

Figure 12–8 *Spot Elevation dialog box*

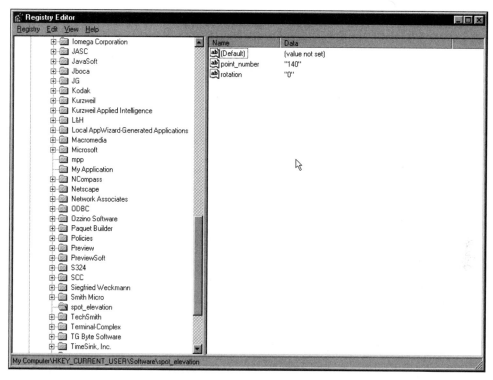

Figure 12–9 *Spot Elevation registry settings*

```
;;;********************************************************************
;;;
;;;Program Name: Spot_elevation.lsp
;;;
;;;Program Purpose: Label spot elevations on an AutoCAD drawing
;;;
;;;Programmed By: Kevin Standiford
;;;
;;;Date: 12/22/99
;;;
;;;********************************************************************
(defun c:spot_elevation   (/ result z_pt field_position elevation pt)
  (vl-load-com)
  (setq cnt 1)
  (setq result 4)
```

```
(while (> result 1)
  (setq dcl_id4 (load_dialog
              "C:/WINDOWS/Desktop/Visual_Lisp/Ch6/LSP/spot.dcl"
         )
  )
  (if    (not (new_dialog "spot" dcl_id4))
    (exit)
  )
  (start_list "layer_pop_list")
  (setq ll  '()
        nl   (tblnext "layer" T)
        idx 0
  )
  (while nl
    (if (= (getvar "clayer") (cdr (assoc 2 nl)))
    (setq cl   idx
          idx (1+ idx)
    )
    )
    (setq ll (append ll (list (cdr (assoc 2 nl)))))
        nl (tblnext "layer")
    )
  )
  (mapcar 'add_list ll)
  (end_list)
  (set_tile "layer_pop_list" (ITOA CL))
  (start_list "style_pop_list")
  (setq ll_s      '()
        nl_s       (tblnext "style" T)
        idx_s      0
  )
  (while nl_s
    (if (= (getvar "textstyle") (cdr (assoc 2 nl_s)))
    (setq cl_s   idx_s
          idx_s (1+ idx_s)
    )
    )
```

```
          (setq ll_s (append ll_s (list (cdr (assoc 2 nl_s)))))
              nl_s (tblnext "style")
        )
     )
      (mapcar 'add_list ll_s)
      (end_list)
;;;~~~~~~~~~~~~~~Registry~~~~~~~~~~~~~~~~~~Registry~~~~~~~~~~~~~~~~~~~
      (setq regInformation
             (vl-registry-descendents
                "HKEY_CURRENT_USER\\Software\\spot_elevation"
               "1"
            )
      )
     (if     (/= regInformation nil)
       (progn
       (set_tile "rotation"
               (vl-registry-read
                   "HKEY_CURRENT_USER\\Software\\spot_elevation"
                  "rotation"
               )
       )
       (set_tile "point_number"
               (vl-registry-read
                   "HKEY_CURRENT_USER\\Software\\spot_elevation"
                  "point_number"
               )
       )
       )
       (progn
       (set_tile "point_number" "100")
       (set_tile "rotation" "45")
       )
     )
;;;~~~~~~~~~~~~~~Registry~~~~~~~~~~~~~~~~~~Registry~~~~~~~~~~~~~~~~~~~
      (set_tile "description" "Spot Elevation")
      (set_tile "style_pop_list" (ITOA CL_s))
      (set_tile "current_layer" "1")
```

```
 (set_tile "current_style" "1")
 (set_tile "donut_marker" "1")
 (set_tile "field_layer" (getvar "clayer"))
 (set_tile "field_style" (getvar "cmlstyle"))
(if    (/= field_position nil)
   (set_tile "field_position" field_position)
   (set_tile "field_position" "No point selected")
)
(if    (/= elevation nil)
   (set_tile "elevation" (rtos elevation))
  (progn
  (mode_tile "elevation" 1)
  (mode_tile "rotation" 1)
  (mode_tile "description" 1)
  (mode_tile "point_number" 1)
  (mode_tile "c1" 1)
  (mode_tile "c2" 1)
  (mode_tile "c3" 1)
  (mode_tile "field_style" 1)
  (mode_tile "field_layer" 1)
   )
)
 (mode_tile "layer_pop_list" 1)
 (mode_tile "create_layer_box" 1)
 (mode_tile "style_pop_list" 1)
 (action_tile
   "current_layer"
    "(progn (mode_tile \"layer_pop_list\"  (atoi $value))
          (mode_tile \"create_layer_box\" (atoi $value))
          (mode_tile \"field_layer\" (- 1 (atoi $value))))"
)
 (action_tile
   "select_layer"
    "(progn (mode_tile \"create_layer_box\"  (atoi $value))
          (mode_tile \"field_layer\"  (atoi $value))
          (mode_tile \"layer_pop_list\" (- 1 (atoi $value))))"
)
```

```
 (action_tile
    "create_layer"
     "(progn (mode_tile \"layer_pop_list\"  (atoi $value))
           (mode_tile \"field_layer\"  (atoi $value))
           (mode_tile \"create_layer_box\" (- 1 (atoi $value))))"
 )
 (action_tile
    "current_style"
     "(progn (mode_tile \"style_pop_list\"  (atoi $value))
           (mode_tile \"field_style\"  (- 1 (atoi $value))))"
 )
 (action_tile
    "select_style"
     "(progn (mode_tile \"field_style\"  (atoi $value))
           (mode_tile \"style_pop_list\" (- 1 (atoi $value))))"
 )
 (start_image "test")
 (setq x (dimx_tile "test")
      y (dimy_tile "test")
 )
 (slide_image 0 0 x y "/temp/donut.sld")
 (end_image)
 (action_tile
    "donut_marker"
     "(progn (start_image \"test\")
           (fill_image 0 0 x y 0)
           (slide_image 0 0 x y \"/temp/donut.sld\")
           (end_image))"
 )
 (action_tile
    "cross_marker"
     "(progn (start_image \"test\")
         (fill_image 0 0 x y 0)
           (slide_image 0 0 x y  \"/temp/cross.sld\")
           (end_image))"
 )
 (action_tile
```

```
        "circle_marker"
         "(progn (start_image \"test\")
               (fill_image 0 0 x y 0)
               (slide_image 0 0 x y  \"/temp/circle.sld\")
               (end_image))"
    )
   (action_tile
      "no_marker"
       "(progn (start_image \"test\")
             (fill_image 0 0 x y 0)
             (end_image))"
   )
     (action_tile "accept"  "(get_info)(done_dialog 1)")
     (action_tile "s_point" "(done_dialog 4)")
     (action_tile "cancel" "(setq result nil) (done_dialog 0)")
     (setq result (start_dialog))
    (if    (and (= result 1) (= pt nil))
       (progn
       (alert "No point Specified")
       (setq result 2
             elevation        nil
             field_position
             nil
        )
       )
    )
    (if    (= result 4)
       (progn
       (setq pt (getpoint "\nPick insertion point : "))
       (setq field_position
             (strcat (rtos (car pt))
                 ", "
                 (rtos (cadr pt))
                 ", "
                 (rtos (setq z_pt (caddr pt)))
             )
             elevation     z_pt
```

```
            )
          )
        )
      )
    (princ)
  )
(defun get_info    ()
  (setq    description     (get_tile "description")
      elevation     (get_tile "elevation")
      rotation      (get_tile "rotation")
      point_number  (get_tile "point_number")
      no_marker     (get_tile "no_marker")
      cross_marker  (get_tile "cross_marker")
      donut_marker  (get_tile "donut_marker")
      circle_marker       (get_tile "circle_marker")
      style_pop     (nth (atoi (get_tile "style_pop_list")) ll_s)
      select_style  (get_tile "select_style")
      current_style       (get_tile "current_style")
      layer_pop     (nth (atoi (get_tile "layer_pop_list")) ll)
      create_layer_box  (get_tile "create_layer_box")
      select_layer  (get_tile "select_layer")
      current_layer       (get_tile "current_layer")
      create_layer  (get_tile "create_layer")
  )
  (if (/= pt nil)
    (progn
      (setq pt (list (car pt) (cadr pt) (atof elevation)))
      (if (= no_marker "1")
      (no_marker_function)
      )
      (if (= circle_marker "1")
      (circle_marker_function)
      )
      (if (= donut_marker "1")
      (donut_marker_function)
      )
      (if (= cross_marker "1")
```

```
          (cross_marker_function)
        )
      )
    )
;;;~~~~~~~~~~~~~Registry~~~~~~~~~~~~~~~~~~Registry~~~~~~~~~~~~~~~~~~
    (if (/= point_number nil)
      (progn
        (vl-registry-write
        "HKEY_CURRENT_USER\\Software\\spot_elevation"
        "rotation"
        rotation
        )
        (vl-registry-write
        "HKEY_CURRENT_USER\\Software\\spot_elevation"
        "point_number"
        (itoa (+ 10 (atoi point_number)))
        )
      )
    )
;;;~~~~~~~~~~~~~Registry~~~~~~~~~~~~~~~~~~Registry~~~~~~~~~~~~~~~~~~
)
(defun no_marker_function ()
  (setq    object_1 nil
       object_2 nil
  )
   (make_object_function)
)
(defun circle_marker_function ()
  (setq    object_1 (LIST (CONS 0 "CIRCLE")
                    (CONS 10 pt)
                    (CONS 40 0.5)
                    (CONS 8 "pt_layer")
            )
       object_2 nil
  )
   (make_object_function)
)
```

```
(defun donut_marker_function ()
  (setq    object_1
       (list (cons 0 "LWPOLYLINE")
             (cons 100 "AcDbEntity")
             (cons 67 0)
             (cons 410 "Model")
             (cons 8 "0")
             (cons 100 "AcDbPolyline")
             (cons 90 2)
             (cons 70 1)
             (cons 43 0.5)
             (cons 38 0.0)
             (cons 39 0.0)
             (cons 10 pt)
             (cons 40 0.5)
             (cons 41 0.5)
             (cons 42 1.0)
             (cons 10 (list (+ 0.5 (car pt)) (cadr pt) (caddr pt)))
             (cons 40 0.5)
             (cons 41 0.5)
             (cons 42 1.0)
       )
     object_2 nil
  )
  (make_object_function)
)
(defun cross_marker_function ()
  (setq   object_1 (list
             (cons 0 "line")
             (CONS 10 (list (- (car pt) 0.5) (cadr pt) (caddr pt)))
             (cons 11 (list (+ (car pt) 0.5) (cadr pt) (caddr pt)))
             (cons 8 "pt_layer")
          )
     object_2 (LIST
             (CONS 0 "line")
             (CONS 10 (list (car pt) (- (cadr pt) 0.5) (caddr pt)))
             (CONS 11 (list (car pt) (+ (cadr pt) 0.5) (caddr pt)))
```

```
                              (CONS 8 "pt_layer")
                    )
        )
      (make_object_function)
    )
    (defun make_object_function ()
      (setvar "cmdecho" 0)
      (setvar "cmddia" 0)
      (if (= current_layer "1")
        (setq layer (getvar "clayer"))
      )
      (if (= create_layer "1")
        (setq layer create_layer_box)
      )
      (if (= select_layer "1")
        (setq layer layer_pop)
      )
      (if (= current_style "1")
        (setq style (getvar "cmlstyle"))
      )
      (if (= select_style "1")
        (setq style style_pop)
      )
      (ENTMAKE (list (cons 0 "block")
                (CONS 2 "point_block")
                (cons 10 pt)
                (cons 70 64)
            )
      )
      (ENTMAKE object_1)
      (ENTMAKE object_2)
      (ENTMAKE
        (list (cons 0 "TEXT")
            (cons 100 "AcDbEntity")
            (cons 67 0)
            (cons 410 "Model")
            (cons 8 (strcat layer "_ptn"))
```

```
      (cons 100 "AcDbText")
      (cons      10
        (list (+ (car pt) (* 1.7143 (getvar "textsize")))
              (+ (cadr pt) (* 1.7143 (getvar "textsize")))
              (caddr pt)
        )
      )
      (cons 40 (getvar "textsize"))
      (cons 1 point_number)
      (cons 50 (* (atof rotation) (/ 3.14 180)))
      (cons 41 1.0)
      (cons 51 0)
      (cons 7 style)
      (cons 71 0)
      (cons 72 0)
      (cons 100 "AcDbText")
      (cons 73 0)
    )
  )
  (ENTMAKE
    (list (cons 0 "TEXT")
        (cons 100 "AcDbEntity")
        (cons 67 0)
        (cons 410 "Model")
        (cons 8 layer)
        (cons 100 "AcDbText")
        (cons      10
          (list (+ (car pt) (* 1.7143 (getvar "textsize")))
                (cadr pt)
                (caddr pt)
          )
        )
        (cons 40 (getvar "textsize"))
        (cons 1 elevation)
        (cons 50 (* (atof rotation) (/ 3.14 180)))
        (cons 41 1.0)
        (cons 51 0)
```

```
                (cons 7 style)
                (cons 71 0)
                (cons 72 0)
                (cons 100 "AcDbText")
                (cons 73 0)
            )
        )
        (ENTMAKE
           (list (cons 0 "TEXT")
                (cons 100 "AcDbEntity")
                (cons 67 0)
                (cons 410 "Model")
                (cons 8 (strcat layer "_des"))
                (cons 100 "AcDbText")
                (cons      10
                   (list (+ (car pt) (* 1.7143 (getvar "textsize")))
                         (- (cadr pt) (* 1.7143 (getvar "textsize")))
                         (caddr pt)
                   )
                )
                (cons 40 (getvar "textsize"))
                (cons 1 description)
                (cons 50 (* (atof rotation) (/ 3.14 180)))
                (cons 41 1.0)
                (cons 51 0)
                (cons 7 style)
                (cons 71 0)
                (cons 72 0)
                (cons 100 "AcDbText")
                (cons 73 0)
           )
        )
        (ENTMAKE (list (cons 0 "endblk")))
        (ENTMAKE (list (cons 0 "INSERT")
                (cons 2 "point_block")
                (cons 10 pt)
           )
```

```
    )
    (loc)
)
(defun loc ()
  (setq
    ent1 (tblsearch "block"
                (cdr (assoc 2 (setq entkev (entget (entlast)))))
      )
  )
  (setq rnam (cdr (assoc -1 entkev)))
  (setq namkev (assoc 2 entkev))
  (setq ent1 (subst (cons 2 "*Unnn") (assoc 2 ent1) ent1))
  (setq ent1 (subst (cons 70 1) (assoc 70 ent1) ent1))
  (entmake ent1)
  (setq ent1 (entget (cdr (assoc -2 ent1))))
  (entmake ent1)
  (while (/= ent1 nil)
    (setq ent1 (entnext (cdr (assoc -1 ent1))))
    (if    (/= ent1 nil)
      (progn
      (setq ent1 (entget ent1))
      (entmake ent1)
      )
    )
  )
  (setq l (entmake '((0 . "endblk"))))
  (setq entkev (subst (cons 2 l) (assoc 2 entkev) entkev))
  (entmod entkev)
  (princ)
)
```

VISUAL BASIC FOR APPLICATIONS AND VISUAL LISP

In addition to the Visual LISP IDE integrated into the AutoCAD environment, Autodesk has provided the AutoCAD application developer with a second IDE for developing applications, Visual Basic for Applications (VBA). While it is not the intent of this book to teach VBA, there are a few Visual LISP functions that are provided to allow the developer to incorporate VBA applications into AutoLISP programs. Although AutoLISP is a powerful programming language for developing AutoCAD applications, VBA has the ability to create GUI interfaces much more easily and quickly than DCL can. Because of Visual LISP's ability to incorporate other programming languages, the developer can create the GUI portion of an application in VBA. To incorporate VBA applications into an AutoLISP program, Autodesk has provided two functions, VL-VBALOAD (vl-vbaload filename) and VL-VBARUN (vl-vbarun macroname), for loading and executing VBA applications. The VL-VBALOAD function, when supplied with a string representing the name and location of a Visual Basic project file, loads the project into memory. The VL-VBARUN function, when supplied with a string representing the name of a loaded macro, executes that macro. This is illustrated in the following example where an AutoLISP application loads and then executes a Visual Basic application.

 Note: The only way variables generated in a VBA macro can be passed to an AutoLISP application is through the use of the AutoCAD system variables USERI1 through USERI5, USERR1 through USERR5, and USERS1 through USERS5.

```
;;;****************************************************************
;;;
;;; Program Name: VBA-EXAMPLE.LSP
;;;
;;; Program Purpose: Demonstrate how Visual Basic and Visual LISP can
;;;                  be integrated into a single programming package.
;;;
;;; Date : 05/19/99
;;;
;;; Written By; James Kevin Standiford
;;;
;;;****************************************************************
(defun c:vba-example ()
  (vl-vbaload
    "C:\\WINDOWS\\Desktop\\Visual_Lisp\\Ch12\\Program\\Project.dvb"
  )
  (vl-vbarun
  "C:\\WINDOWS\\Desktop\\Visual_Lisp\\Ch12\\Program\\Project.dvb!Module1.bend"
  )
  (setq    users1 (getvar "users1")
      dcl_id4
```

```
            (load_dialog
               "/windows/desktop/Visual_Lisp/Ch12/program/vba-example.dcl"
            )
    )
     (if (not (new_dialog "bend" dcl_id4))
       (exit)
     )
     (set_tile "in" "1")
     (set_tile "len" "12")
     (action_tile "accept" "(retreve) (done_dialog)")
     (start_dialog)
     (if (= tog 1)
       (progn
          (if (not (new_dialog "lay" dcl_id4))
          (exit)
          )
          (setq fir1 "Bend allowance = "
               sec1 "Orginal Length = "
               thi1 "New Length = "
          )
          (set_tile "fir" (strcat fir1 all))
          (set_tile "sec" (strcat sec1 len))
          (set_tile    "thi"
               (strcat thi1 (rtos (+ leng (atof all))))
          )
          (action_tile "accept" "(done_dialog)")
          (start_dialog)
      )
   )
    (princ)
)
(defun retreve ()
   (setq
     out               (get_tile "out")
     in          (get_tile "in")
     len               (get_tile "len")
     tog               1
```

```
      all            (getvar "users1")
        thickness (getvar "users2")
        radius      (getvar "users3")
    )
  (if (= out "1")
      (setq leng (- (atof len) (+ (atof radius) (atof thickness)))))
    )
  (if (= in "1")
      (setq leng (atof len))
    )
  )
```

SUMMARY

Traditionally, DOS and early versions of Windows (3.X) applications stored essential startup information in initialization (.INI) files. Initialization files are ASCII-based text files that are limited in length to 64K. This created a problem: because these files were ASCII text files, they were considered low security and could be edited using any text editor. It wasn't until the introduction of Windows NT that Microsoft began addressing these problems by developing a system of database files (collective known as the registry). These files are stored in a centralized location housing important startup information used by Windows and other Windows-based applications. The registry is a system of proprietary binary database files whose primary purpose is to ensure the integrity of the computer's operating system and applications. The registry uses a hierarchical structure to store information. The structure is comprised of six main components called subtrees: HKEY_LOCAL_MACHINE, HKEY_CLASS_ROOT, HKEY_CURRENT_USER, HKEY_USERS, HKEY_CURRENT_CONFIG, and HKEY_DYN_DATA. Each subtree is made up of subkeys. The subkeys contained in each subtree may also contain subkeys or they may contain active keys. It is the active keys that store the actual values associated with a particular application.

Out of the six subtrees located in the Windows registry, just two contain 90 percent of the data used by the registry, HKEY_LOCAL_MACHINE and HKEY_USER. The remaining subtrees (with the exception of the HKEY_DYN_DATA) are actually aliases that contain a copy of the contents of the other two keys. All information stored in the registry by an AutoLISP application should be placed in either the HKEY_LOCAL_MACHINE or HKEY_USER.

In addition to the Visual LISP IDE integrated into the AutoCAD environment, Autodesk has also provided the AutoCAD application developer with a second IDE for developing applications: VBA. Although, AutoLISP is a powerful programming language for developing AutoCAD applications, VBA has the ability to create GUI interfaces much more easily and quickly than DCL can. Because of Visual LISP's ability to incorporate other programming languages, the developer can create the GUI portion of an application with VBA and exchange data between VBA and AutoLISP components.

REVIEW QUESTIONS

1. Define the following terms:
   Windows registry
   initialization file
   subtree
   subkey

2. What are the limitations associated with initialization files?

3. What are the advantages of using the registry as opposed to an initialization file to store important program information?

4. List the six main subtrees associated with the Windows registry.

5. List the three AutoLISP functions that can be used to read information from the registry.

6. List the two AutoLISP functions that can be used to modify the contents of the registry.

7. How are VBA applications loaded and executed using AutoLISP?

! (exclamation mark)
  lists and, 77
  recalling global variables with, 46, 47
!=, DIESEL and, 190
" " (quotation marks)
  arguments and, 30
  attributes and, 201
  DIESEL and, 182
  empty strings and, 224
  lists and, 77
  symbol names and, 27
  variable names and, 54
$ (dollar sign)
  DIESEL and, 181, 182
  menu files and, 188
$Data, 232
$Key, 232
$Reason, 232
$Value, 232
' (single quote)
  arrays and, 375
  lists and, 77, 78
  QUOTE and, 119
'(*), arguments and, 131
'all-documents notification setting, 420
( ) (parentheses)
  argument lists and, 45
  AutoLISP and, 197-198
  color scheme for, 287
  defining functions with, 28
  DIESEL and, 181
  expressions using, 18, 20
  lists and, 41
  matching command for, 293
*/, DCL and, 199, 200
* (asterix)
  classification of, 38

DIESEL and, 190
  multiplication using, 61
  retrieving application names with, 153
  searches using, 308
+ (plus sign)
  addition using, 59-60
  classification of, 38
  DIESEL and, 190
- (minus sign)
  classification of, 38
  DIESEL and, 190
  subtraction using, 60
. (period)
  decimals and, 39
  dotted pairs with, 119-120, 122, 153-155,
    158, 375, 408, 409
/*, DCL and, 199, 200
/= comparisons, 63-64
/ (forward slash)
  classification of, 38
  declaring variables with, 44
  division using, 61, 190
  specifying directories with, 32
// (forward slashes), DCL and, 199
: (colon)
  DCL and, 200
  tiles and, 204
:E selection mode, 163
:N selection mode, 163
:VLA-TRUE, 386
:VLR-AcDb-Reactor, 401
:VLR-Command-Reactor, 402
:VLR-DeepClone-Reactor, 402
:VLR-DocManager-Reactor, 401-402
:VLR-DXF-Reactor, 402
:VLR-Editor-Reactor, 402
:VLR-ERASED event, 433

:VLR-Insert-Reactor, 402
:VLR-Linker-Reactor, 403
:VLR-Lisp-Reactor, 402
:VLR-Miscellaneous-Reactor, 402
:VLR-Mouse-Reactor, 402
:VLR-Object-Reactor, 403
:VLR-SysVar-Reactor, 403
:VLR-Toolbar-Reactor, 403
:VLR-Undo-Reactor, 403
:VLR-Wblock-Reactor, 403
:VLR-Window-Reactor, 403
:VLR-XREF-Reactor, 403
; (semicolon)
    DCL and, 200
    inserting comments using, 29, 199, 200
< comparisons, 64, 190
<= comparisons, 64, 191
= comparisons
    DCL and, 200
    DIESEL and, 190
    mathematical calculations using, 62-63
> comparisons, 64, 190
>= comparisons, 64, 191
[ ] (square brackets)
    modifying elements in, 317
    SETQ and, 46
\", escape codes and, 84
\ (backslash)
    escape codes and, 40-41
    GETSTRING and, 47
\\ (backslashes)
    GETSTRING and, 47
    specifying directories with, 32
{ } (curly brackets)
    DCL and, 197-199
    modifying elements in, 317
    tiles and, 201, 204

A

ACAD* entries, 161
ACAD2000 Help directory, 364-365
Acaddoc.lsp file, 447
Acad.lsp files, 431
Acad2000.lsp files, 179

Accuracy, in problem-solving process, 11-12
Action expressions
    defined, 225
    defining, 228-232
    reactors and, 400
ACTION_TILE, 224, 229, 344
ACTIVE table entries, 158
ActiveX
    accessing entities with, 405
    accessing NameSpaces with, 464
    accessing non-AutoCAD applications
        with, 385-387
    accessing object properties with, 378-380
    application for, 387-393
    AutoCAD and, 360-367
    AutoLISP and, 265, 367-369
    collections and, 380-385
    introduction of, 359-360
    modifying entities with, 413
    releasing objects with, 385
    viewing object properties of, 369-371
    Visual Basic and, 372-378
    Visual LISP and, 360-367
    VLA data type and, 43
ActiveX and VBA Reference manual, 362, 364,
    365
ActiveX and Visual LISP Reference Guide, 369
Add Watch dialog box, 315, 316
Addcircle method, VLA-ADDCIRCLE and, 366
Addition, 59-60
ADD_LIST, 234-235
ADS files, 32
ADSRX applications, 39
Aesthetics, dialog boxes and, 216
ALERT, 387
Alert boxes, callback functions and, 405
Alignment
    DCL statements and, 200
    dialog boxes and, 216
    expressions and, 21
Alphanumeric strings, converting, 85
AND
    described, 191
    test expressions and, 96, 98-99, 100-101
ANGBASE system variable, 51

ANGDIR system variable, 51
Angles
   radian-string conversions for, 90-91
   requesting data on, 50-51
   trigonometric functions for, 65-67
ANGTOF, 90-91
ANGTOS, 91, 191
Animate debugging function, 336-337
ANSI standards, 26
API (Application Programming Interface), 264
APPEND, 78-79
APPID tables
   adding application names to, 149-150
   checking application names in, 150-151
   symbol tables and, 158
Applications
   ActiveX and, 385-387
   checking syntax in, 296-298
   COM-based technologies for, 360
   creating projects for, 337-349
   DIESEL expressions in, 185-187
   generating compiled, 349
   loading third-party, 475
   making functions accessible to, 450-453
   making variables accessible to, 453-460
   modular, 357
   multiple NameSpaces for, 447-448
   NameSpace options for, 449, 450
   ObjectARX, 39, 403
   registering names for, 149-151
   retrieving names for, 153
   reusability of, 358, 359
   standalone, 344, 349-350
   storing execution records for, 319-325
   valid and invalid names for, 150
   viewing registered, 333-334
   Visual Basic for (VBA), 491-494
   Visual LISP and, 303-337
   VLA-Object pointer to, 367
Applications, illustrative
   calculating bend allowances, 67-71
   constructing and attaching Xdata to simple
      gear train, 169-172

constructing simple harmonic motion cam
      displacement diagram, 101-109
determining spot elevations, 238-257
incorporating a reactor, 433-436
spot elevation, 477-491
using ActiveX to write results from
      AutoLISP application to Word,
      387-393
APPLOAD command, 29, 32-33
APPLY, 344
Apropos search engine
   finding symbols using, 306-309
   word completion using, 314-315
   working with results from, 309-314
Arc tangent, calculating, 65, 67
Arguments
   assigning multiple expressions to, 94
   callback function, 406
   checking number of, 297
   defined, 18
   DEFUN, formatting list of, 28, 44, 45
   dotted pairs as, 120
   format styles for, 291
   mathematical comparisons between, 62-65
   prefix notation for, 59
   syntax for, 19-20
   testing condition of, 92-101
   valid, 19
   Visual Basic-to-LISP conversions for, 366
Arrays, 375-378
Artificial intelligence, 6, 76
ARX files, 32
AS INTEGER, 375
ASCII format
   decimal character codes in, 56-57, 58
   dialog boxes and, 193
   saving files in, 21, 22, 26
   security problems with, 468
ASHRAE Fundamentals Handbook, 10
ASSOC
   retrieving dotted pairs using, 153, 154
   retrieving list elements using, 122-123, 131
Association lists
   acquiring entity names in, 124-127

application exercise for, 169-172
checking available memory for, 155-156
complex objects and, 127-129
creating, 119-122
deleting and restoring entities in, 129-130
dotted pairs in, 119-122
entity creation functions and, 135-137
filtering with, 164
linking elements in, 116-119
modifying, 132-135
retrieving elements from, 122-123
returning object data in, 130-131, 167
substituting sub-lists in, 123-124
working with AutoCAD, 131
Xrecords and, 156
Assumptions, in problem-solving process, 10
ATAN, 65, 67
ATOF/ATOI, 85, 209
Atoms, data storage and, 76
Attributes
  listing unknown, in problem-solving
      process, 8-9
  tile
    characteristics of, 201-202
    controlling dialog box alignment with,
        216
    defined, 193
    DIALOG and, 203-205
    syntax for, 200
AUNITS system variable, 91
AUPREC system variable, 91
AutoCAD
  ActiveX and, 360-367
  association lists and, 124-137
  AutoLISP and, 264-265
  AutoLISP variant functions in, 372
  dotted pairs in, 119-120
  editing multiple entities in, 161
  editor reactors identifiers for, 402-403
  entity data stored in, 116
  error function in, 463
  file searches by, 31-32
  help files in, 362
  launching Visual LISP from, 265-266

lists in, 77
MDI in, 285, 442
NameSpace in, 443
object model for, 360-362, 367
obtaining system and environmental data
    from, 53-55
OOP and, 360
reactors and, 400, 408-409
text window in, 21
working with multiple drawings in, 418
Xrecords and, 156
AutoCAD 2000 Command Reference, 54
AutoCAD 2000 Customization Guide, 38, 51
AutoCAD Customization Guide, 119
AutoLISP
  ActiveX functions included in, 365
  converting Visual Basic to, 366
  data types in, 38-43
  DCL compared to, 197-200
  declaring variables in, 45-46
  defined, 6
  dialog box management in, 193
  DIESEL compared to, 181-182
  enhancements to, 265
  file management, Visual LISP and, 298
  limitations of, 148, 264-265, 464
  MDI and, 464
  modifying Windows registry using,
      475-477
  obtaining and modifying reactor data using,
      421-424
  reading Windows registry using, 470-475
  strengths of, 264-265
  using ActiveX with, 367-369
  VBA and, 491-492
AutoLISP Function Catalog, 38-39
AutoLISP Programming Reference, 43
AutoLISP Reference Guide, 404, 405, 406
AutoSAVE time settings, 187, 188-189
AUX menus, values for, 185

B

Backup files, Windows registry, 470
Bearing selection chart, 97

Bend allowances, calculating, 67-71
BGLCOLOR, 236
Binary system, defined, 3
Blackboards, 460-462
Blocks, 136, 137, 330-332
BLOCKS table, 158
Bookmarks, 303-305
Boolean data type, 373
Boxed columns, grouping tiles into, 213-214
Break on Error option, trace stack and,
    323-324
Browsers
    block, 330-332
    drawing table, 328-330
    entity name, 326-328
    selection set, 332-333
    Xdata, 333-334
Build options, compiling source code with,
    342-349
Build Project FAS File option/toolbar button,
    349
Build window, errors displayed in, 298
Buttons, defining, 206-208
Buttons menu, values for, 185

C

C: prefix, use of, 27, 28
C selection mode, 162, 163
C/C++, 200, 298, 403
CADDR, 80, 81
CADR, 80-81
Calculations, in problem-solving process,
    10-11
Callback events
    callback functions associated with, 406-407
    listing, 404-405
    purpose of, 404
    reactor types associated with, 401-403
Callback functions
    defined, 225, 400
    defining, 229-230
    example of, 406-407
    executing, in non-active drawings, 418-419
    guidelines and restrictions for, 405-406

identifying and modifying, 423-424
modifying AutoCAD entities using, 413
object reactor, 407-408
persistent reactors and, 431
syntax for, 408-409
unknown commands and, 406-407,
    408-409, 419
VL-LOAD-COM and, 401
Cam displacement diagram, application for,
    101-109
Caption, inspection window, 317
CAR, 80-81
Carriage returns, writing expressions and, 21
Case
    DCL and, 200
    SET_TILE and, 228
    strings and, 91
    symbol names and, 308
CDATE system variable, 187
CDR, 80-81, 123
Child/children, defined, 193
Child/parent relationship, 202-203, 359
CIRCLE command, 92
CIRCLE_MARKER slide, 240, 242, 243
Class, OOP and, 358-359
Clipboard, copying symbol names to, 312
CLOSE, 57-58
Close command, 302-303
Close toolbar button, 303
Cluster tiles, 194, 212-215
Clusters, 193
CMDECHO system variable, 183-184
CMDNAMES system variable, 183-184
Collections, 367, 380-385
Color attributes, symbolic names for, 236
Color style dialog box, 289
COlumn format style, 291
Columns, grouping tiles into, 212-213
COM (compound object modeling), 360
Comma-delimited expressions, 182
COMMAND
    ACTION_TILE and, 229
    callback functions and, 405
    DIESEL expressions and, 183

modifying AutoCAD entities with, 132
modifying selection sets with, 166
Command prompt
  AutoCAD
    invoking NEW_DIALOG from, 225
    launching Visual LISP from, 265
    recalling variables at, 46-47
    setting MODEMACRO system variable
      with, 178
    writing results to, 55-57
  console window
    clearing text from, 282-283
    launching programs from, 278, 280
Commands
  activating Visual LISP, 272
  callback functions and, 405
  displaying current, 184
  lists and, 76, 77
  modifying selection sets with, 166
  programming and, 3
  retrieving, 280
Comments
  AutoLISP code for, 29, 199-200
  color scheme for, 287
  DCL code for, 199-200
Common LISP, 6, 76
Comparisons, mathematical operations for,
  62-65
Compilation Mode for Build Options,
  343-344
Compiled applications
  creating FAS file for, 349
  creating standalone version of, 349-350
  directories for, 345
  NameSpaces created by, 448-450
  Separate Namespace option for, 450
  standard *vs.* optimized options for,
    343-344
Compiler, defined, 3
Complete Word by Match feature, 306
Complex objects
  accessing entities in, 127-129
  creating, 136
  selection sets and, 163
Compound documents, 359

Computer revolution, 2
Computers, terms and concepts related to,
  2-5
Concatenated text, dialog boxes and, 212
COND, 100-101
Conditional evaluators, 92-93
CONS
  arrays and, 375
  association lists and, 120, 122, 123
  COlumn format for, 291
  creating dotted pairs with, 153, 154
Console window
  characteristics of, 275-285, 290
  copying symbol names to, 312, 313
  executing programs from, 294-295
  invoking NEW_DIALOG from, 225
  searching symbol table from, 306
Constants
  accessing, 386-387
  integers as, 39
  real numbers as, 40
CONTINUOUS table entries, 158
Control characters, 40-41
Coordinates
  dialog box, 226
  image buttons for, 207
  image tile, 236
  obtaining, 51-53
  returning entity names with, 124-125, 127,
    128
Copy to Console option, 313, 314
Copy to Trace/Log option, Apropos,
  310-311
Copy_rotate dialog box, 204-205
COS, 65, 67
CP selection mode, 162, 163
CREATE_LAYER, 231
CROSS_MARKER slide, 240, 241
CURRENT_LAYER, 230
Cursor
  inserting bookmarks at, 304
  repositioning, 305-306
CXXXX functions
  accessing extended data with, 131
  replacing sub-lists with, 123

retrieving association list elements with, 122-123

retrieving list data with, 80-81

D

Data exception errors, 375

Data storage

checking available space for, 155-156

conserving memory for, 385, 387

methods for, 76-77

Data types

arrays and, 375

attribute, 201

AutoLISP, 38-43

DIESEL, 181-182

mathematical comparisons between, 62-63

testing, 79-80

Visual Basic, 372-373

VLA, 43

Database file system, Windows, 468-470

Database reactors, 401

Date, displaying, 185-187

DATE system variable, 187

DBGLCOLOR, 236

DCL (Dialog Control Language)

alignment and spacing in, 200

attributes used in, 201-202

defined, 193

defining dialog boxes using, 203-216

parent/child relationship in, 202-203

previewing files in, 334-336

syntax for, 197-200

tile format in, 200-201

DCL files, 222-223, 298

DDEDIT command, 161

DDL files, 386

DDMODIFY command, 161

Debug menu, Visual LISP, 268, 271

Debug on Entry (De) symbol flag, 326

Debug toolbar, Visual LISP, 273

Debugging

animate feature for, 336-337

Break on Error tool for, 323-324

callback functions for, 407

developments in, 265

global variables and, 44, 46, 47

symbol service and, 325-326

Decimal values, date/time format using, 187

Decision making

COND function in, 100-101

AND function in, 96, 98-99

IF function in, 93-96

OR function in, 96, 99-100

process for, 91-92

Decorative tiles, 194

Default actions, dialog box, 224

Definition data

appending Xdata to, 149, 151-152

retrieving, 135, 152-153

Definition files, dialog box, 222-223, 227

DEFUN

argument list for, 28, 44-45

callback functions and, 405

components of, 27

creating error traps with, 463

DIALOG tile *vs.*, 203-204

launching programs with, 278, 280, 294

Degrees, converting radians to, 67

Demand loading, 475

Descriptive Geometry: An Integrated Approach Using AutoCAD, 68

Design problems

flowchart for, 15

sketching, 9

solving, 6-14

DFGLCOLOR, 236

Diagnostic messages, console window display of, 283-284

Diagrams, in problem-solving process, 9

Dialog box facility, 225

Dialog boxes

action expressions for, 228-232

Add Watch, 315, 316

advantages of, 192-193

application exercise for, 238-257

APPLOAD command, 32-33

Apropos Options, 307, 310

Apropos Results, 309, 312

callback functions for, 228-232, 405
child/parent relationship in, 202-203
color style, 289
components of, 193-195
copy_rotate, 204-205, 214
defining elements in, 203-216
design considerations for, 216
displaying, 225
emergency escape from, 226
event-driven programming for, 358
file format for, 193
Files, 33
Find command, 296
flowcharts for, 195, 197
Go to Line, 305
images in, 210, 235-238
initializing, 223-224
Inspect, 370-371
list boxes and popup lists in, 208-209,
    233-235
loading, 222-223
nested, 226
New Project, 338-339
Open command, 299
preferences, 31
previewing, 334-336
project, 349
Project Properties, 339-343
Registry subtree program, 472
Save-as command, 24, 26, 301
select color, 227
setting and retrieving tile values in,
    227-228
setting mode of, 232-233
spot elevation program, 230-231, 239-243,
    478
style and syntax for, 195-202
Symbol Service, 325-326
terminating, 215-216, 225-227
unloading, 227
window attributes, 288
DIALOG tile, 202, 203-204
DICTADD, 161
Dictionary objects, 160-161
DICTNEXT, 160, 161

DICTREMOVE, 161
DICTRENAME, 161
DICTSEARCH, 160, 161
DIESEL
    AutoCAD menu files and, 188-190
    AutoLISP programs and, 183-187
    defined, 178
    functions associated with, 190-192
    syntax for, 181-183
DIM, 375
Dimension style, displaying, 183-185
DIMSTYLE tables, 158
DIMX/Y_TILE, 236
Directories
    adding, to library path, 31
    file searches in, 32
    specifying, for compiled files, 345
    structure of, 196, 202
Do Not Link option, Link Mode and, 346
Document reactors, 401-402
Documentation, adding, 29
DOCUMENTCREATED event,
        VLR-DOCMANAGER-REACTOR and, 406
Documents
    compound, 359
    making functions accessible to, 450-453
    making variables accessible to, 453-460
    retrieving VLA-Object type, 367
DONE_DIALOG, 225-226, 229
DONUT_MARKER slide, 240, 241
DOS, 24, 26, 468
Dotted pairs
    arrays and, 375
    association lists and, 119-120, 122, 158
    creating and replacing, 153-155
    reactors and, 408, 409
Double-precision floating-point format, 40
Double-precision floating-point number data
        type, 373
Drawing database
    acquiring entities from, 125
    browsers for, 326-334
    filter-list option for, 164
    purging objects from, 129
    recording changes in, 153, 155

storing non-graphic objects in, 157-161
updating entity data in, 151-152
updating object definitions in, 132
Drawing tables, viewing, 328-330
Drawings
listing reactors defined in, 421
making applications accessible to, 447-448
NameSpace concept for, 443-447
obtaining entity names for, 42
regenerating, 133
using reactors with, 409-420
Drive, specifying, 32
Drop-down menus, values for, 185
DWGNAME system variable, 87
DXF group codes, 148
attaching Xdata with, 149
characteristics of, 118
ENTMAKE/ENTMOD and, 158
obtaining list of, 119
scanning association list for, 122-123
Xrecords and, 156
Dynamic memory, NameSpace and, 443
Dynamic text, 211, 344
Dynamic tile values, 228

E

Edit boxes, defining, 209-210
Edit Global Declarations option, Build
Options and, 345
Edit menu, Visual LISP, 268, 269
Editor reactors, 402-403
EDTIME, 185-186, 191
Elements, list
defined, 77
inspection window for, 317-319
linking, 116-119
retrieving, 122-123
ELSE expressions, WHILE and, 138
Embedded expressions, 19, 20
Empty sets, 28
Empty variants, 372
Encapsulation, OOP and, 359
Endblk sub-entities, 136, 137
END_IMAGE, 238

END_LIST, 235
Engineering skills, 2, 50
ENTDEL
deleting dictionary objects with, 160, 161
removing entities with, 129-130
symbol tables and, 157, 158
ENTGET
ActiveX functions and, 369
callback functions and, 405
confirming current settings with, 135-136
dictionary objects and, 161
obtaining association lists with, 130-131,
167
retrieving Xdata with, 152-153
symbol tables and, 157, 158
Entities
accessing, within complex objects, 127-129
acquiring names of, 124-126
creating, 135-137
deleting and restoring, 129-130
drawback related to, 148
modifying, 132-135
naming, 41-42
returning lists associated with, 130-131
selection set
adding, 164
building, 161-165
defined, 41
determining members of, 165
determining number of, 166-167
extracting, 167-168
removing, 164-165
storage space for, 155-156
using Xdata with, 149-157
viewing, 326-330
Entity-modifying commands, 152
ENTLAST, 126, 135-136
ENTMAKE, 131, 135-137, 157
ENTMAKEX, 131, 137
ENTMOD
callback functions and, 405
dictionary objects and, 161
modifying entities with, 131, 132-135
recording database changes with, 153, 155
space limitations and, 156

symbol tables and, 157, 158
updating entity data with, 151-152
ENTNEXT, 125-126, 127
Entry counter, resetting, 159
ENTSEL, 124-125, 127, 128-129
ENTUPD, 131, 133-135, 157
ENTXXX functions, 161, 168, 413
Environmental data, 53-55
EQ, 191
Equalities, mathematical operations for,
    62-65
ERASE command, 129
Ergonomics, dialog boxes and, 216
Error messages, 462-464
Error tiles, 194
Error trace stack feature, 323-325
Error traps, 463, 464
Errors
    Break on Error option for, 323-324
    checking for, 287, 296-298
    data exception, 375
    formatting, 293
    loading programs and, 29-30
    MDI and, 462-463
    object reactors and, 413
    persistent reactors and, 431
    rejected function, 225
Errors and Warnings option, Message Mode
        and, 346
Escape buttons, 207
Escape codes
    inch symbol and, 84
    PRXXX functions and, 56
    strings and, 40-41, 48
EVAL, 191, 344, 456
Evaluator
    AutoLISP, 20, 29
    conditional, 92-93
Event-driven programming, 357-358
Excel, Microsoft, 385, 386
EXE files, 386
Exit buttons
    action of, 207
    tiles for, 194, 215-216
Exit tiles, predefined, 226

Expert mode, Make Application Wizard, 350
EXP/EXPT, 65
Exponential functions, 65
Export to AutoCAD (Ea) symbol flag, 326
Expressions
    action expressions and, 225
    adding comments to, 29, 199
    animate debugging feature for, 336-337
    AutoLISP, 18-21
    clearing, 282-283
    COND tests for, 100-101
    data types for, 38
    DCL, 197-200
    defining functions with, 28
    DIESEL, 181-182
    evaluating, 29-30, 275-278
    examples of, 18
    executing, 294-295
    IF tests for, 93-96
    OR tests for, 96, 99-100
    PRXXX functions and, 55-56
    retrieving, 280
    saving, 21-26
    AND tests for, 96, 98-99
External files
    descriptors for, 42-43
    importing, 386
    reading, 58-59
    writing to, 57-58
External sub-routines, 38-39

F

F selection mode, 162, 163
FAS Directory option, Build Options and,
        345
FAS files, 344-345, 349
Fatal errors, error traps for, 462-463
Fatal Errors option, Message Mode and, 346
File descriptors, 42-43, 57
File menu, Visual LISP, 268, 269, 299-302
Files
    external
        descriptors for, 42-43
        importing, 386

reading, 58-59
writing to, 57-58
finding, 295-296
program
creating and saving, 21-26
loading, 29-33
project
build options for, 342-349
creating, 338-340
FAS files for, 349
opening, 349
selecting, 340-342
source code
closing, 298, 302-303
creating, 298, 300
opening, 298-299
saving, 298, 300-301
Fill area, image tile, 235-256
FILL_IMAGE, 235-236
Filter flag search option, 308-309
Filters search option, 309
Find command, 295-296
FIX, 191
Flags, symbol, options for, 326
Floating menu, 427
Flowcharts, 14-17, 195, 197
FOREACH, 321
Formatter, Visual LISP, 290-293
Formulas
bend allowance calculation, 68
distance between points, 50
Julian format conversion, 187
listing, in problem solving, 9-10
radian-to-degrees conversion, 67
4D arrays, 376
Full Report option, Message Mode and, 346, 347-349
Function Call Frame, trace stack and, 321
Functions
ActiveX, 365
adding comments to, 29
color scheme for, 287
defined, 6, 18
defining, 26-28

exporting, 448, 449, 450-453
finding, 295-296
NameSpace and, 443
nesting, 95-96
prefix notation for, 59
reactor, 400
sub-routines for, 38-39
syntax for, 19-20, 297
valid, 18

G

Garbage collection, 44
GC, 44
GETANGLE, 51
GET_ATTR, 228, 229
GETCORNER, 53
GETDIST, 49-50
GETENV, 54-55, 192
GETINT, 49, 51
GETKWORD, 405
GETORIENT, 51
GETPOINT
callback functions and, 405
obtaining coordinates with, 52
point lists and, 77, 378
GETREAL, 48, 51
GETSTRING, 47-48
GET_TILE, 228, 229
GETVAR, 54, 87, 192
GETXXX functions, 40-41
Global declaration files, 345
Global variables
declaring, 45-46
recalling, 46-47
Go to Last Edited/Go to Line options, 305-306
Graphic entities
creating, 135-137
obtaining space for, 367-368
updating definition of, 132-135
Graphic symbols, 14, 17
Graphics editor
displaying changes in, 133

interfacing with, 132
selecting entities from, 124
slides of, 237
Graphics screen
  deleting entities from, 129
  switching from console window to, 285
  updating, 133, 135
GUI (Graphical User Interface)
  advantages of, 192-193
  basis for, 357
  defined, 3
  operating system for, 5
  VBA for, 492

H

Handles, 41, 158, 160
Hardware, 2
Height, image tile, 236, 237
Help
  ACAD2000, 364-365
  AutoCAD, 362
  Visual LISP, 269, 272, 311-312, 363
History, console, 280-282
HKEY subtrees, 469-470, 477

I

I selection mode, 162, 163
IBM Corp., 298
Identification numbers, dialog box, 222, 223
IDEs (Integrated Development
    Environments)
  VBA, 491-494
  Visual LISP, 265
IF
  DIESEL syntax for, 192
  evaluating multiple arguments with, 93-94,
    96
  integrating AND in, 98-99
  integrating OR in, 99-100
  MODE_TILE and, 238
  nesting of, 95-96
  testing expressions with, 93
  WHILE compared to, 138

IF/IF THEN ELSE statements, 92-93
Image buttons, 207-208
Image tile menus, values for, 185
Image tiles
  constructing lines in, 236-237
  filling, 235-236
  inserting slides into, 237
  markers in, 240
Images, dialog box, 210, 235-238
Implementation inheritance, 359
INDEX, 192
Index numbers, 167
Inequalities, mathematical operations for,
    62-65
Infix notation, mathematical operations and,
    59
Information tiles, 194
Inheritance, 359
INI files, 468
INITGET, 49
Inspect dialog box, 370-371
Inspection tool
  examining reactors with, 421, 424-429
  viewing object properties with, 369-371
Inspection windows
  browser, 327-330, 332
  copying symbol names to, 312
  features of, 317-319
Integer data type, 372
Integer overflow, 39
Integers
  adding, 59-60
  attributes and, 201
  characteristics of, 39
  color scheme for, 287
  date/time format using, 187
  dividing, 61-62
  examples of, 39, 40
  LOOPS and, 140, 141
  multiplying, 61
  string conversions for, 84, 85
  subtracting, 60
Interactive functions, 405
Interface, graphic environment, 132
Interface inheritance, 359

Interface tools, text editor, 334-335
Internal option, Link Mode and, 346
Interpreter, AutoLISP, 20, 294
Itoa, 84

J

Julian format, 187

K

Key values, association lists and, 117
Key word searches, 295
Keys
    text name, 160, 161
    tile, 228
    Windows registry, 469, 470-477
Keyword Frame, trace stack and, 321

L

Labels, edit box, 209
Lambda Form, trace stack and, 321
Layer command, 226, 227
Layers
    changing, 132-135
    listing names of, 234-235
Length, 82-83
Libraries, slide, 237
Library path, AutoCAD, 31
Line inspection window, 328
Linear units, formats for, 84
LINECOLOR, 236
Linelen, 192
Lines, image tile, 236-237
Linetype tables, 158
Link Mode and option, Build Options and,
    346
Linker reactors, 403
List, 77-78, 120, 123
List boxes, 208, 233-235
List processing
    angular string conversions in, 90-91
    association lists and, 116-137
    changing string's case in, 91
    determining string length in, 88-90

manipulating and converting strings in,
    83-85
merging text strings in, 85-87
truncating text strings in, 87-88
Lists
    available events, 404-405
    characteristics of, 41, 76-77
    combining, 78-79
    creating, 77-78
    determining length of, 82-83
    dialog box, 208-209, 234-235
    object element, 317-319
    object properties, 379-383
    problems with, 116-117
    reactor type, 403-404, 421
    retrieving data from, 80-82
    symbol name, 307-314
    testing identity in, 79-80
Load, 29-32, 344
Load_dialog, 222-223
LoadTypeLib API, 386
Local variables, declaring, 44-46
Localize Variables box, Build Options and,
    346
Log, 65
Logarithmic functions, 65
Logic charts
    AND, 98
    color number, 236
    OR, 99
Long integer data type, 372
Loops
    callback functions and, 406
    counting entities with, 166
    creating, 138-141
    described, 137
    extracting entities with, 168
    generating list with, 234-235
    introducing, 357
Lowercase, converting strings to, 91
Lowercase symbols search option, 308
LSP files
    acad, 431
    acad2000, 179
    Acaddoc, 447

defaulting to, 30
saving as, 22
LUNITS system variable, 84
LUPREC system variable, 84

M

Machine language, 3
Make Application Wizard
creating standalone applications using, 344, 349-350
NameSpace options in, 448, 449
MAPCAR, 344
Match by prefix search option, 307-308
Matching, completing words by, 306
Mathematical operations
basic types of, 59-62
comparisons in, 62-65
exponential and logarithmic, 65
trigonometric, 65-67
Matrix, 4x4 transformation, 128
McCarthy, John, 6
MDI (Multiple Document Interface)
AutoLISP and, 464
blackboards and, 460-462
error conditions and, 463
NameSpace and, 443-460
purpose of, 285
SDI compared to, 442
Memory
freeing up, 385
loading dialog boxes into, 222-223
object libraries and, 387
removing dialog boxes from, 227
Menu files
using DIESEL expressions in, 188-190
MENUCMD, 185-186
Menus, Visual LISP
activating formatter through, 290-291
categories of, 267-272
selecting format options from, 292
setting bookmark from, 304
syntax coloring, 288, 289
Merge Files Mode, Build Options and, 344-345

Message boxes
callback functions and, 405
date/time displays in, 187
Message Mode options, Build Options and, 346
Messages
user, 199
writing, 55-57
Methods
accessing, 386-387
determining existence of, 380
objects and, 362
OOP and, 359
VLA- conversions for, 366
Microsoft Corp., 26, 359, 360, 468, 469
Microsoft Excel, 385, 386
Microsoft Word, 24, 26, 387-393
MNC files, 190
Mode settings, dialog box, 232-233
Model Coordinate System, 128
Model space, 163, 367, 368
MODEL_SPACE table entries, 158
MODEMACRO command, 182
MODEMACRO system variable, 178-181, 183-184
MODE_TILE, 229-233, 238
Modify command, 317
MS-DOS, 24, 26, 468
MSLIDE command, 237
Multidimensional arrays, 375, 376, 377
Multiplication, 61

N

N escape code, 40, 41, 48
N> prompt, 20
NameSpace
blackboards and, 460-462
concept of, 443-447
defining reactor and callback functions in, 431
error messages and, 463-464
limitations related to, 464
loading AutoLISP applications into, 447-448

making functions accessible in, 450-453

making variables accessible in, 453-460

VLX applications and, 448-450

NArrow format style, 291

NENTSEL, 127-129

NENTSELP, 127, 128-129

Nested dialog boxes, 226

Nested entities

accessing, 127-129

changing layer of, 133-135

ENTDEL and, 130

viewing, 327-328

Nested expressions, 182, 319-325

Nested functions

concept of, 95-96

loops and, 139-140

trace stack for, 319-325

Nested loops, 139-140

Nested tiles, 201

New File command, 300

New Project command/dialog box, 338-339

NEW_DIALOG

action expressions and, 228-229

caution for, 225

compiling problems with, 344

initializing dialog box with, 223-224

Nil values, comparisons and, 62-65

NO_MARKER slide, 240, 242, 243

Non-graphic entities

accessing, 367

creation of, 135-137

storage methods for, 157-161

updating definition of, 132-135

Non-GUI (Non-Graphical User Interface),
3, 5

Notepad, 22-24

Notification settings, callback event, 418,
419, 420

NTH, 80, 81-83, 192

Numeric data

requesting, 48-51

Numeric data conversions, 83-84, 85

O

OBJ system variable, 312

Object data type, 373

Object element list, 317-319

Object libraries, importing, 386-387

Object line, inspection window, 317

Object model, AutoCAD, 360-362, 367

Object reactors

arguments required with, 406

callback functions for, 407-408

creating, 409-418

disabling, 430

function of, 403

listing, 421

ObjectARX

compilation problems with, 344

functions created by, 39

functions of, 128

linker reactors and, 403

Object-listing commands, 152

Objects

accessing constants and methods for,
386-387

accessing properties of, 317-319, 369-371,
378-380, 386-387

assigning Xdata to, 151-152

attaching Xdata to, 149

collections of, 380-385

creating, 368-369

external, connecting to, 385-386

handles for, 41-42

model for, 360-362, 367

obtaining space for, 367-368

OOP and, 359

reactor owner lists and, 422-423

releasing, 385

retrieving Xdata from, 152-153

selection modes for, 162-163

viewing, 326-328

Visual Basic-to-LISP conversions for, 366

VLA-Object conversions for, 413

OFFSET command, 47
OK_/OK_CANCEL, 215-216
OLB files, 386
OLE (Object Linking and Embedding), 283, 285, 359-360
OOP (object oriented programming), 265, 357-359, 360
OOPS command, 129, 130
OPEN, 38, 57
Open File command, 298-299
Operating systems, 3, 5
Optimize option for compiling, 343-344, 346
OR, 96, 99-101, 192
Overwriting files, 57, 468
Owner list, reactor
   defined, 410
   expansion of, 427
   generating and modifying, 422-423
   modifying objects in, 413

P

P selection mode, 163
Paper space, 163, 367, 368-369
PAPER_SPACE table entries, 158
PARAGRAPH tile, 212
PARC Laboratory, Xerox Corp., 192
Parent objects, 127
Parent/child relationship, 202-203, 359
PDB (programmable dialog box) facility, 234
PDF (Programmable Dialog box Facility), 193
Persistent reactors, 430-433
PLane format style, 291
Planning process
   application exercise for, 67-69
   planning source code in, 14-18
   solving design problems in, 6-14
Pointers, VLA Object
   creating, 367
   removing, 385
Points
   calculating distance between, 49-50
   obtaining coordinates for, 51-53
Polymorphism, 359

Popup lists, 209, 233-235
Predefined functions, 27
Predefined tiles, 194, 226
PREFERENCES command/dialog box, 31
Prefix notation, mathematical operations and, 59
PRIN1, 55-56
PRINC, 55-56, 96, 323
PRINT, 55-56
PRJ files, 338
Problem statements, 8
Problem-solving process, 6-14
PROGN, 94, 225
Programming
   advancements in, 357-358
   planning source code in, 14-18
   problem solving in, 6-14
   terms and concepts related to, 2-5
Programming languages
   characteristics of, 3
   developments in, 264-265
   dialog boxes and, 193
   list-processing, 6
   OOP and, 358
Programs
   animate debugging feature for, 336-337
   bend allowance calculation, 67-71
   creating and saving files for, 21-26
   decision making process in, 91-101
   defined, 2-3
   defining functions in, 26-28
   graphically-oriented, 193
   launching, from console window, 278, 280, 294-295
   loading, 29-33
   reformatting, 290-293
   syntax coloring scheme for, 287-289
Project menu, 338
Project Properties dialog box, 339-343
Projects
   concept of, 337-338
   creating, 338-349
   FAS files for, 349
   menu for, 268, 270

opening, 349
standalone applications for, 349-350
Prompts
constructing user-input, 47-55
flowcharts and, 15, 16
improving readability of, 48
Properties
accessing
ActiveX for, 378-380
inspection tool for, 369-371
support code for, 386-387
AutoCAD object model and, 362
OOP and, 359
Protect Assign (Pa) symbol flag, 326
Protected symbols, 287
Prototypes, 193
PRXXX functions, 55-56
Pseudo-code, 17-18, 69-70

Q

QUOTE, 119, 120
Quoted lists, 291

R

Radians, conversions for, 67, 90-91
Radio buttons
defining, 206-207
grouping, 214-215
Reactors
ActiveX and, 360
application for, 433-436
callback events associated with, 404-405
callback functions for, 405-408
categories and types of, 401-403
creating AutoCAD, 408-409
creating object, 409-418
disabling and enabling, 429-430
identifying type of, 421
incorporating, 265
listing, 403-404, 421
multiple drawings and, 418-420
notification settings for, 420
obtaining and modifying data for, 421-429

persistent or transient, 430-433
purpose of, 400
support code for, 400-401
READ-CHAR, 58
READ-LINE, 58
Real numbers
angular conversions for, 90-91
attributes and, 201
characteristics of, 39-40
color scheme for, 287
date/time format using, 187
loops and, 141
mathematical operations for, 59-61
requesting data on, 48-51
string conversions for, 84, 85
REDO command, 401
Reformatting text, 290-293
REGAPP, 149-151
REGEDIT utility, Microsoft, 468-469, 470
REGEN command, 133, 135
REPEAT, 137, 140-141
Reserved variables, 232
Reserved words, 201
Restricted attributes, 202
Restricted tiles, 194
Results, reading and writing, 55-59
Retrieval functions, 80-82
Rich Text Format, 24, 26
Rows, arranging tiles in, 214
RTOS, 84, 192
Rubber-band effect, 49, 50, 52
Runtime errors, 297

S

Safe Optimize box, Build Options and, 346
SafeArrays, 375, 377-378
Save command, 300, 301
Save time settings, 183-187, 189
Save-As command, 22, 24, 26, 300-301
SAVETIME system variable, 188
Screen, writing results to, 55-57
Screen menus, values for, 185
Scroll bars, 208, 209
SDI (Single-Document Interface), 442

Search engine, Apropos, 306-315
Search menu, Visual LISP, 268, 270
Search toolbar, Visual LISP, 273
Security, INI files and, 468
Selection mode options, 162-163
Selection sets
    adding entities to, 164
    building, 161-165
    counting entities in, 166-167
    defined, 41, 161
    determining members of, 165
    extracting entities from, 167-168
    PRXXX functions and, 55-56
    removing entities from, 164-165
    viewing, 332-333
SELECT_LAYER, 231
Separate Namespace option, 450, 452
Seqend sub-entities, 136, 137
SET, 344
SETQ, 38, 45-46, 457
SET_TILE, 228, 234, 238
SETVAR, 178, 181, 183-184
Shortcut menus
    accessing inspection tool from, 429
    Apropos, 312-313
    project dialog box, 349
Side effects, 44
Signed numbers, 39
Simple option, Make Application Wizard,
        350
SIN, 65, 66
Single-precision floating-point number data
        type, 373
16-bit numbers, 39
Sketches, in problem-solving process, 9
Slide library, 237
SLIDE_image, 237
Slides, 237, 240, 241-243
Software
    defined, 2
    evolution in, 357-358
    standardizing, 468
Solid regions, constructing images from,
        235-236

Source code
    animate debugging feature for, 336-337
    bookmarking, 303-305
    build options for, 342-349
    checking for errors in, 296-298
    developing projects using, 337-350
    dialog box, 193
    format styles for, 291
    managing files for, 298-303
    planning, 14-18
    reformatting, 290
Spacebar, GETXXX functions and, 47, 48, 49
Space-delimited expressions, 182
Spacing
    in DCL statements, 200
    in dialog boxes, 216
    in expressions, 21
Spaghetti programming, 357
Special Form, trace stack and, 321
SPLINE command, 102
Spot elevations, applications for, 238-257,
        477-491
Spreadsheets, 385-386
SQL files, 298
SQRT, 65
SSADD, 164
SSDEL, 164-165
SSGET
    dialog boxes and, 225
    selection sets and, 55-56, 161-164, 168
SSLENGTH, 166
SSMEMB, 165
SSNAME, 167
S::STARTUP, 179
Stacks, trace, 319-325
Standard notation, mathematical operations
        and, 59
Standard option for compiling, 343
STANDARD table entries, 158
Standard toolbar, Visual LISP, 273
Start menu, Windows
    launching Notepad from, 22, 23
    launching Wordpad from, 24, 25
START_DIALOG, 225, 228, 229

START_IMAGE, 235, 238
START_LIST, 234
START-TRACE command, 323
Static text, dialog boxes and, 211
Static tiles, 228, 233
Status bars
    AutoCAD, configuring, 178-181, 182-187
    Visual LISP, 268, 272, 275
STRCASE, 91
STRCAT, 85-87, 183-184
String data type, 373
Strings
    angular conversions for, 90-91
    changing case of, 91
    characteristics of, 40-41
    color scheme for, 287
    determining length of, 88-90
    empty, 224
    merging, 85-87
    numeric conversions for, 83-84, 85
    PRXXX functions and, 55
    requesting data on, 47-48
    truncating, 87-88
STRLEN, 88, 192
Structured programming, 357
STYLE tables, 158
Subassemblies, 193
Subkeys, Windows registry, 469, 470-475
Sub-lists
    defined, 77
    dotted-pair, 119-120
    ENTDEL and, 130
    reactor, 421
    retrieving, 122-123
    substituting, 123-124
Sub-routines
    data type for, 38-39
    defining, 28
    external
        data type for, 38-39
    structured programming and, 357
SUBST
    substituting list items with, 123-124
    switching dotted pairs with, 153, 154-155
SUBSTR, 87, 88, 192

Sub-strings, 87-88
Sub-tables, 158-160
Subtraction, 60
Subtrees, Windows registry, 469-475
Support code
    ActiveX, 367
    application object, 386-387
    reactor, 400-401
Symbol names
    characteristics of, 27
    removing, 343, 346
    variables names and, 43
Symbol service, 312, 325-326, 387
Symbol tables, 157-160
Symbols
    Apropos feature for, 306-314
    NameSpace and, 443
    variables and, 43
Syntax
    checking for errors in, 296-298
    coloring scheme for, 287-289
    DCL, 200-201
    defined, 3
    DIESEL, 192
    expression writing and, 18-20
    Visual Basic, 366
System, obtaining data about, 53-55

T

T variables, 48, 62-65
Tablet menus, values for, 185
Tangent, calculating inverse of, 65, 67
TBLNEXT, 159-160
TBLOBJNAME, 158
TBLSEARCH, 158-159, 160
Temporary files, 136, 345
TERM_DIALOG, 226
Test expressions
    AND for, 96, 98-99
    COND for, 100-101
    IF for, 93-96
    OR for, 96, 99-100
    WHILE and, 137, 138
Tests, callback functions for, 407

Text
  adding dialog box, 210-215
  attributes and, 201
  clearing, 282-283
  displaying current style of, 181-185
  dynamic, 211, 344
  formatting, 290-293
  transferring, 283, 285
Text cluster tiles, 194
Text descriptors, 84
Text Document formats, 22, 24, 26
Text editor
  animate debugging feature in, 336-337
  checking syntax in, 296-298
  executing expressions from, 294-295
  FAS files in, 345
  file searches in, 295-296
  formatting with, 290-293
  matching parentheses in, 293
  security of, 468
  symbol table searches from, 306
  syntax coloring in, 287-289
  tools in, 286, 334-335
Text strings
  association list for, 117
  list boxes and, 208
  merging, 85-87
  truncating, 87-88
Text window, AutoCAD, 21
Third-party applications, loading, 475
32-bit numbers, 39
3D arrays, 376
TILEMODE, 160
Tiles
  action expressions for, 225, 228-232
  attributes of, 201-202
  callback functions for, 229-232
  classification of, 194
  clustered, 194, 212-215
  DCL format for, 200-201
  defined, 193
  grouping, 203-204
  parent/child relationship for, 202-203
  redefining actions of, 226
  setting and retrieving value of, 227-228

  setting mode for, 232-233
  spacing, 216
Time, displaying current, 185-187
TLB files, 386
TLBSEARCH, 150-151
Tmp Directory option, Build Options and, 345
Toggle Bookmarks option, 304
Toggles, 206
Toolbar buttons, Visual LISP
  Close command, 303
  descriptions of, 272, 275
  project dialog box, 349
Toolbars, Visual LISP, 272-275
Tools menu, Visual LISP, 269, 271, 335
Top Form, trace stack and, 321
Trace log file, 310
Trace stack, 319-325
Trace (Tr) symbol flag, 326
Trace window, 285-286, 310
Transient reactors, 430
Tree-structure format, 193, 195, 202
Trigonometric functions, 65-67
Truncation, text string, 87-88, 204
2D arrays, 376
TYPE, 79-80
Typing errors, detecting, 287

U

UNDO command, 129, 401
Unicode Text Document format, 24, 26
Unknown commands
  linking callback function to, 408-409
  notification for, 419
  VLR-EDITOR-REACTOR and, 406-407
UNLOAD_DIALOG, 227
UPPER, 192
Uppercase, converting strings to, 91
User interface, Visual LISP
  components of, 266-267
  menu region of, 267-272
  toolbar region of, 272-275
User-defined functions, writing, 26-29
USERI system variables, 492

USERR system variables, 492
USERS system variables, 178, 492

V

Variables
action expression, 232
assigning arrays to, 375-378
checking syntax for, 297
creating access to, 453-460
declaring local and global, 44-47
inspection tool for, 317-319
listing unknown, 8
monitoring value of, 315-316
NameSpace and, 443-447
naming, 43
resetting, in test expressions, 138
storing data using, 76-77
transferring value of, 460-462
verifying value of, 278, 279
Visual Basic data types for, 372-373
Variants, 372-374, 378
Vector graphic image, 210
Vector_image, 236-237
View menu, Visual LISP, 268, 270
View toolbar, Visual LISP, 273, 274
Visual Basic
data types in, 372-378
reference guide for, 365
syntax for, 366
Visual Basic for Applications (VBA),
491-494
Visual effects, dialog box, 210
Visual LISP
ActiveX and, 360-367
components of
console window, 225, 275-285, 290
menus, 266-272
text editor, 286-298
toolbars, 272-275
trace window, 285-286
data type for, 43
defining SafeArrays in, 375
developing applications with

animate debugging function for,
336-337
Apropos features for, 306-315
bookmarks for, 303-305
browsing database for, 326-334
inspection window for, 317-319
previewing dialog boxes for, 334-336
repositioning cursor for, 305-306
symbol service for, 325-326
trace stack for, 319-325
watch window for, 315-316
word completion for, 306, 314-315
dialog box management in, 193
examining reactors with, 424-429
features of, 265
help files in, 269, 272, 311-312, 363
importing type library into, 386-387
launching, 265-266
manipulating files in, 298-303
projects in, 337-350
VBA and, 491-494
Visual LISP Developers Guide, 202
VLA- functions, 365, 386
VLA-ADDCIRCLE, 366
VLA-DUMP-OBJECT, 379-380
VLA-GET, 365, 379
VLA-IMPORT-TYPE-LIBRARY, 387
VLA-MAP-COLLECTION, 380
VLA-Objects
data type for, 43
reactors and, 410, 413, 420
retrieving, 367, 369
viewing properties of, 369-371
VLA-PUT, 365, 379
VLA-PUT-VISIBLE, 386
VLAX- functions, 365, 405
VLAX-CREATE-OBJECT, 385
VLAX-ENAME->VLA-OBJECT, 413
VLAX-FOR, 383-384
VLAX-GET-ACAD-OBJECT, 367-369
VLAX-GET-ACTIVEDOCUMENT, 367-369
VLAX-GET-MODELSPACE, 367-369
VLAX-GET-OBJECT, 385
VLAX-GET-OR-CREATE-OBJECT, 385, 386, 387

Vlax-get-paperspace, 367-369
Vlax-get-property, 365, 387
Vlax-import-type-library, 386
Vlax-invoke-method, 365, 387
Vlax-make-safearray, 375-376
Vlax-make-variant, 372, 373-374, 378
Vlax-method-application-p, 380
Vlax-name->vla-object, 369
Vlax-product-key, 470, 475
Vlax-property-available-p, 380
Vlax-put-property, 365, 387
Vlax-release-object, 385
Vlax-safearray->list, 378
Vlax-safearray-fill, 375, 377
Vlax-safearray-get-element, 377-378
Vlax-safearray-get-l-bounds, 378
Vlax-safearray-get-u-bounds, 378
Vlax-safearray-put-element, 375, 377
Vlax-3d-point, 378
Vlax-variant-change-type, 372, 374
Vlax-variant-type, 372, 374
Vlax-variant-value, 372, 374
Vlax-vbarray, 373
Vlax-vbboolean, 373
Vlax-vbdouble, 373
Vlax-vbempty, 372, 373
Vlax-vbinteger, 372
Vlax-vblong, 372
Vlax-vbnull, 373
Vlax-vbobject, 373
Vlax-vbsingle, 373
Vlax-vbstring, 373
Vl-bb-ref, 460, 462
Vl-bb-set, 460, 462
Vl-doc-export, 450-452, 450-453
Vl-doc-import, 452-453
Vl-doc-ref, 453-456
Vl-doc-set, 453, 456-457
Vl-exit-with-error, 463
Vl-exit-with-value, 464
Vlide command, 265
Vlisp command, 265
Vl-list-exported-functions, 452
Vl-list-loaded-vlx, 452

Vl-load-all, 447-448
Vl-load-com
    ActiveX and, 367
    reactors and, 400-401, 405
    Windows registry and, 470
Vl-propagate, 453, 458-460
Vlr- functions, 365, 400
Vlr-add, 429, 430
Vlr-added-p, 430
Vlr-beep-reaction, 407
Vlr-commandCancelled, 404
Vlr-commandWillStart, 404
Vlr-data, 422
Vlr-data-set, 422
Vlr-docmanager-reactor, 406
Vlr-editor-reactor, 406, 409
Vl-registry-delete, 475
Vl-registry-descendents, 470-474
Vl-registry-read, 470, 474
Vl-registry-write, 475, 476-477
Vlr-notification, 420
Vlr-object-reactor, 410, 413
Vlr-owner, 423
Vlr-owner-add, 423
Vlr-owner-remove, 423
Vlr-pers, 430-433
Vlr-pers-list, 431-432
Vlr-pers-p, 431, 432
Vlr-pers-release, 433
Vlr-reaction, 423
Vlr-reaction-names, 404-405
Vlr-reaction-set, 424
Vlr-reactor, 421
Vlr-remove, 429
Vlr-remove-all, 429
Vlr-set-notification, 419
Vlr-trace-reaction, 407
Vlr-type, 421
Vlr-types, 403
Vl-unload-vlx, 450
Vl-vbaload, 492
Vl-vbarun, 492
VLX applications
    accessing functions with, 450-453

accessing variables with, 453-460
creating NameSpaces for, 448-450
error-trapping function for, 463-464
VLX files, 344
VPORT symbol tables, 157, 158, 160
VSLIDE command, 237

W

W selection mode, 163
Watch window, 312, 315-316
WCMATCH search option, 308
WHILE, 137, 138-140
WIde format style, 291
Width, image tile, 236, 237
Wild cards, searching with, 153, 308
Window attributes dialog box, 288
Windows
    database file system in, 468-470
    status line in, 178
    text-editing applications in, 22-26
    tree-structure format in, 195-196
Windows Explorer, 202
Windows menu, Visual LISP, 269, 272
Windows registry
    application for, 477-494
    development of, 468
    files and subtrees in, 468-470
    modifying, 475-477
    reading, 470-475
Word, Microsoft, 24, 26, 387-393
Word completion methods, 306, 314-315
Word processing programs, Windows-based,
    22-26
Wordpad, 22, 24-26, 70
World Coordinate System, 128
World matrix, 127-128
WP selection mode, 163
Wrapper function, 386, 387
WRITE-CHAR, 56-57
WRITE-LINE, 56-57
Write-xxx functions, 56-57

X

X coordinates, 236, 237
X selection mode, 163
Xdata (Extended Entity Data)
    application for, 169-172
    assigning, 151-152
    attaching, 149
    managing, 155-156
    modifying, 153-155, 161
    purpose of, 148
    registering application name for, 149-151
    retrieving, 131, 152-153
    viewing, 333-334
    Xrecords and, 156-157
XDROOM, 155-156
XDSIZE, 155-156
Xerox Corp., 192
XOR, 192
Xrecords, 148, 156-157

Y

Y coordinates, 236, 237

LICENSE AGREEMENT FOR AUTODESK PRESS

THOMSON LEARNING™

Educational Software/Data

You the customer, and Autodesk Press incur certain benefits, rights, and obligations to each other when you open this package and use the software/data it contains. BE SURE YOU READ THE LICENSE AGREEMENT CAREFULLY, SINCE BY USING THE SOFTWARE/DATA YOU INDICATE YOU HAVE READ, UNDERSTOOD, AND ACCEPTED THE TERMS OF THIS AGREEMENT.

Your rights:

1. You enjoy a non-exclusive license to use the enclosed software/data on a single microcomputer that is not part of a network or multi-machine system in consideration for payment of the required license fee, (which may be included in the purchase price of an accompanying print component), or receipt of this software/data, and your acceptance of the terms and conditions of this agreement.

2. You own the media on which the software/data is recorded, but you acknowledge that you do not own the software/data recorded on them. You also acknowledge that the software/data is furnished "as is," and contains copyrighted and/ or proprietary and confidential information of Autodesk Press or its licensors.

3. If you do not accept the terms of this license agreement you may return the media within 30 days. However, you may not use the software during this period.

There are limitations on your rights:

1. You may not copy or print the software/data for any reason whatsoever, except to install it on a hard drive on a single microcomputer and to make one archival copy, unless copying or printing is expressly permitted in writing or statements recorded on the diskette(s).

2. You may not revise, translate, convert, disassemble or otherwise reverse engineer the software/data except that you may add to or rearrange any data recorded on the media as part of the normal use of the software/data.

3. You may not sell, license, lease, rent, loan, or otherwise distribute or network the software/ data except that you may give the software/data to a student or and instructor for use at school or, temporarily at home.

Should you fail to abide by the Copyright Law of the United States as it applies to this software/data your license to use it will become invalid. You agree to erase or otherwise destroy the software/data immediately after receiving note of Autodesk Press' termination of this agreement for violation of its provisions.

Autodesk Press gives you a LIMITED WARRANTY covering the enclosed software/data. The LIMITED WARRANTY can be found in this product and/or the instructor's manual that accompanies it.

This license is the entire agreement between you and Autodesk Press interpreted and enforced under New York law.

Limited Warranty

Autodesk Press warrants to the original licensee/ purchaser of this copy of microcomputer software/ data and the media on which it is recorded that the media will be free from defects in material and workmanship for ninety (90) days from the date of original purchase. All implied warranties are limited in duration to this ninety (90) day period. THEREAFTER, ANY IMPLIED WARRANTIES, INCLUDING IMPLIED WARRANTIES OF MERCHANTABILITY AND FITNESS FOR A PARTICULAR PURPOSE ARE EXCLUDED. THIS WARRANTY IS IN LIEU OF ALL OTHER WARRANTIES, WHETHER ORAL OR WRITTEN, EXPRESSED OR IMPLIED.

If you believe the media is defective, please return it during the ninety day period to the address shown below. A defective diskette will be replaced without charge provided that it has not been subjected to misuse or damage.

This warranty does not extend to the software or information recorded on the media. The software and information are provided "AS IS." Any statements made about the utility of the software or information are not to be considered as express or implied warranties. Autodesk Press will not be liable for incidental or consequential damages of any kind incurred by you, the consumer, or any other user.

Some states do not allow the exclusion or limitation of incidental or consequential damages, or limitations on the duration of implied warranties, so the above limitation or exclusion may not apply to you. This warranty gives you specific legal rights, and you may also have other rights which vary from state to state. Address all correspondence to:

Autodesk Press
3 Columbia Circle
P. O. Box 15015
Albany, NY 12212-5015